The Evolution and Function of Cognition

The Evolution and Function of Cognition

Felix E. Goodson
Greencastle, Indiana

LONDON AND NEW YORK

The cover art depicts stone age men armed with bows and arrows preparing for battle or the hunt. In the world of primitive magic success in either was sought by formal representations such as these. From a cave painting at Teruel.

First published 2002 by Lawrence Erlbaum Associates, Inc.

Published 2014 by Psychology Press

Published 2021 by Routledge
2 Park Square, Milton Park, Abingdon, Oxon OX14 4RN
605 Third Avenue, New York, NY 10017

Routledge is an imprint of the Taylor & Francis Group, an informa business

Cover design by Kathryn Houghtaling Lacey

Library of Congress Cataloging-in-Publication Data

Goodson, Felix E., 1922– .
The evolution and function of cognition / Felix Goodson.
p. cm.

Includes bibliographical references and index.
ISBN 978-0-8058-4216-6 (cloth : alk. paper)
ISBN 978-0-8058-4217-3 (pbk : alk. paper)
1. Genetic psychology. I. Title.
BF701 .G62 2002
155.7—dc21 20020022509
CIP

ISBN 13: 978-0-8058-4216-6 (hbk)

For my children

Jane, James, John, Holly, Felix, Boyd and Brad

In the distant future I see open fields for far more important researches. Psychology will be based on a new foundation, that of the necessary acquirement of each mental power and capacity by gradation.

—Charles Darwin, *The Origin of Species*, 1859.

Contents

Foreword

As a third party writing a foreword to any volume, one should prepare a potential reader for its contents and, hopefully, inform the potential reader of the value of the volume, honestly portraying both its merits and its limitations. This is done with the intention of helping the reader decide whether they are going to: (a) buy the volume, (b) bother to read the rest of the volume, and (c) use it as a textbook for a course. For that reason, it is important to clarify in advance both what the volume is and what the volume is not.

I start with a few words on what the volume is not. This volume does not construe evolutionary psychology as either a new science or as some kind of speciality within psychology with its own theories, methods, and content area. The volume is about understanding all of mainstream psychology from an evolutionary perspective. I find that to be its major strength. The volume is rife with references that are decades, and sometimes even centuries old. Some readers may be surprised to find that it includes no comprehensive review of the current research in this area, but that is not what this volume is about. Whereas many other evolutionary psychologists emphasize novelty, Goodson emphasizes historical continuity and shows how evolutionary psychology fits in with the mainstream of thought in psychological theory. Judged on this basis, Goodson incorporated the most important and relevant work of all time. His treatment is extremely accurate and superbly well-integrated.

Goodson essentially builds up the human mind from scratch, literally constructing it step by step from the primaeval slime. He builds principle upon principle in a way no longer seen in contemporary psychology, rivaled only by the axiomatic systems of Hull and Spence in the 1950s. Prediction is risky, but I suspect that this volume might become to evolutionary psychology what Euclid was to geometry. However, it should be noted that all such systems of organization are necessary subjective and creatively constructed by the author, and some might disagree with the interpretive scheme used by Goodson. Personally, I have doubts about the centrality ascribed to homeostatic mechanisms, but I can see the value of the use that Goodson makes of it. Some partisanship towards certain ideas seems to me to be unavoidable because an integrative synthesis requires a conceptual scheme to crystallize around. Homeostasis might be as good as any other.

What Goodson does is take us on a personal journey of discovery that includes many tangents and historical allusions, delving into detail where delightful discoveries are made and inspiring insights are revealed, and then returning us to the main narrative without missing a step. The volume is written more in the style of the traditional humanities than that of the contemporary hard sciences, and recall a bygone era when scientists were also scholars of some culture and refinement. There is little here of the technical bottom line and more of the process of understanding ourselves. Those who enjoy that journey will find the volume riveting. Students who want to know exactly what material will be on the final examination will doubtless become exasperated. Although the writing style is interesting and engaging for the truly curious mind, the sentence structure and vocabulary are relatively sophisticated. The volume abounds with historical examples and analogies of all the principles covered, but these are pitched at a pretty high level and not intended as simple parables for the intellectually lazy or feeble-minded. The level is appropriate for use as an upper-division (junior or senior) undergraduate textvolume, or for a graduate textvolume, but not for first-year introductory students unless they can pass an IQ test for fairly high verbal ability. Pedagogically, this volume is an excellent exercise in higher-level thinking about psychology, evolutionary and otherwise, and in the training of the mind to more complex and sophisticated modes of thought. I would give it to my students to teach that skill, as well as to convey the basic principles of evolutionary psychology. The logical framework provides the foundation for a mighty edifice of ideas, for those who can conceive of grand schemes. There are no short-and-sweet sound bytes here or simplified take-home messages. This volume is no easy exercise for the casual student. I must acknowledge here that I am typically very demanding of my students. Nevertheless, they rarely disappoint me when so challenged.

Certain merits of the volume should be emphasized. These noteworthy characteristics are those of Niceness, Maturity, and Integration.

Niceness

It has been said that in the physical sciences one advances by standing on the shoulders of one's predecessors, whereas in the social sciences one advances by stepping on their faces. This unflattering characterization is too often true, and contemporary evolutionary psychology is no exception. In many recent works in the field, the introductory opening arguments are marked by the firing of an initial salvo of critiques of mainstream psychology, aimed at distinguishing what we do and think from the nefarious practices and numerous errors of a pathetic and moribund social science paradigm. Goodson instead follows Darwin himself in emphasizing historical continuity over discontinuity. He does this while also unabashedly following Darwin in basing all of psychology on evolutionary theory. This shift in emphasis cannot be sufficiently commended.

Maturity

I grew up with evolutionary psychology, and I would venture to guess that Goodson did also. The span of my life has witnessed its contemporary beginnings in the 1960s and 1970s to its gradually breaking into the mainstream in the 1990s. As an unflinching supporter and active participant in this movement, I hope to survive to witness the triumphant completion of the Darwinian Revolution in the 21st Century, climaxing with the universal recognition of the hegemony of evolutionary theory as the central unifying principle of the social and behavioral sciences, a process which has already unequivocally occurred within the biological sciences. That being said, I see that much of what has been written to date by many evolutionary psychologists (sometimes even by myself) smacks more of adolescent rebellion than mature scholarship. Frankly, I was waiting for this volume, or something like it, to come along, and I was very much afraid that I might have to write it. Goodson might have saved me the trouble, and for that I am thankful. This is a work of maturity, reflecting pride in our illustrious past as well as hope in our magnificent future. Goodson makes it known early on that he treats the psychology of the past with due respect. He does more than this, he treats it with love. This does not mean that he remains mired in the errors of the past out of blind reverence. Everyone gets a slap on the wrist when they have gone astray, from Freud to Gould, the way a loving parent might correct a wayward child. Yet Goodson makes it clear that we no longer need to kill our fathers, if we ever did. With this volume, evolutionary psychology has truly come of age.

Integration

Goodson's most impressive service is in the integration of the evolutionary perspective with those of the past intellectual movements within psychology. Again, he emphasized scientific continuity over the more fashionable Kuhnian revolutions, even between past movements that are still widely believed to have been in total contradiction with each other. The breadth and depth of scholarship required to accomplish this task is impressive. As a result, the solid groundedness of evolutionary psychology is revealed in all its grandeur. The point is made emphatically that evolutionary psychology is not a new science. It is arguably one of the oldest, if not the oldest and most venerable, movements within psychology, soundly initiated by Darwin and Galton decades before Wundt supposedly began it all in Leipzig. True, it has evolved over the years, with no pun intended, but the continuities are multifarious and manifest. Goodson's work brings that out gently, without the polemical tone that someone like me might have taken.

In summary, even with these various caveats for those who might be looking for something else, I cannot sufficiently endorse and recommend Goodson's volume. This volume represents an entirely different creature for which there is currently no adequate comparison. This volume definitely breaks new ground and hopefully represents the beginnings of the evolutionary psychology of the future, now finally come of age. I invite you to enjoy it as I did and to share it with your students.

30 August 2001
—*Aurelio Jose Figueredo*
Tucson, Arizona

Prologue

And man built a machine and then wondered
if he had been built in the image of it.
—*Morrison W. Crawley*

When Descartes was young, he and his friends spent much time in the public gardens of Paris, conversing about man's nature and man's condition. Descartes was particularly intrigued with the hedge mazes where at certain turns or blind alleys mechanical figures of human beings and animals, activated by hidden springs and hydraulic tubes, would appear with frightening suddenness, to be greeted by the shrieks of the young ladies and the opportune protective embraces of the young gentlemen.

Descartes was inspired by these mechanical figures and with them as models developed the first physiological explanation of how the body works as a lawful mechanism:

> It is to be observed that the machine of our bodies is so constructed that all the changes in the motions of the spirits may cause them to open certain pores of the brain rather than others ... *causing the spirits* (italics mine) to be conducted into the muscles which serve to move the body ... in the same way that the movement of a watch is produced by the force solely of its mainspring and the form of its wheels. (1892, part 1, pp. 291–292)

Descartes, of course, was a dualist, believing in both the mechanical body and the immortal soul, but that is not the issue here, which is man's tendency to construe himself in the image of the machines that he has built.

With the development of computers we have come full circle. Once again we gaze with hope and inspiration at machines in our search for an explanation of how humans work as lawful mechanisms. A colleague who has thought long and written much in psychological theory even maintains that when we build a computer that can do everything a human can do it will have the same cognitive processes that humans have. I do not accept this extrapolation, believing that information can be processed in many different ways. We share certain obvious operations with computers; input, storage, and retrieval are common to us both; but we are made of much different stuff, and we have far different attributes. We bleed, get hungry, weep, fear death, and smell bad. Indeed many of the characteristics that are most definitive of humanness are missing in computers altogether. Anxiety, hope, despair, awe, jealousy, and envy are not found in computers; as methodical, uncaring, plodding, and invariably precise as they are.

Computers, like humans, are information processors, but the processing that takes place in humans is considerably more complex; outcomes are always the result of a dynamic interplay of a multitude of factors that perpetually interact to occasion action, where a seemingly trivial encounter from the past may momentarily intrude. Prufrock asked, "Is it the perfume from a dress that makes me so digress?" In an organism the input is perpetually shifting in information cogency as the conditions within and without the organism change; information processing is influenced not only by its accumulated past both as species and individual, but also by its present state, a state in perpetual and momentary flux as needs, emotions, threats, change, etc.

In passing it might be noted that to build a computer that could do everything a human can do would require that we knew how the model (the human) worked, which is what we are trying to accomplish, not vice versa. If we truly understood the human being the use of computers in an explanatory effort, except as an arcane exercise, would obviously be moot.

The recent efforts to bring psychology into the evolutionary synthesis are marked by a return to a machine model. Such domain-general categories as learning, and such historical dichotomies as nature versus nurture have been replaced, in certain cases, with an emphasis upon specific adaptive biologically based mechanisms called modules that reflect adaptive solutions to particular survival problems and, perhaps not coincidentally, with the use of computers as an explanatory metaphor. Much research during the past few years has involved the search for modules in language (grammar and syntax predispositions) social interchange (cheater detection mechanisms), sex differences (in jealousy), resource allocation (meat distribution), and many others (see Barkow, Cosmides, & Tooby, 1992) for an update of research and theory.

In this search for modules there has been a corresponding rejection of, if not contempt for, historical categories, with the view that such domain-general (i.e., inclusive), explanatory involvements obscure specific biological survival-reflective propensities. These domain-general categories are often designated by the dismissive rubric standard social science model—while the virtues of what is called computational theory, are extolled. Computational theory is used to designate the various steps involved in determining the evolutionary origin and function of a particular attribute, an activity with which I completely agree, and one that is central to my own development in the following pages. The word computational apparently symbolizes their commitment to a computer analogue, but their use of the word theory in this context remains unclear to me.

It is in their emphasis upon the term module that recent evolutionary psychologists most clearly signal their dependence on computers as analogues; thus computers and evolutionary theory become conjoined in an explanatory effort. A module, as the term is used in computer terminology, is a self-contained assembly of electronic components and circuitry designed to perform a particular task. This is an attractive definition and seemingly analogous to what may obtain in nervous systems. Yet in the writing of many contemporary evolutionary psychologists there are presumed modules, (biologically based, highly specific determinants) that account for all kinds of activities and propensities. Indeed, in their rejection of domain-general classifications, they seem to view the mind of human beings as entirely made up of a multitude of such specific determinants tenuously tied together, if at all, by a hypothetical integration mechanism.

The term module may be useful in explaining humans, but there are often domain-general solutions for general adaptive problems, as I indicate at length in the following chapters. These general solutions may have areas of particular sensitivity, precision and appropriateness, but such specifics inevitably reside within a broader adaptive context. For instance, organisms may be more adept at solving learning problems reflecting their species history, but such particularized facility always resides within a more general capacity for learning. In other words, the term module may be useful in our explanatory efforts, but organisms, particularly humans, do not consist of a large bundle of isolated and independently operating components. The observation by Symons (1987) that humans have great adaptive flexibility because of the number of domain-specific mechanisms we possess (i.e., the more modules, the greater the plasticity) overlooks two salient points: (a) such mechanisms (hard wired as they are) inevitably reduce flexibility and (b) where general problems are faced (and there have been a number of them, as I describe at length) general solutions resolving such problems provide parsimony (reduced complexity) rather than burdening biology (and theory) with a proliferation of mechanisms.

A problem arising from the use of computers as explanatory models may be called the metaphor or simile error, the offering of a seems like or works like concept in an effort to explain, and then confusing the concept with the factor being explained, so that in certain cases the model or analogue takes on a life of its own, with misleading and ever-widening repercussions.

I have no objection to the use of computer terms and will use them myself when it is helpful and appropriate; we have to get our language from somewhere. We can all agree that computers are very good at what they do. They are much better than humans at mathematical manipulations, facile storage and rapid retrieval, and they reduce the tedium of literature search and preparation. As was the case with Descartes' frog, computers suggest a commitment to a lawful and naturalistic interpretation, and they provide hints about the nature and operation of information processing mechanisms. Concepts from computer theory and operation, such as the term module, may help us develop an appropriate language. However we must always keep in mind the possibility of metaphor or simile error, which occurs when such terms as module lead us into an erroneous segmentation of cognitive processes, or when spandrel, a recent metaphor from evolutionary theory (Gould & Lewontin, 1979) picks up a life of its own and confuses the very process it is designed to clarify.

Recently there has been a tendency to look upon contributions (empirical research and theory) from other schools of thought as suspect and inappropriate. I do not share in the neglect of alternate positions. There have been blind alleys, willful distortions, and tedious fixations on outmoded paradigms, but there is much accumulated wisdom reflecting sincere contributions by dedicated searchers, contributions that should not be dismissed without respectful perusal. In the following pages many of those contributions are considered; many facts and insights from the past are relevant to our efforts to understand the evolutionary product that we are. A fact about some aspect of our mentation, behavior, or physiology—be it the phi-phenomena of Gestalt psychology, the Weber-Fechner function of psychophysics, the laws of reinforcement of the behaviorists or the findings on jealousy of the contemporary evolutionary psychologists—is a statement about the nature of organisms, and as such fits into an evolutionary perspective. Facts have no personal or theoretical agenda; they are nonpartisan.

Further, I try to salvage some of the historical categories and contrasts that have recently been questioned by a number of writers (see Barkow, Cosmides, & Tooby, 1992) in the field. The accumulated wisdom of the past, although sometimes burdened by distortions and omissions, cannot be summarily dismissed because of counter-enthusiasms. The present trend towards particularization (modules, analogues) does not negate such historical divisions of cognitive processing as sensation, perception, attention, learning, etc., nor does it obscure the fact that in the course of evolution relatively few progressive steps in information processing facility have emerged to account for the remarkable adaptive potential of organisms, particularly of human beings.

In the following pages these progressive steps (which often mirror the historical categories of mentation) in information-processing facility are traced, while the insights and research from all persuasions are incorporated into a comprehensive explanatory system reflecting the nature of human beings as evolutionary products.

Acknowledgments

My greatest debt is to three enduring friends and colleagues: George Morgan, for his support and counsel, both at the very beginning of my efforts toward this synthesis and in its final version. Melvin Marx, my major professor and steadfast friend over the years, provided detailed constructive feedback on each chapter. Al Hillix made innumerable corrections and suggestions, also chapter by chapter, ranging from changes in style to the way basic ideas were presented and integrated.

A. J. Figueredo, as expressed in the foreword, immediately grasped the essence, implications and scope of the integration and with his timely and astute review inspired me to complete it. Steve Gangestad made many suggestions for improvement leading to my reevaluation of many sections, particularly those treating evolutionary theory, language and dreams.

Thanks are due to Keith Opdahl for our many productive discussions on emotion as language, Ernest Henninger and Mike Hildreth for insights into the laws of physics, Roger Roof for alerting me about the energy utilization of plants, Louis Smogor and David Maloney for their suggestions about dreams, Margaret Berrio for our talks on Piaget, Bob Garrett for insights into physiological processes, Terri Bonebright for our discussion of cognitive psychology, Kenneth Wagoner—a great phenomenologist—for looking and listening, Cheryl Sorenson who transcribed and contributed to my early notes, Holly Goodson-Hildreth for our discussions on the behavior of microorganisms, and Felix Goodson for helping me see the universality of dynamic resolutions.

And thanks to my students for the give and take in classrooms and seminars that inspired and tested me over many years on the implications of the development: Dick Weigel, Tom Tombaugh, Jo Wood, Judy Diehl, Jane Middleton, Norene Goode, Joe Danks, Al Rosenquist, Phil Dunham, Bob Rudolph, Gurnot Doetsch, Ed Crossman, Fritz Knarr, Brian Pope, and many others. Robin Smoger, Emily Hunteman, and

Glenn McCain, I could not have proceeded without your diligent and continuing search of pertinent abstracts and other sources. Thank you, Brad Goodson, who not only did the indexing but provided detailed proofreading and evaluation of the final typescript, and Maxine Davies who was always there when I needed help.

I deeply appreciate the help given by people at Lawrence Erlbaum Associates. Thank you Bill Weber for immediately seeing the significance of my work and for finding three astute reviewers, Erica Kica for handling the details of initial evaluation, Phoebe Ochman for hundreds of corrections and constructive changes and Marianna Vertullo who with remarkable tact and insight during many phone calls and letters guided it through to its completion.

1 Background

EVOLUTION AND COGNITION

More than two decades ago it was argued that "psychology should be concerned with determining how living organisms work in the same sense that a watch specialist is concerned with how a watch works or an automobile specialist is concerned with how an automobile works" (Goodson & Morgan, 1976a, p. 406). It was also urged that all three levels of data that have traditionally been of interest to psychologists (the experiential, the physiological, and the behavioral) should be included in a comprehensive view. These opinions are still affirmed and they are strengthened by trends that, although they have had long historical significance, have picked up particular momentum since 1991. The first of these trends is the resurgence of cognitive psychology, a branch of that is making progress in unraveling the various stages involved in information processing; the second is the maturing of evolutionary theory and its rapid dissemination through most disciplines in the life sciences. Until recently psychology resisted inclusion in this evolutionary synthesis, but now this inclusion is rapidly taking place.

Since this volume represents an effort to integrate the facts and principles of psychology into a unified structure based upon both evolutionary and information-processing theory and research, this chapter is a summary of the developments in these two converging, often overlapping, fields. It is a review for scholars in psychology and the life sciences and provides the interested layman with an introduction to these topics.

Cognitive Psychology

Cognitive psychology, as it is treated today, has two major divisions: (a) social cognitive psychology, which is concerned with the way the individual's interaction with others influence his or her conduct and (b) Information processing psychology, which is concerned with how the cognitive machinery works to produce behavior. Although the two approaches occasionally overlap, my emphasis will be on the latter; that is on the evolutionary origin and function of the information-processing machinery that provides the basis for the moment-by-moment adjustments essential for living.

Historical Antecedents. When psychology was formally established as a separate discipline in 1879, there was only one bone of contention: Whether its subject matter consisted of the contents of consciousness (the position of Wilhelm Wundt, [Boring, 1950] the founder of the first official psychology laboratory) or the acts whereby such contents come into consciousness (the position of the philosopher, Franz Brentano, 1874). In the statement "I see red," the red was the subject matter according to Wundt, while the act of seeing the red was the subject matter according to Brentano. Wundt was interested in such contents of mind as sensations, memory images, and feelings, while Brentano was concerned with the processes whereby such contents come into consciousness.

Thus psychology was concerned with information processing at its inception. The reemergence of interest in cognition, after a 40-year hiatus caused by the ascendancy of behaviorism, owes a considerable debt to both Wundt and Brentano. With the advantage of aftersight, it is clear that Wundt and his followers provided the experimental methodology and categories (i.e., structures) for the resurgence (beginning about 1950) of cognitive psychology, while Brentano and his followers provided the direction. Brentano insisted that psychology should be concerned with such acts as sensing, judging, discriminating, wishing, and remembering. Yet it was Wundt and his followers who developed the equipment, refined the measurement of such acts, and provided the conceptual structure within which such acts occurred.

Wundt and his followers, the structuralists, adopted the basic categories of the mind—sensations, images, and feelings—from the English Empiricists, and then proceeded to evaluate these categories in an effort to determine the basic structure of the mind. For example, sensation was broken down into the various modalities of vision, hearing, olfaction, etc. Then each of these was further broken down. Vision, for instance, was broken down into hue, saturation, and brightness, and then each of these was further analyzed into just noticeable differences (JNDs) along a single dimension. So it was with all the modalities. The J.N.D. became the smallest brick in the structure of the mind. Their research, all inspired by Gustav Fechner, produced some of the most reliable facts in psychology, and, as we shall see, they fit precisely into an evolutionary perspective.

Mental Processes and Their Measurement. Both apparatus and design problems began with astronomy. As every psychologist will remember, Kinnebrook was fired by Maskelyne, because of pronounced differences in their observations of a star crossing a marker in a telescope. The differences obtained inspired Bessel (1823) to study what came to be known as the personal equation, that is, personal differences in reaction time. With the development of the chronometer (a clock that recorded time in seconds with a pointer on a kymograph) in 1850 and the Hipp chronoscope (a stop clock that measured time-intervals in thousandths of a second) in 1860, the investigation of the intervals required for different mental processes could begin.

In 1861, Wundt (Boring, 1950) constructed the first complication apparatus: It consisted of a pendulum that moved across a measured scale, with a click sounding at certain predetermined points. Wundt and his students performed many experiments with various modifications of what came to be known as the complication clock. Boring stated: "The determination of the times of various mental processes by the reaction method and the subtractive procedure was one of the outstanding activities of the new psychology and of Wundt's laboratory during the 1880s ..." (1929/1950, p. 147). Von Tchisch used a modification of the complication clock to study what is now seen as the single-channeled capacity of the apperceptual (attentional) process. When subjects were told to report which number on the scale the pointer indicated when a click sounded, they tended to respond with the previous number rather than the one the pointer was actually over; apparently apperceiving (attending) the sound, jammed out the visual input.

However, it was Donders (1862), a Dutch physiologist, who originated and first explored mental chronometry, the basic technique of modern cognitive psychology. Donders, using the subtractive procedure,[1] measured the latency of such cognitive activities as choice, dis-

---

[1]There were hundreds of these experiments, and the subtractive procedure remains as one of the most precise designs for exploring mental processes. In simplest form such research involves a subject sitting in front of a stimulus board reacting as fast as he or she can when certain stimuli are presented. The baseline of a subject's reaction time is first determined by having him or her press a button as fast as he or she can when a light is turned on. After this simple reaction time is determined, various kinds of stimuli and response complications can be studied. Whatever the complication, the time required for it to occur can be determined by subtracting (thus the name of the procedure) the baseline reaction time from it. In this fashion, many mental processes such as discrimination and choice have been studied.

crimination, association, cognition reaction, sensorial reaction, muscular reaction, and judgment.

James McKeen Cattell worked in Wundt's laboratory and later at Columbia to explore some of the problems that are still central to the study of information processing. With a chronoscope for measuring reaction times, Cattell (1885) found a relationship between stimulus intensity (both light and electricity) and shorter reaction latencies. Using the subtractive technique of Bessel and Donders, he measured the time taken for discrimination and for will (Cattell, 1886). He also measured the time it takes to see and name objects, finding, not surprisingly, that the rate at which a person reads a foreign language is directly related to his familiarity with it. He went on to suggest that a reaction time test might be developed (similar to Hunt's notion, as we shall see later) for determining a person's facility with a language. He also found that colors and pictures can be recognized in a shorter period than words.

In another study reminiscent of Galton's association experiments to be discussed later, Cattell (1886) found that the time taken for a particular mental operation increased as a function of the complexity of retrieval; thus, judgment took longer than choice and choice longer than simple retrieval. He also found that it takes longer to go from whole to part than from part to whole; thus, given a stimulus term such as Hamlet, it is easier to think of Shakespeare than to think of one of his works when given the name Shakespeare.

Wundt and his students also determined the optimal interval for associations to develop to be around 750 milliseconds. Many studies searching for the optimal interval for associations to be established (see Blumenthal, 1977, for an excellent review of these experiments) have been completed in such diverse areas as classical conditioning, instrumental conditioning, verbal associations, and reaction times, with Wundt's finding being substantiated again and again. Thus contemporary cognitive psychology owes Wundt and his laboratory a considerable debt, for both design and apparatus. It also owes a debt to Brentano and his followers, for direction and theoretical emphasis.

Brentano (1874) insisted that psychologists should be concerned with the acts whereby contents come into consciousness. This certainly has a modern ring: When cognitive psychologists measure the span of attention, immediate memory, and short term memory, the development of schema, the use of mnemonic devices, chunking, retrieval, etc., they are purportedly measuring some aspect of the processes whereby contents are either initially brought into consciousness as sensations or returned to consciousness as memories. The acts Brentano was talking about. Although Brentano did not believe in formal experimentation, he defined the direction of modern experimental cognitive psychology. This direction was also emphasized by such early British psychologists as Stout and Ward (Brett, 1921), who were inspired by Brentano. According to Ward, psychology deals with the mental life of people as it affects behavior. In his view there is a sensory continuum that is automatically and irresistibly imposed. Attention (or, as he termed it, cognition) is continually and nonvoluntarily taking place relative to components in this sensory continuum. As a function of such attending the person is either pleased or pained (in his terms feeling results), which in turn brings about changes in behavior (in his term conation).

Thus in Ward's formulation, as in Stout's who followed him, emphasis is placed upon the acts of sensing, attending, and reacting; thus, they not only brought the act psychology of Brentano into central focus, but also directly anticipated the direction and subject matter of modern-day cognitive psychology.

There was not as much difference between the content psychology of Wundt and the act psychology of Brentano as adherents of the time and as historical pigeonholers would have us believe. Wundt insisted that his was a psychology of process, while Brentano included contents as necessary precursors to the demonstration of his acts. Not only were the acts of associating and attending studied at length in Wundt's laboratory, but also the notion of short-term memory, as providing a buffer function, was generally accepted. Boring, speaking about the atmosphere in 1900, states: "... psychologists were ready to accept the principle that the latent times for perception vary so greatly that attentive predisposition may cause an incoming impulse to mill around in the brain waiting for the attention to be ready to receive it" (1929/1950, p. 147).

Thus one of the most important research involvements of modern cognitive psychology, that is, the measurement of the reaction time for various mental processes, finds its roots in the very first schools (act and content) of psychology. It should not be surprising that another major focal point of interest and research of modern cognitive psychology, the nature and function of mental imagery, also has a long history.

Galton (1880) found that there are remarkable individual differences in the illumination, definition, and coloring of mental images. To his astonishment, he found that the eminent scientists he questioned maintained that they were almost entirely devoid of mental images, while other people, particularly women and children, stated that they experienced images of great clarity, in full color. Such people described their images in minute detail, so much so indeed that Galton hesitated to believe them. He felt that he would have spoken as they did if he were describing a scene that lay before his eyes in broad daylight, to a blind man who persisted in doubting the reality of vision

(Galton, 1883). Galton came to the conclusion that scientific men have feeble powers of visual imagery. The perception of sharp mental pictures is antagonistic to the acquirement of habits of highly generalized and abstract thought and that the highest minds are probably those in which visual imagery is subordinated by word usage that in effect dilutes the clarity of the visual image.

In experiments similar to those of Cattell, already described, Galton (1883) used the reaction time method to study the accessibility and variety of associations. His approach, which may seem crude by today's standards, was to write stimulus words on small cards, shuffle the cards, and then expose them one by one to himself, taking care that he did not know which word would appear on each trial. As soon as he saw a word he would start a small chronograph (stopwatch) and stop it as soon as an associated word popped into his mind. In general, he found that the more inclusive or abstract the word the longer was the reaction time. There was much less variety in associations than he expected, and there is a perpetual and rapid reminiscence over stored ideas that serves to renew them. This mumbling over its old stores, as he called it, could be a factor in dreams, where old scenes and acquaintances are revived by reiteration.

Thus, two major foci of modern studies of information processing, the nature and function of imagery and the measurement of reaction times for different mental operations, have had a long history. Yet for many years, from 1900 until 1950, many of the issues central to modern cognitive psychology were disregarded, but not entirely. Studies during the twenties and thirties (as reported in Paivio, 1971) seeking to relate brain wave type to different indexes of visual imagery produced conflicting results. However in direct support of Galton's findings, Roe (1951) found a relationship between type of memory representation and the person's profession, with biologists and experimental physicists falling into the visual imagery group and theoretical physicists, psychologists, and anthropologists falling in the verbal group.

During the twenties and thirties (overlapping and in many cases representing the ascendency of behaviorism), psychology branched in many directions. Psychologists became interested in animals, children, abnormal people, and testing. The problems of emotion, thought, will, memory, and intelligence were occasionally treated, but not as issues in information processing, until the resurgence of an avowed cognitive psychology in the 1950s.

The Reemergence of Cognitive Psychology. Whether or not one views the computer as an adequate analogue for human information processing, the growth of computer technology since 1981 has directed our attention to the question, "how do information-processing systems work as lawful mechanisms"? The computer operations of input, storage, processing, and retrieval are parallel to, if not accomplished in the same manner as, those in human beings. Computers have also (I believe this contribution will ultimately prove more important than the analogue implications) provided a reliable and versatile apparatus for presenting stimuli for brief intervals, the technique that has long been basic to the measurement of different mental operations through the subtractive process.

As is often the case, the emergence of a new trend, or the recurrence of an old one, may not depend as much on the appropriateness of the trend as on the demise of competing positions. The fractionation of both psychoanalysis and Gestalt psychology and the ebbing influence of behaviorism created slack for change, a growing intellectual vacuum that became translated into an accepting environment for the reappraisal of and involvement in cognitive issues.

Certain seminal ideas and experiments, which were both expressions of this new interest in cognitive psychology and catalysts for its increasing importance, should be mentioned. The shadowing design invented by Cherry (1953), demonstrating that subjects missed obvious information being introduced in the shadowed or unattended ear, inspired many replications. The research by Broadbent (1954) on auditory localization and attention initiated a long series of experiments on the single versus multi-channeled nature of the attention process. The paper by Newell, Shaw, and Simon (1958), emphasizing the computer as an analogue to human information processing, initiated the emerging tendency to use computers as explanatory models. Piaget's (1924) pioneer work on developmental stages in the formation and use of concepts by children also continues to influence contemporary psychology. Tolman (1948), while ostensibly a behaviorist, developed a position that was compatible with cognitive psychology and that, ironically, helped loosen the bonds of the behavioristic commitment. Chomsky's (1965) book emphasized the cognitive analysis of language, is still influential. Miller, Galenter, and Pribram (1960), anticipating and hastening the demise of behaviorism, offered a cognitive alternative. Neisser (1967) produced a landmark textbook that emphasized a cognitive model consisting of a series of memory stores and processes and then, belatedly it would seem, Paivio (1971) reaffirmed the importance of the mental image in both memory and verbal processes.

Studies that should also be mentioned include Sperling's (1960) experiment on the span of immediate visual memory and its extrapolation by Bliss, Crane, Mansfield, and Townsend (1966), Darwin, Turvey, and Crowder (1972), Norman (1969), and Treisman (1960), to include immediate memory for tactual and auditory input. All of these experiments are replications or extensions of the Wundt-Donders design, and all of them agree with remarkable consistency that immediate memory (echoic, iconic, sensory memory, specious present, field of awareness, etc.) lasts approximately 750 milliseconds, as Wundt had previously stated.

Other pivotal experiments include: Miller's (1956) study of short-term memory span and chunking, Shepard and Metzler's (1971) investigation on the mental rotation of three-dimensional objects and Hunt, Frost, and Lunnebork's (1973) demonstrations (supporting Cattell's previously mentioned suggestion) that the latency for various mental operation can be used to measure intelligence. More recently, a longitudinal study by Chase and Ericcson (1980) not only clarified the function of chunking and mneumonic devices in retention and recall, but also provided important data on the nature of meaning.

This brief overview is but a sampling of the historical antecedents, theoretical perspectives, and experiments important as incremental achievements in the accumulating impetus toward modern cognitive psychology. Two basic themes are apparent: (a) a resurgence of interest in the nature and function of imagery, and (b) a renewed effort to probe the nature of cognitive processing by measuring the latency (using the subtractive procedure) of different mental operations.

Both historically and at the present time, a number of corollary issues, complementing and overlapping these two basic concerns, have often emerged: What happens when memories become processed? Is attention single or multi-channeled? How does affect fit into the stages of information processing? How do the various stages of processing finally culminate in reaction? Where, if at all, do the notions of self and will fit into the picture? How do inherited and learned mechanisms interact to influence the various processing stages?

Very recently, with the belated inclusion of psychology into the evolutionary syntheses, other research has appeared on such topics as human sexual strategies (Buss & Schmitt, 1993) the adaptive function of color (Hardin, 1990), the effect of relevance on attention and learning (Gelman, 1990b), the origins and functions of emotion (Ekman, 1992), the origin and function of moral systems (Alexander, 1987), and language instincts (Pinker, 1994). These, along with many other studies on the evolution of information-processing refinements, will be considered where appropriate in the following chapters.

Evolutionary psychologists during the past few years have focused on the search for biological (evolutionarily derived) predispositions in human culture. So one of the oldest and murkiest issues in the field, that of heredity versus environment (nature vs. nurture, innate vs. learned, call it what you will) has again resurfaced. An unusual amount of contention, at times even personal invective, has resulted, but, as the following pages will demonstrate, the tension has been the inspiration for much research and healthy reappraisal.

Although the development of new apparati and designs may render the following warning inappropriate, I terminate my brief overview of the history of information processing in cognitive psychology with a statement made many years ago by Boring. Commenting on the early work of Donders and the later experiments of Watt and Ach, Boring states:

> As a matter of fact the subtracting procedure never worked well. The times were too unreliable, and the differences between them were more so. It was, however, Kulpe who in 1893 argued successfully that these total processes are not compounded of elements with separate part-times ... The change in the task and in the attitude which induces action alters the whole process instead of merely adding an additional part. (1929/1950, p. 140)

Continuing research should tell us whether this admonition was prophetic or premature.

Evolutionary Theory

Early Developments. We are so immersed in evolutionary theory at the present time that it seems incredible that this view of the earth and its inhabitants is so recent. Up until 1800, just a few generations ago, even the most enlightened scientist in the western world accepted the notion of creation as given in the book of Genesis. There were some anticipations of evolutionary thinking; Erasmus Darwin, Charles Darwin's grandfather, indulged in evolutionary speculations without postulating the mechanism in 1794. In 1795, Goethe vaguely hinted at evolution in his claim that there is a theological archetype that is gradually being achieved as modification takes place through reproduction, and Matthew (1831), as we shall see, clearly stated a theory of evolution through natural selection almost 30 years before the publication of *On the Origin of Species*. These were the exceptions. Most of the great scientists of that time, including both Cuvier and Linnaeus, who, although they generally accepted the notion of cross-species similarities, as implied in Aristotle's *scala naturae*, were convinced of the fixity of species.

Yet in the latter part of the 18th century, evidence was coming in from a new quarter. Paleontologists, both amateur and professional, were digging up bones that resided in different layers of rock, suggesting that the earth was indeed very old, rather than being recently created only 6,000 years ago, as Bishop Usher insisted. The bones, many of which were petrified, kept appearing, and as they accumulated both theologians and scientists were faced with the problem of explaining them.

Explanations, some of which seem ridiculous at the present time, were immediately forthcoming. One was the jealous God argument. As is generally known, the scriptures suggest that God is not without certain anthropopsychic idiosyncracies, jealousy being one of the attributes most often mentioned, as evidenced by such episodes as the tower of Babel, the dispersion of the Jews, and the sudden salinity of Lot's wife. Thus certain thinkers of the late 18th century reasoned that God might have salted the different strata of rock with pre-aged bones, just to befuddle presumptuous man.

Another argument, perhaps not so fanciful, indeed one which was accepted by the great Cuvier, was that of multiple catastrophe. According to this view God both created and destroyed the earth, not once but many times—each time leaving another layer of bones. Apparently God first had a great affinity for trilobites, then for ancient fishes, then for dinosaurs, then for mammals, and finally and most recently, for primates. The theory of multiple catastrophe is not completely dead even among modern paleontologists. In disguised form it is present in the bone record of mass extinctions and may, at a more particular level, be involved in the punctuationalism of such modern evolutionists as Gould (1994). Generally speaking, however, catastrophism takes its place in the wastebasket of discarded theories, which found their appeal largely to the degree that they were expressions of man's enduring penchant for squaring the facts with his preconceptions.

On May 11, 1800, it happened, the appearance of the first comprehensive theory of evolution; one that has never been surpassed in terms of its simplicity of expression or the plausibility of its logic. On that fateful day, Lamarck registered in *Discourse* his shift, almost overnight it would seem, from a view of fixed and unchanging species to a theory that life demonstrates a trend towards increasing complexity through gradual evolution (Mayr, 1959).

Lamarck's theory, though it is now almost entirely discounted, is so simple and so seemingly obvious that I will use it as a vehicle for explaining evolutionary thinking. It also had profound historical significance, being used by Charles Darwin to buttress his own theory, and by the Russians from 1938 until 1965 as the rationale for agricultural decisions that are still having negative repercussions, much to their chagrin and to the benefit of our farmers, who sell them grain.

Lamarck believed that, to some slight degree, whatever an organism acquires during its own lifetime as a function of its own effort can be transmitted to its offspring. Thus the giraffe (his favorite example) gradually achieved its long legs and neck because for thousands of generations its progenitors stretched to reach the leaves of the Eucalyptus tree, their favorite food.

Why are there so many different species on the earth? The answer is obvious according to Lamarck. There are millions of different species because there are millions of different environments (we would call them niches), each of which makes the organisms within it exert themselves in different ways. Thus, to fulfill their needs organisms struggled in different ways until the tree of life (Lamarck's concept, borrowed by many others) with its expression of diversity and increasing complexity became an aptly descriptive metaphor.

Why was Lamarck's theory not more influential? There were many reasons. The zeitgeist (spirit of the times) was not quite ready. Very little specific empirical evidence was given by Lamarck to support his position. Also, Lamarck was a Frenchman, and the English have often been both suspicious and hostile to any offering from the mainland. Whatever the causes, the implications of Lamarck's theory and of evolutionary thinking in general had little influence in the life sciences (zoology, biology, taxonomy, etc.) until the publication of Darwin's theory of evolution through natural selection in 1859.

Both Darwin and Wallace (the co-discoverer of the theory) were influenced by the *Essay on Population* by Malthus, who, as a minister, was influenced by the description of the fourth horseman of the apocalypse given in the scripture: "I saw, and, look! a horse; and the one seated upon it had the name Death. And authority was given ... to kill with a long sword and with food shortage and with deadly plague and by the wild beasts of the earth" (Revelation 6:8). Malthus was also influenced by our own Benjamin Franklin. Indeed on the first page of the *Essay*, Malthus quoted Franklin to the effect that "there is no bound to the prolific nature of plants or animals but what is made by their crowding and interfering with each other's means of subsistence" (1826, p. 1).

Malthus saw selection pressures purely as instruments for the inexorable carnage that the imbalances between the population (increasing geometrically) and food supply (increasing only arithmetically) impose upon human populations. He did not see, as did Darwin, that the terrible ravages of the rider of the pale horse were actually the crucible for creation; that death, that most individually terrifying and terminal of all events, was the arbiter of evolutionary change. Yet it was Malthus' vivid account of the constraints on population growth that planted the seeds of a theory of evolution through natural selection in Darwin's mind. It happened, as he himself admitted, in one glowing moment of insight on September 28, 1838, when he read from Malthus "It may safely be pronounced, therefore, that population, when unchecked, goes on doubling itself every 25 years, or increases in a geometrical ratio ... it at once struck me that under these circumstances favorable variations would tend to be preserved, and unfavorable ones to be destroyed" (Darwin, 1958, p. 120).

If Darwin's sudden illumination upon reading Malthus appeared remarkable, it seemed astonishing that Wallace, the co-discoverer of evolution through natural selection, also had a sudden illumination with the same source as inspiration. While on an expedition as a naturalist at Ternate in the Moluccas in February, 1858, he came down with an intermittent fever. "As I was lying in bed something led me to think of the 'positive checks' described by Malthus in his *Essay on population*" (Wallace, 1891, p. 20). Wallace got up from his sick bed and in a few hours wrote a paper describing his revelation. Realizing its importance, he sent his paper to the leading zoologist in the world who, of course, was Darwin. Darwin read it and was dumbfounded. In a letter to Lyell he agonized, "... if Wallace had my ms. sketch written out in 1842, he could not have made a better abstract! Even his terms now stand as heads of my chapters ... so all my originality, whatever it may amount to will be smashed" (Mayr, 1982, p. 423). Darwin's anguish was premature; Lyell and Hooker simultaneously presented Wallace's paper and extracts from Darwin's writing to the Linnean Society on July 1, 1858, where both were greeted with surprising indifference. However with the publication of the *Origin* on November 24, 1859, both interest and sales were considerable, and it was obvious that Darwin's monumental volume gave him priority over Wallace, who most generously agreed, maintaining that his three-day effort could hardly be compared with Darwin's 20-year involvement.

To a student of the history of ideas, it must seem astonishing that two men could come up with the same all-encompassing point of view at approximately the same time, and that both achieved their inspiration from the same source, the writings of Malthus. Did they really go far beyond the thinking of Malthus? Yes, in a simple, yet most profound way. Where Malthus saw the rider of the pale horse of the apocalypse in terms of suffering and decimation, both Darwin and Wallace turned the coin over and saw creation!

I do not mean to imply that Darwin and Wallace were in perfect agreement. They were not. Wallace never accepted the notion that human beings came into being through evolution, as did Darwin. Wallace soon rejected the Lamarckian theory of evolution through the inheritance of acquired characteristics, something that Darwin not only failed to do, but leaned more heavily upon in each succeeding edition of the Origin. But in their fundamentals—the view that the primary engine of evolutionary change, and speciation, was the gradual accumulation and deletion of variations through natural selection—they were in perfect agreement. I now give a simple exposition of the theory of evolution through natural selection, as it was understood by both Darwin and Wallace in 1859.

Evolution Through Natural Selection

The basic ideas of the theory are very simple (almost as simple as Lamarck's). Three principles, which are logically interdependent, and when taken together have an almost syllogistic self-evidence, are of central importance. Indeed these principles are so well documented and so logically interlocked that to understand them is almost sufficient for the acceptance of the theory.

Chance Variation. (The first principle). During his trip aboard the Beagle, and during many forays into the hinterlands of South America, and particularly during his study of finches on the various islands of the Galapagos, one salient fact about living things stood out: All creatures vary. If one examines any attribute of any species, one will find variability. Thus if one measures the length of elephants' trunks, rats' tails, or human penises, variability prevails. Variability is omnipresent, whether we are measuring weight, length, strength, width, quickness, or even intelligence. It is true that Darwin did not understand what caused the variability (we have achieved that insight only recently by combining the work of Mendel with that of Crick and Watson), but the fact of variability could not then, nor can it now, be denied.

Struggle for Survival. (The second principle). Darwin observed (as had Malthus before him) that every creature has built into it an urge to survive, and to help its offspring survive. Regardless of the circumstances or the odds, creatures will struggle to continue to exist. What do they struggle against? Darwin was clear in his statement about this; there are five major horsemen of the evolutionary apocalypse, which we now call selection pressures.

First, each individual must compete with his or her own peers and with other interacting species for the available requirements (such as food, water, space, etc.) for life to continue, and, as Malthus contended, these requirements are often in short supply.

Second, since life feeds upon itself, every individual must struggle against the inexorable toll of predation. During the evolution of most, if not all, species, some predator has been waiting in the shadows, in the high grass, water, trees, and air, for a chance to gain sustenance so that its existence could continue.

Third, every individual resides in a perpetually dangerous environment, where it can freeze, fall, run over, or be terminated by flood, fire, earthquake, or lightning.

Fourth, every species has evolved within a context where disease is a perpetual possibility, and millions have succumbed to "the thousand ills that flesh is heir to."

Fifth, since death is implicit in all life examples the replication and inheritance of the operative variability is fundamental so that every individual must compete for mates.

These are the selection pressures that have tested the functionality of the myriad of variations that have been present in every attribute of every species since life began.

Survival of the Fittest. (The third principle). Darwin did not like the statement of this principle very well. He borrowed it from Spencer, puzzled about and then finally accepted it because he could not think of anything better. Given that every attribute of every species varies, and given that all creatures struggle to survive, it seems obvious that those variations that abet survival (i.e., the fittest ones) will gradually be incorporated into the species in question, and conversely those variations that are detrimental to survival will tend to be eliminated. Thus evolution becomes, according to Darwin, a blind, relentless process where variations are continually being subjected to the selection pressures encountered by each individual and are consequently either selected in or out.

This is the theory as both Darwin and Wallace understood it, or at least the way scholars in the life sciences understood them. The teleological explanations that were rampant even among the scientists of the time were totally rejected, to be replaced by attributes that mirrored particular environments because they were gradually adapted to such environments through natural selection. Indeed, it was Darwin's view, as it had been Lamarck's before him, that the millions of different species were reflections of the fact that there are millions of different environments on the earth. But here the similarity ends. Lamarck believed that different environments caused individuals within them to exert themselves in different ways, and that the characteristics acquired because of such exertions could, to some slight degree, be transmitted to their offspring, while Darwin believed that different environments impose different selection pressures on the variations among individuals, which are consequently either selected in or out.

In passing, it is interesting to note that there were anticipations of the Darwin–Wallace theory of evolution through natural selection. Most of them were vague and inconclusive, but one published by Matthew in 1831 was stated so clearly that one wonders why he was not given more credit for his insights. The following excerpt from Matthew not only gives recognition to his precedence, it also provides a brief summary of the theory of evolution through natural selection just given.

> The self regulating adaptive disposition of organized life may, in part, be traced to the extreme fecundity of Nature, who ... has, in all the varieties of her offspring, a prolific power much ... beyond what is necessary to fill up the vacancies caused by senile decay. As the field of existence is limited and pre-occupied, it is only the hardier, more robust, better suited to circumstance, individuals, who are able to struggle forward to maturity from the strict ordeal by which Nature tests their adaptation to her standard of perfection and fitness to continue their kind by reproduction (1831, p. 385).

This review of evolution through natural selection should suffice to acquaint or refresh the reader with its basic essentials and to provide some insight into the atmosphere that prevailed during early considerations of the topic.

In retrospect it seems cruelly ironic that Mendel's work on what caused the variation was not brought to Darwin's attention before he died. Nevertheless Darwin believed that empirical support of evolution through natural selection was abundantly available. He relied heavily on the evidence from man's ability to bring about fortuitous changes in species through unnatural selection and published two massive volumes on the topic (Darwin, 1897). Yet the empirical support that was perhaps most gratifying to Darwin came from the naturalist Bates. "The discovery of mimicry by H. W. Bates (see Mayr, 1982) came as a godsend, and Darwin at once wrote a joyous and highly laudatory review of it" (Mayr, 1982, p. 522). What Bates had discovered was that palatable butterflies tended to mimic unpalatable or poisonous ones, thus fooling their predators. Even more striking was the fact that when the nonpalatable butterflies varied in appearance from one location to the next the palatable mimics underwent the same changes in appearance. Both Bates and Darwin believed that Batesian mimicry, as it is now called, could only be explained through natural selection. As documented in many places in the following chapters mimicry (camouflage) is a widely used survival gambit in many different species, and it still provides some of the most satisfying empirical evidence in support of evolution through natural selection.

What happened to Darwin's theory after the publication of *The Origin*? The presses of 1859 were hardly silent when rents began to appear in Darwin's argument, with the largest gap in the basic premise. What caused the variation? Disturbed, perhaps even desperate, Darwin sought to mend the hole with Lamarck's theory, but the doctrine of acquired characteristics was a poor patch, containing within it as it did the notion that use or disuse could somehow change the sex cells in the gonads. For a time, from 1890 to about 1930, the principle of natural selection lost its influence among scientists, even as it was becoming understood and assimilated by the public. The saltationist interpretation, the view that species could be established by gross mutation, captured the imagination, and natural selection, although not totally disregarded, became eclipsed by this more dramatic explanation of change. Yet the developing science of genetics, the labors of much research, and the application of population statistics all conjoined to suggest that Darwin wrote better than he knew. Mayr (1959) suggested, in the opening address of the symposium honoring the 100th anniversary of the publication of *On the Origin of Species,* that we have completed a full circle and that we are closer to Darwin's original concepts now than we have been at any time during the intervening periods.

Evolutionary theory has become sophisticated. It is now realized that mutations are typically minute and, as such, may account for the variation that Darwin talked about, that many such genetic changes must occur before a noticeable alteration takes place within a species, and that the adaptive contribution of any genetic change must be evaluated in terms of its integration into the total synchrony of a complex living system. A genetic change can be fortunate or foul depending on its accompanying attributes, and one that could be beneficial in one genetic context might be lethal in another. Evolution is a game of chance in which every card that is dealt changes the value of all the other cards, and where selection pressures may momentarily shift the set of rules. Yet natural selection serves as the ultimate judge of which hand wins, and so sets the course that all evolution takes.

Update of Darwin's Theory

Beginning about 30 years ago and gathering momentum there have been remarkable changes in evolutionary thinking. It is now realized that, from an evolutionary standpoint, the sole function of the individual is to transmit his or her genetic material to the next generation. Any variation, whether behavioral or morphological, which increases the likelihood that this transmission will take place will tend to be incorporated into the species.

Postulate of Evolution. *Any inheritable variation that increases the probability of its own sexual (or asexual) transmission (replication) relative to alternative variants will tend to increase in frequency in the population.*

The following list includes most, if not all, the variables that participate in the evolutionary process. These categories are not mutually exclusive. There are always varying degrees of overlap and interaction. In general, the following analysis represents the contemporary view of evolution.

Individual Survival Variables. Sexual activity (whatever its nature) requires a minimal level of maturation. This means that any factor that increases the chance that the individual will survive until sexual maturation is achieved will tend to be incorporated. There are many such factors, all of which interact, but in general they relate to the relative efficiency with which the individual can retain, regain the equilibrium necessary for life to continue, or both. These include ease in finding food and avoiding predators, in coping with environmental hazards, effectiveness in peer group competition and fighting off the hazards of disease, and so forth.

Individual Attractiveness Variables. It typically takes two to reproduce (unless the organism is unisexual) and an individual who cannot attract a mate makes no evolutionary contribution. There are many sexual attractors, as Darwin documented at length; (e.g., sounds, smells, movements, plumage or coloration, big buttocks, large breasts, etc.) that increase the chance of sexual activity and, thus, tend to be incorporated.

Sexual Facilitator Variables. These are obviously related to the previously mentioned variables, but no matter how attractive, strong, or available the individual is, he or she makes no evolutionary contribution if something interferes with sexual activity. Conversely, any variation that increases the probability of consummation will tend to be incorporated. This explains why organisms have such an impulse to copulate, and why orgasms are so pleasurable. Sexual activity is the ultimate activity in evolution. Any variation that makes it a little more likely (i.e., more pleasure, a stronger drive, etc.), which facilitates the coalescence of the sperm and ovum (the ultimate evolutionary event), will tend to be selected.

Sexual Mutuality Variables. This relates to both of the previously mentioned variables, but in a slightly different vein. It also relates to something called outlocking. Since sexual activity is so important, any variation (as repeatedly mentioned) that increases its probability will tend to be incorporated. This means that certain trends, or tangents— call them what we will— toward mutuality in behavior and structure would be predicted to emerge between males and females of a given species. Thus, the morphological and behavioral synchrony between penis and vagina, lordostic posturings and movements, and the often complex and stylized mating dances and rituals observed in many creatures, particularly birds. Once such a *tangent into mutuality* has developed, any variation that lessens an individual's capacity to participate will tend to lock out his or her genetic material. Conversely, any variation that increases an individual's capacity to participate will tend to be incorporated and to further abet the tangent. Of course, all such tangents, as well as any attribute for that matter, must abide in dynamic conjunction with many other attributes, so there is always a functional limit imposed upon their extension. For instance, a courtship ritual can continue in its complication only up to a point, the point where fatigue, energy depletion, predator arousal, or hazard, begins to chop if off at the top.

Territory Variables. Territoriality is critical, not only for survival (i.e., getting food for both individual and offspring), but also for finding and protecting a mate, and we find many examples of territorial behavior (i.e., singing, fighting, bluffing, pheromone marking, etc.) in many species.

Kin Selection Variables. Nothing underscores the relative unimportance of individual survival, as against germ plasm survival, more than these variables. All altruistic behaviors apparently fall here, and it is here that we finally (since Hamilton, 1964) understand how such altruism came into being and became incorporated into a species. If a given individual's behavior (such as feigning death, disability, the indulgence in distraction antics, or even literal self-sacrifice) increases the tendency for the critical genetic material to survive and to be transmitted (thus the kinship factor) that behavior will be incorporated in the species (read kinship group). This explains not only the present functionality of such behaviors, but also how they developed in the species in the first place. Any variation that increases the survivability of the kinship group can thus be incorporated, even though that variation may decrease the probability of individual survival (see Mueller & Feldman, 1987).

There are other variables that may enter into the natural selection process, but these are the major ones. It should also be remembered that a number of circumstances or conditions may affect the relative contribution of any one of the previously-listed variables. For instance, a species that is less vulnerable to predation or environmental hazards might be expected to develop more elaborate courtship displays or sexual signals because, with survival variables less operative, these more subtle influencers on sexual activity would have a greater relative influence.

Empirical Support for Darwin's Theory

Although most contemporary thinkers in the life sciences would insist that even a casual appraisal of the various forms of life provides ample evidence for the generality of Darwin's theory, it seems strange that there is such a paucity of direct empirical support, and only during the past few years have there been actual experiments designed to test specific predictions. The most famous and by far the most convincing is that of Kettlewell (1959) on *industrial melanism*. As is well known, a certain species of moth (Biston betularia) grew rapidly darker during a period in which the trees of England were being blackened by industrial soot. Kettlewell surmised that this change was due to a natural selection occasioned by the fact that birds that normally feed on this moth could not see the darker members of the species. Kettlewell, in a clear demonstration of Batesian mimicry, documented the recovery rates of both light and dark moths released in industrial and nonindustrial areas and found that the proportion of moths reclaimed after various elapsed times supported the notion of directional selection. Kettlewell photographed birds picking conspicuous individuals off trees, while protectively colored moths remained unharmed nearby. He also demonstrated the application of selective principles by including photographs in which dark moths meld into darker background, while light moths become almost invisible in the lighter context. Cook, Mani, and Varley (1986) strikingly extended Kettlewell's work by showing that industrial melanism decreases in predictable frequency as pollution controls become effective.

Another example of natural selection in process reported by Johnston and Selander (1964) involved the ecogeographical adaptation of house sparrows. They found that there was a marked inherited body weight increase in birds living in colder climates and higher northern latitudes. There were also changes in color and tarsus length. All such changes took place within 100 generations after the sparrows were imported into North America from England in 1852.

The fourth piece of evidence was recorded by Fenner and Sobey, as reported in Hamilton (1967). In 1859, an auspicious date for evolutionary theory, a few rabbits were brought to Australia. By 1950 the population had increased to hundreds of millions, so that rabbit inundation seemed imminent. To combat this rapidly expanding population explosion, a virus peculiarly effective against rabbits was introduced. Yet as the years passed the proportion of rabbits that would succumb to the virus kept decreasing. Apparently rabbits with a slightly higher resistance to the virus were surviving and transmitting this attribute to their offspring. High death rates, up to 99.8% in the beginning, followed by slower rates as time passed, provided a circumstance for genetic selection of virus-resistant rabbits, and this adaptation took place in approximately 10 years.

Recently, in an area that is both appropriate and prophetic, an avenue of research has been undertaken that bears directly on the applicability of Darwin's theory. In the Galapagos Islands, Darwin's finches are being used as subjects for a series of observations and experiments designed to test Lack's (1944) theory of speciation. It should be emphasized that such experiments do not question (nor does Lack) the validity of natural selection as the arbiter in species emergence. Rather, they examine the mechanisms whereby new species do emerge, within the structure of Darwin's explanation. Since 1983 Peter and Rosemary Grant, with their students and colleagues, have been studying Darwin's finches focusing on the adaptive shifts in beak and body size as a reciprocal reaction to climatic shifts.

There are 13 species of finch on the Galapagos, each with different equipment (primarily the beak) reflecting the species' particular method of getting food: cactus dwellers, tool users, leaf eaters, iguana groomers, blood drinkers, and vegetable strippers, and some specialization even within these categories.

Why decide that there are 13 different species? Not only because their morphology (beaks and bodies) and behavior (eating habits) differ but, and this is the primary criterion of speciation, because they do not interbreed.

The Grants' research (with time off for information processing and writing) has continued unabated since 1973. They have caught, measured, and branded thousands of birds, through many generations. Their work is one of the finest achievements in the annals of evolutionary biology and the book *The Beak of the Finch* by Weiner (1995), a well-deserved winner of the Pulitzer Prize, is an eminently suitable and timely presentation of this work.

Among the Grants' findings: (a) evolution is always taking place, (b) it (at least in the short term) is not directional or linear but a series of oscillations (seesaws) in response to climatic shifts, (c) great change occurs during periods of maximal stress, with little change occurring during periods of plenty, giving direct supporting evidence for Gould's (1994) principle of punctuated equilibrium, and (d) speciation occurs when divergence is accelerated by separation (genetic drift) and by hybridization, which increases competition between slightly different groups.

Genetic drift is not a purely random phenomenon, particularly where migration into new territories is involved. Kinship divergence when coupled with what Mayr (1982) called the *founder effect* probably enters into variations in gene frequencies between groups and may provide an initial push toward speciation. The founder effect represents the fact that when a number of individuals migrate they automatically have gene frequencies that vary from those of the larger population. This is the case because a few individuals cannot carry with them the full genetic diversity of the larger group. Such founders thus are different even before the selection pressures of the new environment have had a chance to operate. The founder effect is enhanced by the fact that migrating individuals tend to be interrelated (the kinship factor), which further reduces the degree to which the founders represent the gene frequencies of the population. That both kinship divergence and the founder effect were operative in the development of the several species of finches on the Galapagos, and in the genetic differences between human groups, seems probable. In his defense of the founder effect as an important determinant of human genetic diversity, Diamond stated: "Virtually every isolated human population that has been examined turns out to have its own 'private' genes—ones that are common in that population but rare elsewhere" (1988, pp. 12–13).

In the last few years, inspired by the Grants, a number of experimenters worked to demonstrate evolution taking place during relatively brief time frames. Research on what has been termed the *Lincoln effect* (see Weiner, 1995) demonstrates the subtle influence of single variables on the shifting characteristics of a particular attribute. Somebody once asked Lincoln how long a man's legs needed to be. "Long enough to reach the ground" was, of course, his answer. There is a bug, the soapberry bug, that gains its sustenance through a long (probocis) tube. Carroll and Boyd (1992) studied the beak of the soapberry bug in various regions throughout the United States, with a diligence reminiscent of the Grants study of the beak of the finch in the Galapagos. Where the soapberry bug eats from its traditional source, (a relationship lasting thousands of generations), the soapberry tree, its beak averages 6 mms in length, just right to reach the seeds. Yet where soapberry bugs have switched to the heartseed vine (which has occurred since 1970), which has seeds that average about 9 mms from the outer peel, the beak has increased in length to just under 8 mms. Conversely, where soapberry bugs have switched from the seeds of the balloon vine, where seeds are deeper in the interior, to the seeds of the golden rain tree, which has seeds that are near the surface, the beaks have shortened correspondingly. This research by Carroll and Boyd demonstrated the rapid adaptive response (within a few hundred generations) of an attribute to a particular environmental shift. It also demonstrates the law of least complexity that any attribute will only be as complex as the adaptive circumstances require; that is that excess baggage is lethal over time, since complexity without corresponding adaptive efficiency increases the likelihood of breakdown. (Incidentally, this law obtains throughout all systems, mechanical, electrical, biological, and psychological.)

Sometimes the most impressive research on the origin and function of a particular characteristic comes from a careful examination of a single attribute in just a few individuals. Benkman and Lindholm (1991) studied a group of finches whose beaks were crossed as though by some genetic anomaly. However, they believed that the crossing was functional and decided to find out. They captured a number of finches and trimmed out the crossing. Such birds could still efficiently eat the seeds from open tree cones but they could no longer open closed cones. Crossed bills were cone openers. It was a very functional attribute; as the crossing grew back, the birds were able to open closed cones with their original efficiency.

Darwin's finches in the Galapagos have been long studied (primarily by the Grants) as pristine examples of speciation. Ironically, it is also in finches where, rather than speciating, learning to adapt to different opportunities is dramatically demonstrated. Werner and Sherry (1987) intensively studied the finches on Cocos Island, a single isolated spot of land 500 miles from Costa Rica. Much to their astonishment and mine, they found that, although there was but one single species of minimally variable finches on the island, many of the ecological niches were filled. Different groups did different things. Some were branch probers, some were leaf scanners, some were nectar sippers, etc., and each finch practiced the same trade as its parents and retained its particular method of foraging throughout its life. Rather than speciating (behavioral, morphological, and sexual separation), these birds were learning their different survival techniques, apparently through a very sensitive kind of imprinting (observation and action learning). The researchers, on many different occasions, observed the juveniles following adults around, imitating their behavior.

By far the most dramatic example of evolution in process, and man's greatest experiment, although unpremeditated, on the adaptive plasticity of DNA comes from a relentless war that is presently being waged on this planet. It is man's war against the countless legions of small things. It is being fought primarily against insects, viruses, and bacteria. As in all wars it involves the repeated introduction of novel weapons of offense and the activation of reciprocal countermeasures of defense. There have been many weapons of offense but only one of defense: rapid adaptive change. A new weapon (pesticide or antibiotic) is deployed against a particular invader; the invader is decimated, almost wiped out. Almost but not quite. There are always a few left over, and these few, that is, those with some genetic anomaly that makes them resistant to the weapon, expand to fill the ecological space, and the war goes on and on. New weapons, new variants.

Hopefully, the ingenuity of man can keep ahead of the seemingly inexhaustible flexibility of the genome, but the outcome is in doubt. Prototypical of this war is the prolonged seemingly endless struggle against AIDS—where weapon after weapon, the natural legions of antibodies plus the adjunct battalions of drugs, sometimes singly, but often in concerted array, have been countered by the adaptive flexibility of the virus. Yet this is only one example. The Heliothis moth, which attacks the cotton plant, has become almost totally resistant to pyrethroids within 60 years; sheep ticks have developed effective resistance to HCH dieldin in just over 30 years; flies have evolved a 100-fold resistance to a presumably resistance-proof poison derived from their own hormones within five years, and the list goes on.

How do insects evolve defenses so rapidly? Weiner (1995) listed the ways:

1. They can simply avoid the pesticide, that is, evolve mechanisms for staying away from it.
2. They can evolve barriers to keep the poison out.
3. They may evolve an antidote to the poison.
4. They may alter their achilles heel through shrinkage, replacement, or by getting along without it.

Whatever the extermination technique employed, the insect evolves a counter maneuver, and this remarkable demonstration of evolution in action, the cycle of ploy, and counter-ploy, continues.

The same story at an even more intimate level is occurring in our efforts to control disease. Tuberculosis, bacillary dysentery, streptococcus, salmonella, and gonorrhea, to mention but a few, are becoming more and more resistant, in some cases almost completely.

All of these examples would have been predicted by Darwin. Where selection pressures are imposed, the pace of evolution quickens. Variation itself is the basic arbiter of change and provides the ultimate defenses again the application of weapons of all kinds. Are humans doomed to lose this war? Not necessarily; it simply may go on and on, an eternal rhythm of measure and countermeasure. Or it may be that in certain cases the variability potential of an enemy may be exceeded. Extinction does occur.

An informal and unpremeditated demonstration of Darwin's theory has been going on for a long time. Selective breeding or what may be termed *unnatural selection*, provides indirect support, as Darwin himself extensively documented (1897). Horses have been bred for centuries, some for strength, beauty, and speed. Dogs have been bred for many purposes: the greyhound for speed, hounds for hunting, German shepherds for police and seeing eye dogs—their uses are almost as numerous as the separate breeds (Scott & Fuller,

1965). Near the turn of the century, a special type of pigeon, the trumpeter pigeon, was bred for an unusual type of cooing (Levi, 1951). Cockerels have been selectively bred for variations in the frequency and vigor of aggressive and sexual behavior (Wood-Gush, 1960). The roller canary has been bred for so many generations for a better song that now its song is completely different from that of wild canaries (Marler, 1959). Everyone is familiar with the different types of cattle we now have, some like the Jersey and Guernsey for milk, others like the Angus and Hereford for beef.

Animals have also been bred selectively for experimental reasons. Manning (1961), working with the fruit fly (Drosophila melanogaster), within seven generations of breeding for fast and slow mating speed, developed one line with a mean mating latency of three minutes, while the other latency was 80 minutes. Bruell (1962), working with mice, bred them for speed in wheel running. McClearn (1959) bred mice for readiness to explore in an open field situation. Tryon (1942) selectively bred rats to learn to run a maze faster than controls. Working with turkeys, as well as mice, Hall (1951) reported the breeding of the quality of wildness in these animals. Anywhere one turns, Darwin's theory can be put to man's purposes. No one needs to be reminded of Hitler's eugenic plan for an Aryan superrace.

PROBLEMS AND PUZZLES

This section deals with a number of issues that are arousing interest and argument at the present time. I treat these issues only briefly here, in a few pages, but try to make a clear statement of each and, where there is controversy, state my own position where I have one.

"It is almost unfair," one of my most perceptive colleagues (Hillix) once complained about Darwin "... by doing everything, he left the rest of us but very little to accomplish, except filling in the trifling supporting rafters for his major structure." There is much truth in this complaint, I believe, but this has not prevented thinkers in evolutionary theory from occasionally staking a small claim, either through rebelling against particular aspects or making a refinement. Of course, bringing psychology into the evolutionary synthesis is no small task, as the remainder of this volume amply demonstrates. Yet here let me briefly present some areas of contention and refinement that are being dealt with by contemporary thinkers.

Group Selection. A fundamental adaptive outcome of evolution (examined at length in chaps. 6 and 7) is the capacity for learning, which is most developed in the human being. This capacity was both the outcome and the occasion for more refined stratagems of group living, which accumulated over time to be represented by the term culture.

With culture, the adaptive situation for humans is dramatically altered. We get our food at supermarkets, travel in machines, live in air-conditioned environments, solve our problems with computers, dull our pain with morphine, watch movies, listen to Mozart, and fight our wars with smart bombs. In short, for many of us, the selection pressures that brought us into being are essentially nullified. Culture serves to accommodate the adaptations already achieved and to radically alter the conditions that might influence the development of new ones.

Culture (i.e., the accumulated influences of the past) impacts on our efforts to understand the evolutionary (biological–innate) roots of our behavior. To understand the evolutionary origins and functions of action we must peel back the obscuring tangles of cultural overlay in order to expose the biological underpinnings.

Culture (particularly as it encompasses the outcomes of group living) has from time to time, again quite recently, entered into our assessment of the very foundations of evolutionary theory under the rubric group selection.

The central notion in group selection theory is that (see Wilson & Sober, 1994 for an extended review of the issues involved) some characteristic (variously called vehicle, interactor) arising from living in groups can become (indeed often does become) a major factor in the evolution of organisms, particularly of human beings.

The proponents of group selection typically reject the view that the gene is a necessary and foundational concept in evolutionary theory. According to the gene-centered view (I accept this position) evolution takes place when certain of the attributes that result from genetic blueprints (genes) increase the chances that replication of that blueprint will take place. Selection pressures perpetually test the functionality of such attributes. The existence of groups certainly has influenced, and continues to influence, selection pressures. This is particularly true of human groups where culture in all of its manifestations alters the selection pressures and indeed, in many cases, negates the operation of at least some of the selection pressures that brought humans into being. Culture does profoundly influence selection pressures but, according to gene-centered theorists, selection, whenever and wherever it occurs, inevitably and always, takes place at the individual level. To say that it occurs at the group level is to obscure the replication process—genetic duplication—at the only level and in the only manner in which it can occur. This is the case because the variability that accounts for differences in the attributes that are being perpetually tested by selection pressures are carried by genes in individuals. It is each particular version of an attribute—one version per individual—that is either selected in or out.

It is in the gene that the operative mechanism of evolution resides, both in terms of the variability that underlies the changes which occur and in the method for the replication and inclusion of such changes.

The higher-level outcomes that the group selectionists emphasize may alter the course of evolution after their appearance, as culture has certainly done, but they are not the central arbiters of evolution. They are a component of the adaptive matrix (which always includes the organism's capacities and the environmental circumstances) within which evolution takes place.

The notion of group selection, however defined, exposes us to an endless welter of complications and to an ever-expanding burden of terms to represent them. What is the vehicle, interactor, conceptual unit? Call them what you will. They can be invented endlessly. Group selection, as an addition to evolutionary theory, is not a trivial flaw. The issue here is the relative simplicity of the conceptual structure involved. Science strives for simplicity. All other things being equal, it accepts the position that explains the most with the least. The macro-evolutionists and group selectionists are not only wrong, but also they introduce an intractable and unnecessary complication.

Let me emphasize that this is still an issue in much debate. Gould, one of the preeminent thinkers in the field and in my view its most articulate writer, is moving away from an individual (gene-centered) position. "I believe that the most portentous and far-ranging reform of Darwinism in our generation has been the growing attempt to reconstruct the theory of natural selection as a more general process, working simultaneously at many levels of a genealogical hierarchy" (Gould, 1994, p. 6769).

Speciation. How did we get all the species on the earth? Darwin's general position (see Weiner, 1995 for an excellent review of Darwin's thinking here) was that different species resulted from the sculpting action of the selection pressures of myriad environments. Each environment was like an evolutionary island where unique survival demands and isolation (genetic drift) would, over time, work their magic. Then Darwin visualized some of the specifics of this process. He realized, in a burst of insight, what would happen if two groups were just beginning to separate. Being similar in behavior and requirements, they would be thrown into extreme competition. It is this competition between similar groups that serves as a fundamental selection pressure that drives them apart. Thus competition between like groups, in his view, is the basic arbiter of cleavage. All this was observed by the Grants in the Galapagos and was dramatically demonstrated when hybrids, perhaps separated on different islands (genetic drift), came together again on the same island; intense competition, rapid change, and separation took place.

Reproductive separation is the defining condition of speciation, but the details of this happening are still far from clear. Separation may occur at many levels and in many complex ways, as any casual observation (this is particularly true in birds) will demonstrate. In order to participate in the ultimate evolutionary event, replication, the individual must look just right (plumage, etc.), act just right (mating dances, etc.), smell just right (pheromones), be in the right place (territory), and the list goes on. A tendency not to measure up in any of these factors, seemingly peripheral to the main event, means that the deficient individual and his or her genetic material is pushed away, and when this happens on a mass scale over long periods speciation results. Through mutation or genetic drift the DNA coding for gamete compatibility becomes sufficiently altered to prevent conception from occurring even if sexual activity does take place. In most cases, barriers preventing sexual intercourse probably arise long before gamete incompatibility occurs.

At the present time divergence through competition, accelerated by hybridization and augmented with genetic drift, appears to be the most widely accepted explanation of speciation. Yet the actual manner in which genetic isolation is finally achieved varies from species to species (see Butlin & Tregenza, 1997).

Exaptation. Another effort to escape the bonds of Darwin's encompassing insight comes from Gould (Gould, & Vrba, 1982) under the term exaptation, a new name for a process that has long been accepted, at least in part, by many others including Darwin. It is the observation that an attribute that became established because of one set of selective circumstances can be coopted for a different function. Feathers, which initially functioned as protective covering, shifted in function to become an integral part of flight. Very few people in evolutionary biology believe that such exaptations (Gould's term) are exceptions to Darwin's explanation. It is when the term is extended, and Gould extended it to a remarkable degree, that the inclusiveness of Darwin's theory is placed into question. The exception occurs, Gould believed, when certain attributes that initially had no adaptive significance, and which developed as a function of fortuitous circumstance, begin to play (at times remarkably pervasive) functional roles. He offered as an explanatory metaphor the spandrels of St. Marcos (Gould & Lewontin, 1979).

In the dome of this church, a number of arched beams meeting at the apex to top off the structure are separated by curved pie-shaped segments that fill in the spaces between the beams. These segments, or spandrels, serve no supporting function in the architecture of the church. They are, Gould suggested, simply filler material without load-bearing function. But these spandrels are covered with art work of great beauty. They have become intrinsic to an artistic contribution of great merit, but one which has no real architectural function. It is at this point that the metaphor becomes unclear. Is the exaptation factor analogous to the spandrel per se, or to the art work on it? A minor point, perhaps, except for compulsive hair splitters. I would not be splitting except for the fact that the spandrel notion has received such usage.

If the exaptation factor is due to the spandrel per se, the assumption of nonfunctionality seems inappropriate. If one defines the function of a building as one of protection from the elements, spandrels make a critical contribution. If such a building is defined as a monument to the glory of God, then the artwork makes an important contribution. It is only if one defines the functions of the components of the church in strict architectural terms that spandrels, or the art upon them, can be viewed as unrelated happenings that can be coopted for some other purpose. My point here is simply that the metaphor is misleading, not that the term exaptation is inappropriate.

If Gould restricted the term exaptation to the cooption of attributes from one function to another, there is no disagreement. Yet he expanded the usage to include attributes that are seen as emergents, completely unrelated to any prior adaptive function. As a final expression of his desire to escape the restrictive confines of Darwinian microevolution, he excludes the human brain. He even gave a nod of appreciation to Wallace, who was also unable to accept such an inclusion. What I have called bonus factors Gould saw as the outcome of macroevolution, that is attributes not reducible to elementary genetic (gene) blueprints, which can become functional at a completely different level.

No evolutionist would deny that a combination of human intellect and the accumulated derivatives of technology and culture have resulted in levels of accomplishment that are the outcomes of attributes that evolved to solve simple problems, and they would certainly not deny that the changes wrought have profoundly altered the circumstances of man's evolution and the selection pressures that will be influential as (or if) he continues to evolve. Certainly western man has achieved all of these bonuses in technology and culture but their achievement has been the outcome of intellectual capacities (learning, foresight, integration, etc.) which are explicable at the microevolutionary (gene) level. Not only is optimal simplicity maintained, but also resort to higher and more complex levels of explanation is unnecessary. The same attributes required to accurately hurl a rock at a rabbit may allow us to hurl a rocket to the moon, but bonus factors are also present in the washing of sweet potatoes by the Japanese macaque and the cracking of nuts by chimpanzees. Indeed with the capacity for learning (and related capacities) a multitude of novel activities and artifacts happened and continues to happen, up and down the phylogenetic scale.

This suggests that present adaptive attributes (i.e., the human brain, human culture, science, math, and music) become explicable at a microevolutionary level. The occurrence of bonus phenomena should not be surprising to anyone who has an appreciation of the capacity of learning and intellect. Such bonus factors are simply demonstrations of the remarkable flexibility implicit in these evolved capacities which, after all, developed because increased flexibility of response to complex situations conferred survival value. Let us retain the term exaptations for those attributes that have made major shifts in function and have, as such, made a significant contribution to evolution.

There is an interesting parallel, if not overlap, between Gould's extension of the term exaptation and group selection, as treated in the previous section. In both cases efforts are made to: (a) make a contribution in the small space for creative maneuver left within the inclusive confines of Darwin's contribution, and (b) escape from (as they see it) the narrow and distorting limits of micro- (gene-centered) theory. The greatest objection to these extensions is a lack of parsimony. In science, to reiterate a well-worn dictum, we strive to explain the most with the least. Both group selection and the expanded version of exaptation give licence to an ever-burgeoning complication of levels and concepts. As such they are unacceptable extrapolations (as well as being wrong, I believe) particularly since Darwinian evolutionary theory (i.e., microevolution based upon adaptive change to selection pressures) so beautifully subsumes all attributes, regardless of level or complexity.

Junk DNA. What a puzzle it is! The percentage of it varies from species to species (from puffer fish, which have almost none, to salamanders where the junk component accounts for more than 99% of the genome) with an average of around 90%. The puzzle is this, since such DNA is so prevalent, and often so lethal, sometimes jumping into other vital DNA sequences to occasion mutations many of which are harmful (accounting for such diseases as Huntington's disease and fragile X syndrome), why is it still around? Indeed, not only is it still present, it comprises most of the genome of most species!

One of the oldest canons in evolutionary theory holds that complication without adaptive compensation is lethal. Junk DNA not only increases complexity, but also makes a deleterious contribution by jumping into, that is inserting itself into functional DNA sequences. The conclusion? It must be making (or at least once made) an adaptive contribution that compensates for these negatives. Yet what? There has been lots of speculation and many hypotheses.

1. It is simply a rider or parasite which makes no contribution.
2. It provides spacing in the folding of other DNA segments.
3. It provides a source of spare parts that can be used for genome repair.

4. It constitutes a hidden language that is responsible for the timing and integration of cell development.
5. It serves as buffer zones between genes.

And the list goes on. A number of these hypotheses may (singly or in conjunction) turn out to be correct but the major contribution of junk DNA apparently comes at a different level, a level that may have little to do with the immediate activity of the genetic machinery but with its literal existence in the first place.

Variation is a fundamental requirement for evolution to occur. The greater the variability, the more rapidly adaptation to selection pressures can take place. So called junk DNA may serve as a variability reservoir, which ensures that an evolving life form will tend to survive. That is, a greater capacity for mutation would abet the survival of an evolving life form and also the tendency for this capacity to be incorporated.

The problem of selection for increased variability potential is related to sex. Here the issue becomes, why is there sex at all? Asexual creatures have done very well. They were the first forms of life on earth, and they still exist in limitless numbers. One may wonder why sex (which requires that two creatures meet at the same time and place, indulge in behavior that is fraught with predator vulnerability, excess energy expenditure, peer competition, etc.), ever started happening and why, once it happened, it has not been selected out.

The answer of course is that increased variability potential allowed greater facility in adapting to different environmental conditions. The appearance of sex provided a quantum increase in variability, which contributed the adaptive plasticity required for the development of complex organisms, from salamanders to humans. In similar fashion, I believe, the increased variability potential provided by junk DNA accounts not only for its occurrence but also for its retention in nearly all species. The story of junk DNA is still unfolding. It may play a single fundamental role or a number in concurrence (evolution is always opportunistic) but the canon of functionality demands that it make a contribution.

That junk DNA contributes to the variability of the genome is supported by recent findings by Lambowitz (1998) and his colleagues. Using common bakers' yeast (a much used and ever convenient substance) they discovered that interons (junk DNA) can insert themselves at various points along a DNA strand, transferring genetic information as they do so. Such interons seem to function as separate organisms (parasites) that surreptitiously intrude and alter the genetic code. It is known that all body cells include vital organelles with their own genetic material. The Lambowitz research (which is still in process) suggested that the gene also may be comprised of symbionts.

In summary, it may be that genetic blueprints with a greater potential for adaptive changes would tend to be replicated. The (junk DNA) interons may provide (and continue to provide) this potentiality for variability and for more flexible adaptation. Indeed, interons appear to work as a mutation reservoir, intruding randomly into the operative genome (as McClintock demonstrated with corn) to occasion mutations, most of which are, as are all mutations, detrimental. Yet an occasional intrusion, depending on the total adaptation matrix of biological and environmental variables, is positive, and this advantage has insured both the survival of the individuals involved and the variability reservoir that the junk DNA constitutes.

Learning (Cultural) Overlay. A perpetual and inevitable complication of the search for the evolutionary origin and function of certain human attributes is that of learning (cultural) overlay. Not only is the capacity for learning itself an evolutionary derivative, but also the sensitivity, subject matter, and timing of its expression are heavily influenced by evolved predispositions. In order to get at the adaptive contribution that a given attribute makes (or once made), the cultural overlay must be peeled away, sometimes layer by layer.

Most contemporary research by evolutionary psychologists involves attributes that are obscured by cultural (learning) overlay. A look at laughter here illuminates the difficulties inherent in all such research. There appears to be little doubt that laughter has an evolutionary origin. It is universal in humans. It appears early in infants, and it seems to be present in related species such as chimpanzees, gorillas, orangutans, and macaques. It can be produced by brain stimulation. It has been observed in people who are deaf, dumb, and blind.

There have been many studies on the circumstances that produce it. Incongruity or upended expectancies appear to produce laughter in all ages and societies (Haig, 1988; Rothbart, 1977; Shultz, 1977). Upended expectancy appears in most jokes, puns, peek-a-boo, riddles, pratfalls, pie in the face, and the list goes on.

Now we come to the difficult part. Where volitional control is a factor, even to some slight degree, learning or cultural overlay confounds the picture. For instance, there is certainly a biological basis for such activity as belches and farts, but they are soon under social control. Farting is generally curtailed, but may become an expression of hostility or contempt. Belching is unacceptable in western society, but is often seen as an expression of appreciation in China. The same cultural inclusion is present in laughter, only more so. Laughter may be an indication that something is funny, but it may also be the result of a remarkable diversity of circumstances. People may laugh to register scorn, to hide embarrassment, to ingratiate ones self to a superior, to give the appearance of conviviality, and the list goes on and on. Indeed, in human society laughter is an integral part of language, with nuances and subtleties far removed from its early origin and function.

What were these origins? How did laughter function in the early days when humans existed in small groups, living in caves, huddled over the primordial fire? We can never know for sure, but we can make some inferences.

In human evolution, group cooperation was a major (perhaps the major) survival gambit. The requirements of group action underlie the emergence and refinements of many human attributes, and it was central, I believe, in the emergence and function of laughter. Laughter, which in contemporary society contributes subtle nuances of communication, had this function in emerging human nature, and, perhaps even more important, it also served as a vital and perpetually overlapping bonding mechanism. In order to survive as a group we had to communicate, but we also had to get along. There are many positive factors in group living in confined spaces: protection, warmth, and companionship. Yet there are many negative ones: crowding, irritating habits, competition for sex and food, bad smells, and the hazards of hunger, danger, and disease. In such a context laughter made us feel better. It was an "I'm o.k., you're o.k." statement that helped bond us into a unity of care and concern. Echoes of this still reside in our nature in the contagiousness of laughter, in its reaction to incongruity, in its capacity to make us feel better, and in children's (and some adults') hilarity at farts.

I treat laughter at some length here at the outset to emphasize a complication and paradox that is intrinsic in any effort to determine the origin and function of human attributes. Such search is inevitably complicated by cultural overlay. Let me emphasize: The capacity for learning, which itself is one of the most important of all adaptive attributes, often overlaps and obscures many biologically based (evolutionary, innate, inborn, and inherited) aspects of human nature.

STATEMENT OF PURPOSE

In this volume the coalescing, often overlapping, trends represented in evolutionary and cognitive theory and research provide the basis for the development of a comprehensive theory of how living organisms work as lawful entities. Evolutionary theory and supporting data serve as a source for the derivation of explanatory principles and as a structure for subsuming the many facts and principles, regardless of data level, of psychology. Although all levels of psychological data (the experiential, behavioral, and physiological) are considered, the research and theory from the experiential (primarily as represented by the cognitive psychologists) are given priority. Emphasis on the cognitive approach is justified, I believe, because workers in this area have been primarily concerned with the manner in which each of the stages of information processing (i.e., input, encoding, retrieval, and reaction), separately and in conjunction, contribute to the efficient operation of the individual. Evolutionary theory as it relates to psychology has been, and continues to be, concerned with the very same issue. Organisms have survived and evolved to the degree that they have developed efficient information-processing capacities, from the simplest forms of life to the most complex.

Combining evolutionary and cognitive perspectives provide a conceptual framework for integrating much seemingly disparate information. This, after all, is the function of theory: to render complexity into simplicity, to subsume the seemingly incompatible and disconnected into an integrative structure that will allow understanding and insight where previously there was a clutter of particulars. (See Goodson & Morgan, 1976b for an extended discussion of the nature and function of theory in science.)

As we bring the overlapping lenses of evolution and cognition to bear on the facts, speculations and hypotheses that relate to living systems, it is my job to convert this complexity into a simplifying and unifying conceptual structure. In this effort, I often make a formal statement about an area of investigation, usually in brief form, with special centering on the page. The reader should regard these statements as hypotheses, as indeed they are. Yet they are studied hypotheses, derived from examining the implications of evolutionary theory and the data available from psychological research. All such statements, taken together, provide the girders of the theory, tied inferentially to an evolution–cognition theoretical foundation on the one side, and empirically to data from both areas on the other.

The theory that emerges in the following chapters is not just a conceptual system locked within a sterile postulational structure. It is a statement about the way living systems actually process information so that appropriate reactions can occur. As such, it is either correct or it is incorrect, true or false. Is it still provisional? Yes, it is. It is designed to be responsive to new data and criticisms. Yet the coalescing, overlapping, mutually supporting contributions from the evolutionary and cognitive perspectives, when combined with the integrative efforts of the writer and the vigilant critical powers of the reader, powerfully suggest, I believe, that this theory provides an outline of the manner in which the information-processing capacities of living systems actually work.

2 Foundations

THE EMERGENCE OF LIFE

We do not know when or how life first appeared on the earth, but we do know that it happened a very long time ago. Samples taken from the chert beds of Western Australia and from the Swaziland system of South Africa all contain microfossils that are between three and four billion years old. Such microfossils are widely distributed throughout the oldest rock strata, regardless of where they are located (see Stewart, 1983). So apparently the occurrence of life was not an unlikely event, as some have supposed (Wald, 1959), but a common outcome of the conditions that were present only a billion years after the formation of the earth. The belief (held prior to Pasteur) that a mix of water and vegetation was all that was required for the spontaneous generation of life has passed, but life must have been generated somehow at some time.

Are new examples of life still emerging? Perhaps, but the search for its emergence has thus far been fruitless. Though many efforts have been made to find life coming into being, there has been no success. Of course, it is possible that the process is too obscure to be captured by our blunt techniques, or perhaps the existence of life provided conditions inimical to a repetition of the event. The stamp of hunger with which nature marks its creatures may ensure that all the primitive prototypes at the threshold, even as they struggle toward life, are devoured or destroyed by those already on the other side. It is even possible that a flight of meteorites in those early times may have provided the seed from which all life as we know it has spread. More likely, and this is generally agreed upon today, (Folsome, 1979; Oparin, 1953, 1968) the appearance of life, its evolution, and its wide dispersion have helped change the environmental conditions necessary for new life to emerge. Three billion years ago there was very little oxygen on the earth. Its almost universal presence now not only demonstrates the way life can change its own environment, but also shows how life might create conditions that make its new appearance impossible.

Although we do not know the details of the environment in which life first emerged, we can make some general inferences that seem indisputable. On the earth, then as now, many different energies were in perpetual display. Vast ranges of the electromagnetic continuum came ceaselessly from the sun, molecules permeated the air and water, and the molecules were in continuing flux. Anything that moved or vibrated produced pressure fronts that fanned out on every side. The earliest examples of life (like all examples of life since that time) were surrounded by this pressing, vibrating, radiating, chemically reacting manifold. Life arose in such a context and continued to exist only because some kind of accommodation to these shifting energies was possible.

One universal characteristic of energy, whatever its type or source, provided the basis for such accommodation. Without this characteristic, this ubiquitous expression of the way nature works, evolution would have been very different, if not impossible. Briefly stated, there is always an inverse relationship between the distance from a particular energy source and the amount of energy striking a given point. As a simple example of what Newton named the inverse square law, as the distance between a hand and a hot stove is increased, the amount of heat striking the hand will decrease proportionately, but not in a linear fashion. The inverse square equation

$$\mathbf{E = K/d^2}$$

states that as the distance from an energy source increases arithmetically, that is, 1,2,3,4,5, etc., the amount of energy impacting on a point will decrease as a function of the square of the distance, that is, 1, 1/4, 1/9, 1/16, 1/25, etc. Obviously, if some particular energy is harmful or dangerous, it would be very functional to move away from it, or to develop some technique to cope with it.

The decreasing influence of energy as a function of increasing distance has been fundamental in evolution, and the specific effects of the nonlinear relationship stated in the inverse square equation, as emphasized at length later, has made its impact on all life forms.

ORGANISMS AS REFLECTIONS OF THEIR ENVIRONMENT

If one of us were suddenly transported to Andromeda and the scientists there were to examine us, they could infer a great deal about the planet on which we evolved: from our legs, a fairly extensive gravity; from our eyes, a particular range of the electromagnetic continuum; and from our ears, the common occurrence of pressure fronts from vibrating objects. The same is true for all the organisms on the earth. Each is a reflection of the selection pressures that have been present during the evolutionary history of its species. There are differences, of course, or different species would not exist—differences depending upon the particular niche that was the unique evolutionary home in which each species happened to evolve. Yet there are common features as well. Gravity has had its impact on all of us. Eyes are universal expressions of the influence of the sun. Ears are ubiquitous reflections of the survival significance of pressure fronts in various media; receptors on the surface of bodies give testament to the evolutionary importance of trauma, pressure, and molecular motion. The homeostatic negative-feedback principle (to be discussed later) implicit in all life forms is a reflection of the inverse square law that describes the impact of energies of all kinds.

This functional interrelationship between organism and environment is dramatically emphasized when humans are exposed to an alien context. Consider the difficulties of surviving in outer space. Suddenly humans are placed in a situation in which one of the major forces that have structured their bodies over the course of evolution is no longer present. Gravity is missing, and, without its perpetual tug, orientation and coordination become difficult. Atmospheric pressure is also missing, and the vulnerable tissues of the body must be protected by creating sustaining pressures in special suits or in the ship itself. Of course, space travelers must take along the most vital of all energy sources, that which sustains life so effortlessly and exists so abundantly on the earth: oxygen.

In outer space, not only are many of the molding forces such as gravity and atmospheric pressure missing, but humans are also exposed to new energies that, because of their novelty, they have no receptor systems to detect. For instance, humans did not develop sensory organs that would allow them to respond differentially to radiation or to oxygen depletion. Radiation was rarely a hazard on the earth because of the protective cover of the atmosphere, and the nearly universal presence of oxygen in the atmosphere made differential behavior (other than breathing) unnecessary. In space, then, humans must build extensions of their sense organs to allow adaptive responses to these alien energies and conditions, or must armor themselves against their effects. In short, humans must create conditions that approximate those in which their species evolved, and that, because of the selection pressures in operation, they essentially represent.

FUNCTIONAL INTERPRETATION

The preceding discussion suggests the plausibility and usefulness of a functional interpretation. If we wish to understand a particular characteristic, we can try to infer the contribution that it makes (or made) to the adaptive facility of the species. Yes, description is also necessary. Yet from a functional viewpoint, we must go beyond description and ask what the characteristic is for. Description is always necessary, but it is not enough.

Consider the red spot on the bill of the adult herring gull. A descriptive statement would emphasize the fact that the adult herring gull has a red spot on the lower side of its bill, that this spot has a certain diameter, and that it appears when the herring gull is approximately two years old. A functional interpretation would go beyond this and ask how this red spot contributes (or contributed) to the survival of the species. Tinbergen and Perdeck (1950) showed that the red spot serves a vital function. When the herring gull parents return from a fishing expedition with their crops full, the red spot serves as a signal that helps give access to this food. When a chick pecks at the spot, regurgitation is triggered and the young are fed.

To give another example, consider the hand of the aye-aye. This little nocturnal lemur from Madagascar has the four fingers and thumb characteristic of all lemurs (and also of all primates), but one of the fingers is remarkably different (Ankel-Simons, 1983). To give a descriptive statement, it is almost twice as long as the other fingers, is very thin, and has an unusual wiry appearance. However what is its function? In order to answer this question it would be necessary to move at night into the forests of Madagascar with infra-red observation equipment. If this were done with great stealth and much luck, one might have the opportunity to observe the function of the remarkable third finger. One might see the small lemur with its head pressed against a rotten log and its posture fixed in the aspect of listening. Then one might observe the third finger in action, see it probing into a hole in the wood and emerging with a grub stuck on the elongated fingernail. There has been (as is the case with all species) a functional interrelationship between the aye-aye and its food supply for thousands of generations. Gradually, as a function of the selection pressures imposed by the requirements of grub hunting, the long wiry finger of the aye-aye has emerged.

The red spot on the bill of the herring gull and the long wiry finger of the aye-aye seem to be dramatic examples of the interplay between selection pressures and the development of a particular attribute. Yet perhaps the most remarkable thing about them both is that they are not remarkable. Most, if not all, characteristics that have been within a species for an enduring period can likewise be understood in terms of their adaptive contributions.

Functional Causality

At least three different levels of causal explanation can be involved in the analysis of any functional attribute. The first, and most important for our development, may be called selection pressure causation. Going back to the example of the spot of the herring gull, we might be concerned with the survival contribution that the development of such a signal makes to the herring gull species. The selection pressures here relate to any factor that might bring about a more adaptive relationship between the adult and its offspring in terms of the sustenance of the young. Accordingly, the red spot, with all of its functional significance, became gradually selected in because it allowed a more efficient interchange of food.

A second kind of causation involves the genetic blueprint that accounts for the appearance and operation of a characteristic. In the case of the red spot, this type of causation has to do with the way in which the genetic information that results in both the appearance of the red spot and in the interplay between pecking and regurgitation has been encoded in the DNA.

Third, we might be concerned with the contemporaneous events that influence the pecking of the young and the regurgitation by the parent. These might include such factors as time since last feeding, health of parents and chicks, time of day, temperature, the presence of predators, etc.

Teleology

Evolutionists occasionally have been accused of indulging in teleological explanations, and perhaps there is some justification for this accusation. Paradoxically, some writers offer a disclaimer of such a scientific sin, and then talk as if such modes of thought were still permissible. Teleological explanations have no place in science, and certainly not in evolutionary theory. No attribute ever developed to fulfill a particular design or plan. The future does not pull us toward some evolutionary goal. Every attribute of every species emerged as a function of the shaping influences of the selection pressures encountered. Genetic blueprints that represent adaptive attributes are preserved from one generation to the next. All attributes emerged without plan or purpose as a function of the continuing process of selecting in and out. So when I suggest that a given attribute such as teeth can be understood in terms of its adaptive contribution, I mean simply that it can be understood as having been gradually shaped into organisms because it contributed to successful predation, fighting, and the preparation of food for digestion. Teeth did not appear in order to provide more efficient digestion, but the survival contribution of more efficient digestion provided the basis for the evolution of teeth.

Anthropomorphism

Another scientific sin that evolutionary thinkers are accused of committing is anthropomorphism, and again there is perhaps some justification for this accusation. But genes are not selfish, nor did Dawkins (1976) really think they were, although he often talked as though they were. Genes are particular genetic blueprints that have become stable through time because the attributes that they code for are more adaptive than others. Thus, it is the functionality of the attribute that determines the stability of genes, not vice versa.

The preceding discussion suggests that every attribute of living things can be considered from two frames of reference. First, it can be described as objectively and as precisely as possible, and second (and here is where the functionalist places his or her emphasis), one can infer what the attribute is for. From a descriptive point of view, the lens of the eye can be defined as a concave, oval-shaped piece of cartilage. A functionalist would agree but would also point out that the lens is an organ that contributes to the focusing of a clear image upon the fovea. The functionalist would deny the circumlocution of teleology. The lens of the eye did not develop for organisms to see better, but the adaptive contribution of better sight was a selective factor that gradually produced the lens. Nor is anthropomorphism admissible. The genes responsible for the development of the eye did not contrive to produce a better lens; rather, the adaptive advantage of better lenses both increased the survival potential of organisms having them and caused the wider distribution of the genetic blueprints (genes) accounting for them.

It should be noted that the term functionalism has a long history in psychology. Indeed one of the major early schools was called functionalism. Yet this earlier functionalism was concerned with the immediate contribution of attributes to mental activity, not with evolutionary origins, although such origins were sometimes implied, as in James's often quoted statement that the mind could not have come into being without adaptive contribution.

Thus, there are two uses of the term functionalism in psychology, corresponding to the first and third kinds of functional causality discussed previously. The third, which may be called immediate functionalism, is concerned with the contribution that a given attribute makes to an ongoing process. Thus the lens of the eye functions to project a clear image on the fovea. The first has to do with the adaptive contribution that both occasioned the origin of an attribute and determined its present contribution. The lens of the eye has the immediate function of projecting a clear image upon the fovea to ensure better vision, but it was the survival advantage of better vision that occasioned the gradual evolutionary emergence of eyes, with the lens serving as a crucial component. Thus evolutionary origins and immediate contribution are profoundly related; taken together they constitute the focus of modern evolutionary psychologists. The genetic coding of such attributes is, of course, important, but this issue is in the province of genetics rather than psychology.

PRIMITIVE LIFE AND THE PRIMAL ENVIRONMENT

Early Life Forms

This volume represents an evolutionary journey, beginning with the first emergence of life and continuing to include such complex organisms as human beings. Remarkably, the same principles that prevailed during the first appearance of life are now demonstrated in its most complex examples. Although we know little about the specific constitution of life's most primitive forms, we can make certain inferences about their general nature.

Early life forms without doubt had the capacity to replicate. Indeed, there are recent pictures of microfosssils (approximately 900 million years old) apparently showing the process of mitosis taking place (Schopf & Oehler, 1976). Also, it is a universal aspect of life (whether in simple cells or in complex organisms) that without replication, death of the species inevitably occurs. There is no exception to this inexorable rule: without replication, extinction. It is true that the death clock runs more rapidly in some species than in others. Certain insects live but a few hours, while humans and elephants live for 60 years. Yet, regardless of such variation, replication is necessary, or extinction inevitably results. An amoeba, in a sense, is immortal, as long as it can continue to divide. However if such fission is prevented, the amoeba rapidly withers and dies. So life and replication are inextricably bound together.

A second inference about early life is that it was able to obtain energy from some source. All life processes that we know about require energy, so it seems reasonable to assume that this characteristic was also present in life's earliest editions. Such energy use may have involved simple passive ingestion, but even ingestion or non-ingestion of assimilable particles requires a certain amount of movement, at least of some part of the organism in question.

Thus, the capacity for movement becomes the third inference that we can make about early life, and such movement was not purely random. It was selective, or at least directional. These behaviors were probably similar to those found in contemporary amoebas, which are able to discriminate assimilable material from particles that cannot be used as energy (Jennings, 1906), or in more complex forms as found in certain bacteria where flagella rotation is counterclockwise to attractants and clockwise to repellents (Carlile, 1975).

Fourth, we can say, without too much chance of error, that very early editions of life were made of vulnerable material. All examples of life that we know about today are so constituted, and there seems little reason to suppose that early life forms were different.

There doubtless were characteristics intrinsic in early life forms other than the ones just mentioned, but these were fundamental ones and the ones that are critical to our development.

The Early Environment

Now it is necessary to ask a similar question about the context in which life first appeared. What can we say about it with some degree of certainty? First, it seems reasonable to assume that the early environment consisted of fluctuating energies, energies that could vary from minimal to high intensities. Energies are in perpetual flux around us today, and the same seems likely three or four billion years ago, except that the shifts were probably more extreme.

Second, it seems reasonable to assume that energy source material was present, probably widely distributed in this primitive environment. If this were not the case, early forms of life could not have survived from one replication to the next.

We are now in a position to infer some general properties of all early life forms. If they were constructed of vulnerable material, and if they did exist in a context of fluctuating energies (some extremes of which could harm or destroy them), then all such life forms must have been homeostatic mechanisms. A detailed explanation of this statement is given later, but in general it implies that a certain balance in the process systems of such vulnerable creatures must have been maintained until replication could occur, or these early examples of life would have perished.

THREE MAJOR SURVIVAL TECHNIQUES IN EVOLUTION

The entire course of evolution has involved the development of techniques (survival gambits) for ensuring that (in a context of energy flux) the vulnerable stuff of life can maintain the balance necessary for existence until replication can take place. As complicated as this process might seem, it appears that only three basic techniques have been involved, and that every example of life, whether at the time of its origin or at present, represents varying combinations of these three major adaptive trends. In every individual of every species that we might examine, all three techniques are found, with varying overlap and emphasis. Let us consider these three techniques in turn.

Armor

First, the vulnerable stuff of life may be protected from the dangerous flux of energies by the development and use of armor, and this has happened in abundance. Wherever we look, we see many different kinds of armor making this vital adaptive contribution: the fur of animals, the bark of trees, the turtle carapace, the shells of seeds and many sea animals, the scales of fish, the rib cage around the vulnerable portions of our bodies, and the skull that protects the most vital and important organ that we possess. Indeed, armor is ubiquitous in living things; the balance necessary for the vulnerable stuff of life to continue has ensured its appearance in its many different forms. Even the amoeba has an outer surface that to some degree protects it from the hazards in its environment.

In every case, of course, armor developed slowly. Those individuals in a particular species that were even slightly more effectively protected from the effects of harmful energy were more likely to survive, while those less well endowed were selected out. Thus, the many forms of armor gradually emerged; the adaptive tangent that in some fashion and to some degree is found in all of us, whether the amoeba, armadillo, or man.

Although all species have developed armor to some degree, certain species have moved into armor as their primary adaptive gambit. In extreme cases, this has resulted in a kind of living death. The more that armor has been utilized, the more heavily burdened the organism becomes, making it less capable of making movements of any kind. Yet, the greatest structure ever created by living organisms, the coral reef off the coast of Australia, is a testament to armor's effectiveness and massive utilization.

Replication

The second major survival gambit in evolution is replication. As previously mentioned, replication is essential to the very existence of life; it has been, and remains, a universal event in every species. Yet in certain species, replication has been emphasized as a major technique allowing the vulnerable stuff of life to continue. Some species have the capacity to reproduce in such vast numbers, that in spite of intensive depredation there are always enough individuals left over for the survival of the species to be assured. Millions of frogs are hatched, but only a few reach adulthood. For every bass that achieves maturity, tens of thousands of fry are lost along the way. Generally speaking, the more vulnerable the individuals of a species, and the shorter the life span, the more rapid replication will be involved as a survival gambit. This is even the case among mammals. The gestation period of mice is only a month, the average litter size is eight, and the life expectancy is two years. For humans or elephants, where life expectancy is long and vulnerability is less extreme, replication is a slower process.

Differential Behavior

It is the third survival technique that is of greatest interest to psychologists. Species cannot exist without the other two, but differential behavior is the survival gambit that our own species has emphasized more than any other, and it is this gambit, in its more complex expressions, that constitutes the essence of what we are.

As previously suggested, even the earliest examples of life were homeostatic mechanisms. That is, an essential balance must be retained or regained or the life process will be disrupted. Also as previously stated, there is an inverse relationship (the inverse square law) between the distance from an energy source and the amount of energy striking a particular point. Thus, moving away from provided a basic maneuver that allowed the vulnerable stuff of life to regain the balance necessary for survival, and the ability to move away became a major survival gambit in the course of evolution. If an energy source became too intense, such differential behavior ensured that tolerable levels would be regained. By differential, I simply mean that the organism could respond differently; that is, move away from harmful energy levels or harmful circumstances and not move away from those that were beneficial or benign.

Yet moving away is but half the picture. All forms of life must utilize energy, and this implies that life forms must have been able to move appropriately to energy source material. So differential behavior not only involves away from responses, it involves toward behavior as well. Even in plants and other static organisms parts of the body respond to deficit and harmful energy.

The inverse square law enters at this point in a profound way. It has insured that organisms are not only homeostatic mechanisms, but that they are also negative feedback mechanisms: that is, there is a relationship between the degree of imbalance and the speed of compensatory reaction. As an example of negative feedback, the more a rubber band is stretched, the greater will be the energy required to stretch it further. That is, the more it is removed from its balanced relaxed state, the greater will be the tendency for it to return to that state. The same is true for the process systems of all organisms.

The inverse square law provided the environmental impetus for the expression of the negative feedback principle in every organism. However another characteristic implicit in the organism itself has helped ensure the universality of the negative feedback principle. Any organic system can tolerate lower energy levels of imbalance for longer periods than it can tolerate higher levels. Thus the tendency to respond with increasing rapidity to higher levels of imbalance has been a critical factor in survival and has conjoined with the survival implications of the inverse square law to ensure that the negative feedback principle will be universally represented in every process system of every organism.

Gambits of Defense and Offense

Are there only three basic tangents in evolution? I believe that this is the case, but from time to time I have considered a fourth possibility—gambits of defense and offense. Certainly, wherever we may look, weapons are represented in marvelous variety. Moreover, species have an intriguing propensity to mimic other species and situations where such imposturings abet, even though slightly, the individual's survival chances. There are plants that look like rocks and bird feces. Certain insects, themselves nontoxic, mimic those who are. There are many examples of species that mimic the environment. Some, as in the case of the chameleon, change on a moment's notice, while others, such as the arctic hare, change more gradually to meld with the tundra or the snow.

What a variety of weapons mark the perpetual interplay between prey and predator! The remarkable array of teeth and horns and claws, from the wart hog's tusks through the horns of the impala to the talons of our nation's symbol. Chemical weapons, both of offense and defense, range from the poisons of plants, fungae, snakes, and toads through the repellent odors of skunks, beetles, and shrews. Nor has electricity been slighted. The electric eel can stun a man with 600 volts, while the knife fish in the muddy tributaries of the Amazon uses rapid bursts of electrical pulses as a kind of radar for navigation and prey detection. The use of projectiles is present in abundance. The archer fish can strike an insect five feet above the surface despite the diffraction of the water. The chameleon can fire its tongue farther than the length of its own body, and the bolas spider can hurl its bolas accurately even in the dead of night.

There are even combinations of chemical and projectile weapons. The nasute (nose) termite, although blind, can fire a thread of entangling glue at intruders from a distance of six inches. The bombardier beetle can fire a boiling solution of abrasive quinones at its enemies, and the whip scorpion has a long tail that can rotate rapidly toward intruders and fire a deadly stream of repellent acid. Yet perhaps the most remarkable of all weapons is that of the grenade ant. It positions itself in the middle of an invading multitude and blows itself up, covering those near its suicidal mission with entangling glue.

These are but a few of the seemingly limitless gambits of offense and defense that are represented among the creatures of the earth. (See Alcock, 1984; Edmunds, 1974; Street, 1971, for a sampling of the literally hundreds of defensive and offensive strategies that are used by living species.) Such gambits are not included as a basic survival tangent in evolution. They are not given equal status to reproduction, armor, and differential behavior simply because they are expressions (in varying proportion) of these more fundamental modes of equilibrium maintenance. But in certain species they are of vital importance and are so considered at appropriate places as the development proceeds.

We might pause here to take note of a striking statement by Darwin, which comes, incidentally, at the end of his great book. "There is grandeur in the view of life ... *that* from so simple a beginning endless forms most beautiful and most wonderful have been, and are being evolved."

Yes, it is most remarkable that the plasticity of the genetic material, under the relentless sculptings of selection pressures, could have resulted in so many diverse and subtle offensive and defensive systems. Yet it is perhaps even more remarkable that the environment not only provides the arena for their deployment, but also (along with interacting species) has provided the occasion for their emergence and the wherewithal for their expression.

STATEMENT OF FUNDAMENTAL PRINCIPLES

I have argued that the existence of life in a context of fluctuating energies presupposed the demonstration of certain principles. Since these early editions of life were constituted of vulnerable organic material, differential response to hazardous energy levels was essential for existence, and more rapid response to more intense energy applications was a critical adjunct. As such responses, and indeed all life processes, required the use of energy, such organisms must also have responded differentially to energy sources, and the greater the energy depletion, the more rapid must have been the compensatory behavior. In short, a condition of minimal variation or balance must have been maintained or regained, and the greater the deviation from this condition, the more rapid must have been the alleviating response. Therefore, the first motile life elements must have been simple, homeostatic, negative-feedback mechanisms, and, considering the context in which they emerged, they could not have been otherwise.

The earliest forms of life were homeostatic negative-feedback systems, and every example of life since that time, regardless of complexity, can be described in the same way. The vulnerable stuff of life can tolerate only so much deviation from an optimal state of balance without perishing, and it is the differential behavior that is instigated by the deviation from such norms that allows an optimal condition to be maintained or regained. Such differential behavior has been a fundamental requisite for survival and for evolution.

The Postulate of Process

This discussion leads directly to a statement of the Fundamental Postulate of Process. This postulate is fundamental because it is presumed to apply to every life process from the simplest to the most complex.

The Postulate of Process. *All overt or covert activity serves the immediate function of impelling the organism toward equilibrium.*

Since life, regardless of its particular expression, can exist only within a certain equilibrium or balance range, the selection pressures encountered down through evolutionary history have ensured that more facile ways for equilibrium maintenance will emerge.

Objections might well be raised at this point. One criticism suggests that organisms could not be homeostatic mechanisms as stated because once equilibrium has been reached, stasis would result, and, obviously, stasis is not descriptive of living things. Yet the postulate states that all activity impels the organism toward equilibrium; it does not imply that equilibrium is necessarily achieved, except perhaps for very short periods. Indeed, life is a process of continuing and perpetual flux, a moving away from and toward equilibrium as imbalances occur and as compensatory responses take place. Nothing is static in life. It is a dynamic and ever-fluctuating process. Death is the only stasis.

Another objection involves a partial if not total denial of the validity of the postulate, where complex organisms are concerned. During the Vietnam war, a number of Buddhist priests poured gasoline over themselves and burned themselves to death. Surely, the critics might say, this behavior cannot be viewed as equilibrium-trending. But even in such extreme examples the postulate applies. The postulate states immediately equilibrium-trending, and at the moment the match was struck, that behavior can be so construed. In human beings, many variables are perpetually interacting (as discussed at length in later chapters), each with varying emphasis. The momentary addition of any new variable may change the weighting of all the others. The striking of the match, in terms of the many coalescing variables influencing the priest, was at that moment the most equilibrating behavior available. To jump from the frying pan into the fire seems to be a disequilibrating action, but at the moment the jump is made, it moves the individual toward relative equilibrium, even though at the next moment, a situation of even greater imbalance is encountered.

A third criticism that might be brought against the postulate has to do with the meaning of the word equilibrium. If we take ourselves as models, it can be seen, although the explanation is difficult, that the term equilibrium is both applicable and reasonable. Each human being is comprised of vulnerable material, continually undergoing the processes of moving away from and toward balance relative to many coalescing and interacting systems. If the individual is healthy, a state of equilibrium exists relative to any body system or to the body as a whole when no compensatory activity is taking place. It should be noted that a range of tolerance exists in every case of equilibrium. This range includes the state of any system from a point where compensatory activity ceases to the point where compensatory activity is again initiated.

How narrow is this range of tolerance? This, of course, varies from one system to the next. As a matter of fact, no system can be considered in isolation. Each system interacts with all the others affecting the point, and thus the range of tolerance, at which compensatory action begins. After eating, for example, as far as that particular aspect of our energy system is concerned, optimal equilibrium has been achieved and a tolerance range of equilibrium holds until the depletion of energy is sufficient to instigate compensatory action. Yet the point at which this compensatory activity occurs depends on the amount of thirst, pain, fatigue, etc., which is also present. The same is true of all equilibrium systems within all organisms, whether few or many interacting systems are involved.

A more general criticism has to do with the testability of the postulate. Can an experiment or series of experiments ever be devised that will unconditionally prove or refute it? Probably not. But postulates by their very nature are, at least in most cases, general statements about nature rather than specific hypotheses. A postulate, however, if it is to be useful, must have a powerful integrating and explanatory function, and it is always open to acceptance or rejection on the basis of this more general, but no less important, requirement. (See Goodson & Morgan, 1976b for an extended treatment of this issue.) The integrating and explanatory function of the previously mentioned postulate should be judged in terms of its effectiveness in subsuming the many diverse principles and facts of psychology. The remainder of this volume consists of my effort to realize this goal and, of course, the reader will and should indulge in a cumulative critical appraisal and make a final assessment of the extent to which this goal has been achieved.

The fundamental postulate of process is derived from the observation that a condition of minimal variation must be maintained if life processes are to continue undisrupted. It is thus derived from the most rudimentary conditions essential to even the most primitive life, and yet it is equally applicable to life's most complex expressions. There is a fundamental corollary to this postulate.

Negative Feedback Corollary: *Within the limits imposed by structure and available expendable energy, the more an organism deviates from equilibrium, the greater its compensatory activity will be.*

Although the manner in which the negative-feedback corollary enters into the continuing adjustment of organisms is considered at length in many sections of the theory as it emerges, a brief account is given here. The interaction between predator and prey has been one of the abiding selection pressures instrumental in the evolution of many species. Every organism signifies its presence with a halo of molecules, by reflecting certain energies of the electromagnetic continuum, and whenever it moves by pressure fronts in a medium. All of these energy dimensions follow the inverse square law. Thus as the distance between predator and prey diminishes arithmetically, the cues signifying the presence of both increases geometrically. When the survival significance of the inverse square law is coupled with the implications of the fact that any vulnerable organic system can withstand lower levels of imbalance longer than higher levels the universal applicability of the homeostatic negative-feedback corollary becomes clear.

This corollary is often emphasized as the development proceeds, but a few examples here at the outset might prove helpful in explaining how universal its application can be. The longer an individual has gone without food, the greater will be the hunger and the more rapid the consuming behavior. The longer an individual has gone without fluid, the greater will be the thirst and the more rapid the process of drinking. The longer an individual has gone without urinating, the greater will be the bladder distension, and the greater the initial output of urine, the output of which will diminish gradually as equilibrium is being achieved. Indeed the negative-feedback principle holds for every cell in living tissue. Thus, the cells in the human body are not only homeostatic systems, but also negative-feedback systems. If an essential ingredient of a cell drops below an optimal balance, the greater the deviation from this balance, the greater will be the compensatory change in the membrane.

The homeostatic negative-feedback corollary describes not only the behavior of complex organisms but also that of microorganisms. Recently there has been a remarkable upsurge of interest in the sensing and response capacities of microorganisms, particularly of bacteria. In 1985, Eisenbach compiled a reference list of research on the topic that included 857 entries, most published during the preceding 20 years. Many of these studies centered on the specific biochemical changes accounting for both the sensing and motility of these animals (a topic beyond the scope of this work), but certain behavioral findings are not only directly pertinent to the negative-feedback corollary but also to an increased appreciation of how seemingly complex processes are demonstrated, albeit at a rudimentary level. Bacteria, for instance, can respond directionally to both attractants and repellents. They respond more rapidly to higher concentrations and to steeper gradients than to lesser ones, as predicted by the negative-feedback corollary.

Bacteria are also capable of memory, discrimination and adaptation:

> The bacterium senses a spatial gradient by detecting the temporal gradient created as it swims. Specifically, the concentration during the current second is compared with the concentration over the previous three. This means that the cell has a way to measure current concentration, a means of storing a record of past concentration, and mechanism for comparing the two values. However, after a period ranging from seconds to minutes, depending on the compound and the magnitude of the gradient, the cells resume their initial behavioral patterns of runs and tumbles, even though the attractant or repellent is still present. Thus bacteria, like most sensory cells, adapt. (Hazelbauer, 1988, p. 466)

Directional movement to chemical gradients (chemotaxis) has been observed not only in a wide range of bacteria, but also in many eukaryotic cells, including free-living microorganisms, white blood cells, sperm and slime mold amoeba, to name a few. Where such movements have been examined in detail, gradients of behavior (as predicted by the negative-feedback corollary) have been demonstrated. That is, the steeper the chemical gradient, whether a repellent or attractant, the more rapid the response. Since the gradient becomes steeper as the cell gets closer to the chemical source (the inverse square law) the influence of the chemical is proportionally greater.

There have been a number of recent efforts to isolate the physicochemical events that not only account for the direction of movement but also occasion the increase in speed as the animals get nearer the attractant (or repellent). For instance, both white blood cells and slime mold amoeba have thousands of receptor sites distributed on their surfaces. There are different classes of receptor sites, each class responding to a particular kind of chemical, so that the same cell may be reactive to a number of different chemical gradients. When a given chemical, say a chemoattractant, is introduced into a medium near an amoeba, this chemical diffuses toward the amoeba until those receptor sites peculiarly attuned to it become occupied, stimulating pseudopod formation and movement in that direction. As the amoeba gets closer to the chemical source, more sites are occupied and the amoeba moves with correspondingly greater speed.

If two or more chemicals (attractants or repellents) are presented to different areas on the surface of an amoeba, pseudopods, each pulling in a different direction, are formed. However, the structural integrity of the amoeba insures that it will not be split, but will move in a direction that best summates the influence of the various chemicals. Although the specific physicochemical processes underlying such directional movement remain to be completely clarified, much progress is being made. To quote Devreotes and Zigmond in their statement of the clearest alternative of three different models they examine, "Each occupied attractant receptor activates a contractile element that pulls the cell in that direction. The resulting tug-of-war produces a net mechanical force in the direction of the highest receptor occupancy. The force is proportional to the difference in occupancy ..." (1988, p. 653).

Thus the negative-feedback corollary not only describes the behavior of complex organisms, but also that of one-celled animals. The tug-of-war and force resolution resulting in the directional movement of the amoeba can also be viewed as a primitive prototype of the singularity implicit in apperception (to be explored at length in chap. 6), where hundreds of inputs from both sensation and memory interact so that the one most critical for adaptive behavior at a given moment can become focalized to insure that human behavior, for instance, will be unified and directional.

The Postulate of Inference

The postulate of process is often be cited throughout this volume because of its explanatory and subsumptive power, but from a theoretical point of view, another postulate is equally important. The example of the aye-aye dramatically portrays the manner in which a given attribute evolves to reflect a particular adaptation circumstance. Yet, as previously mentioned, the truly exceptional aspect of the aye-aye's finger is that as an adaptive device, it is not exceptional. Any organ or process that has been retained by a species for an extended period can also be understood in terms of its survival contribution. The curious grassy material on the human head, the flappy pieces of cartilage called ears, the white protrusions from the gums, are all functional characteristics that can be understood in terms of their survival contributions.

At this point, we again confront a dramatic change that has taken place in evolutionary theory during the past fifteen or twenty years. When we speak of survival, we no longer mean just the survival of the individual, but the survival of the genetic material. It might be helpful to repeat the statement made during my review of evolutionary theory given in chapter 1. "Any inheritable variation that increases the chances of its own replication will tend to be incorporated into the species." This general statement suggests that the emphasis should be moved away from the survival of the individual to the survival of the genetic material.

The Postulate of Inference. *Every attribute that has remained characteristic of a species for an enduring period contributes (or once contributed) to the survival of the genetic material.*

This postulate seems obvious when viewed in the context of neo-Darwinian theory since natural selection is selection according to relative adaptive function. Yet in spite of common lip service to the principle many theorists apparently find it difficult to accept the universality of survival value as the only guiding force in evolution. Cronquist (1968), for instance, stated that there have been few public challenges of the neo-Darwinian position, but then goes on to suggest that certain evolutionary trends in plants seem to be selective, others more doubtfully so, while still others appear to be selectively neutral or even mildly counter-selective. Can this be the case if the basic tenets of Darwin's theory (namely mutation, selection, and incorporation) are correct? In a word, no, simply because the survival value of an attribute is the ultimate determinant of its incorporation. It is the sole arbiter of natural selection. To assume that there is some other factor involved is to abrogate neo-Darwinian theory as it is presently understood.

However if one holds to a strict neo-Darwinian interpretation, how can he or she explain the fact that so many attributes seem neutral while others may even appear to be negative in terms of survival contribution? There are a number of possibilities, each with a number of ramifications.

First, it is possible that an attribute may be in the process of passing. The great majority of mutations are harmful, yet they may linger in a species for a while before they go. If one happened to look at one of these attributes and tried to surmise its adaptive contribution, he or she would obviously be stymied. Or an attribute may be passing because it is no longer functional. The horns of the caribou once were essential weapons against the great wolf packs of the north, but now most of the wolves are gone, and those horns, as heavy as they are and as attractive as they seem to be to hunters, have probably become negative in their survival contribution. In certain areas of Africa many elephants never develop tusks, demonstrating the selective pressure of selective killing for ivory. And so any attribute that once was functional and incorporated because of its survival value may become detrimental if the effective environment changes. Furthermore, an attribute that once was functional may become detrimental because it has been superseded by some other attribute that has made it extraneous or even negative. To be functional, an attribute must be synchronized with all other attributes, and a change in one may alter the contribution of another. Somewhere down the evolutionary path the ostrich moved toward legs and away from wings, but this does not mean that wings were not once its major instrument of locomotion, even though a candid evaluation now might judge them negative.

Second, there is always the possibility that the investigator has made an error in observation or inference, that an attribute may actually have adaptive function but he or she fails to see it. An attribute may be so hidden, subtle, or ephemeral in its effects that its contribution cannot be easily inferred. Consider the pineal gland. Scientists have puzzled over its function since Descartes mistakenly concluded that it was the organ of mind-body interaction, and had almost decided that it had no adaptive function. Now there is a growing body of evidence that it may respond to the amount of activity in the optic nerve and occasion compensatory changes in the production of sex hormones, which in turn may influence such diverse activities as migration in birds and the menstrual cycle in human beings. The pineal gland gives evidence of another complication that may obscure the particular function of an attribute: an attribute may not just pass away, it may literally shift its function. The pineal gland functions as a third eye in certain primitive amphibia, as a circadian pacemaker in birds, and as a governor of sexual development in humans.

Also the function of an attribute may be difficult to determine because it makes little or no contribution in itself but has effects (catalytic or dampening) on other attributes, sometimes a number of steps removed. The function of the pituitary gland was difficult to determine because it seemed to make no contribution in itself. Now we know that it is the master gland that determines the pace of the process in certain other glands. Indeed, every organism is a complex system with thousands of attributes conjoined in perpetually shifting dynamic synchrony. As such, the function of many attributes can never be inferred in isolation, but must be evaluated in terms of their contribution to the operation of others. Attributes are so interdependent in their effects that an alteration of any one may shift the adaptive contribution of the rest. Of course the degree of contribution, whether positive or negative, is always shifting in response to changes in the effective environment. A slight alteration might move an attribute from the plus to the minus column, and vice versa.

Third, it is possible that the investigator may be observing an intrinsic rather than a basic attribute. If one were to try to determine the functionality of the whiteness of an elephant's tusks it is doubtful that he or she would ever arrive at a reasonable explanation. Every aspect of an organism, whether we are speaking of a bit of bone or a hank of hair has to be constructed of some kind of material with certain characteristics. We may designate such characteristics intrinsic attributes. They are intrinsic simply because without them no attribute could exist at all. They, in essence, comprise the basic material from which attributes must be constructed. To try to determine the function of the whiteness of teeth (or elephants' tusks), the softness of flesh, or the breakability of bones is a futile activity, simply because such attributes are intrinsic. Yet it is often difficult to determine whether a given attribute is intrinsic or functional. Brownness in eyes happens to be functional because it provides a greater resistance to the injurious effects of sunlight than other colors, but if it were the only color available, we would classify it as intrinsic. So many characteristics are intrinsic simply because every attribute has to be made of something, have a certain color or lack of it, a certain weight, a certain length, etc. If we pick out such an attribute and try to infer its functions, we are doomed to failure.

Fourth, in certain cases a characteristic may become genetically tied to some other more fundamental trait, and so may endure through many generations as riders of this more basic trait. For example, consider the nipples on the males of many species. They are not in themselves functional, but are tied to a much more fundamental attribute that makes its functional contribution only in the females of the species involved. Why do such rider characteristics endure? Even though the nipples of the male may slightly hinder male survival, the genetic complication required for them to appear in females but be absent in males has apparently been too formidable to have taken place.

It is also possible that variations within an obviously functional attribute may in themselves make no adaptive contribution. Hemoglobin, for example, is vitally important for the distribution and use of oxygen, but there are over 140 different kinds of hemoglobin. Is there an adaptive function for each of these types? This would appear unlikely. Apparently there is a core of mutation-resistant genetic material with blueprints for oxygen utilization, and this one is linked to other mutation-prone genes. Thus, the basic function of hemoglobin can be preserved while the previously named variation can take place. It is also possible that the great number of hemoglobin types may make an adaptive contribution by providing a variability pool that allows more flexible development of disease-resistant genetic combinations, as indeed is the case with sickle cell hemoglobin and malaria.

A number of investigators (Ayala, 1983; Dobzhansky, 1964; Stebbens, 1982) have defined evolution as a change in gene frequencies. In my view, this definition is inadequate. Certainly a shift in gene frequencies is involved whenever a new attribute is incorporated, but something more than a shift in gene frequencies per se is involved. Random shifts are perpetually occurring in populations with no discernible evolutionary change taking place. For evolution to occur the change in gene frequency must become stable through time, and such stability typically occurs only when the attribute makes an adaptive contribution.

Also, of late, there have been suggestions that genetic drift alone, when abetted by the founder effect (see chap. 1), may be responsible for certain evolutionary changes. This appears doubtful to me. Genetic drift, particularly when small groups become isolated, may result in more rapid alteration if the genes code for attributes function in the new environment (or an identical environment for that matter), but it is natural selection that determines the functionality (or lethality) of such genes. Of course, the greater the neutrality of a given attribute, the more genetic drift may be responsible for variation, as is the case with hemoglobin types (see Dillon, 1978 for an extended discussion of this issue).

Another factor that complicates this postulate is preadaptation. It is possible that the characteristic we are examining may have shifted function, as some believe was the case in the development of feathers. Some theorists maintain that feathers initially had the primary adaptive function of protection from the cold (Gould & Vrba, 1982). Only much later, when other attributes had emerged, did feathers become a major factor in flight. Preadaptation does not rule out the applicability of the postulate. It just makes it more difficult to trace the gradual emergence of a trait simply because, where such shifts in function have taken place, there appears to be a sudden leap in the evolutionary process.

A real problem arises in the postulate when we consider the concept enduring period. How long is an enduring period? As a matter of fact, no particular time span can be given as applicable to even a few attributes. Some attributes may have come into being quite rapidly, perhaps over the period of a few hundred generations, while others may have emerged at a much slower pace. When the possibility of preadaptation (exaptation) is brought into the picture, the situation becomes even muddier. To determine the amount of time required in any particular case, an intensive study must be made involving data from population statistics and measurements of attribute alteration, both in contemporary organisms and in the paleontological record.

Perhaps the most serious objection to selection pressures (i.e., the classic Darwinian position) as the ultimate determiner of the attributes of organisms comes from Gould with his view that certain exaptations may, without any relationship with prior selective circumstances, come into being and assume remarkable adaptive importance. (See chap. 1 for an extended discussion and rejection of this view.)

Finally, we should always remember that a number of variables (see chap. 1) may have contributed to the emergence of certain attributes, while only one variable may have been primarily responsible for others. For example, powerful legs and sharp teeth are more likely to be a function of the operation of survival variables, while mating rituals and ornate physical characteristics are more likely to be a function of sexual facilitator variables.

The postulate of inference is so important in my development that I must briefly reconsider (also see chap. 1) a common misapprehension, the notion that evolution somehow reflects some larger purpose or design. The statement that an attribute could not have come into being if it had not made a survival contribution is correct after the fact, but no attribute has ever appeared in order to fulfill the requirements of some teleological plan. Attributes come into being by slow increments, and each such increment accumulates to provide the attribute's functional contribution at a given time. Each attribute is a composite of many small increments that have occurred through chance and have been integrated into the species because they have increased, even if only slightly, their possessors' chances for survival.

Regardless of these limiting and cautionary statements, the postulate of inference reflects one of the most widely accepted ways (by evolutionary thinkers) of viewing organisms and their attributes. As such, this postulate is foundational to the psychological theory to be presented in the following pages. It, along with the postulate of process and the negative-feedback corollary, provides an inferential base that allows the many seemingly disparate facts and principles of psychology, particularly cognitive psychology, to be integrated into a meaningful conceptual structure.

In the following chapters, as we evaluate the various steps toward more facile modes of adaptation (and increasing complexity), I try to indicate the manner in which both fundamental postulates and the fundamental corollary are demonstrated in each attribute.

I begin with a discussion of primitive motile life as a simple homeostatic negative-feedback system and then explore more advanced stages of increasing adaptive facility. Although something very like the various stages must have happened, no claim is made that the actual process of evolution occurred precisely in the manner to be described. Rather, a consideration of evolution in terms of progressive steps provides a logical framework that allows the integration of the many processes and behaviors of the human being into a unified theory.

This statement of intent has a problem that, since it will permeate the development, should be clarified at the outset: *The Intrinsic Dilemma*. Every organism is a complex of dynamic interacting processes so related that a change in any one has immediate repercussions throughout all the rest. Yet, we must deal with such perpetually interdependent processes with explanatory techniques that are both sequential and static. Thus, the very effort to explain a complex dynamic system imposes an intrinsic distortion. We must begin at some particular point and sequentially describe each process as if it were separate from the rest, but in human beings, for instance, such processes as sensation, perception, apperception, motivation, and behavior are all dynamically and perpetually interdependent. Yet I must use separate words in sequence and organize this volume into separate sections in sequence. There is no escape from this dilemma. Fortunately, the reader's memory permits an accumulation that may be melded into a total view. Also, as much as possible, I refer the reader backward and forward to points related to the section he or she is reading. And, of course, Darwin's theory of evolution through natural selection should always be kept in mind. It is, in spite of the paucity of data directly supporting it (see chap. 1), the source of the inferences given in this chapter and the framework for the chapters yet to come.

3 Overview

Evolution may be viewed as the gradual emergence of increasingly refined techniques for maintaining, regaining the balances necessary for vulnerable life to continue, or both. This has involved the development of organic systems of greater sensitivity to imbalance trends and more efficient modes of compensatory reaction. This chapter presents an overview of the progressive steps in adaptation refinement that have taken place.

The notion of progressive steps does not imply that more recent increments are better or worse, in any value sense, than more primitive ones. Rather, they are cumulative developments that in each case allowed more facile processing and reaction. From our discussion the skeleton of the theory that is presented in greater detail in later chapters emerges. Thus, this chapter should serve both as a basis for deriving and integrating a number of important principles and as a summary statement of the general theory. The Postulate of Inference carries a heavy burden in this development and should be kept in mind by the reader.

Within our frame of reference, a progressive step in evolution is an attribute that constitutes a major leap in adaptive facility by providing a more effective, although in many cases, more complex readout and resolution of the imbalances that are perpetually occurring. A progressive step cannot be defined in terms of any particular organ or structure. Indeed, the workings of natural selection produce many different creatures for each progressive step that appears. Although there are many different kinds of organisms, the number of progressive steps that have been achieved during the evolutionary process is remarkably few.

It is noted as we proceed that these progressive steps in increasing information-processing efficiency occur in a sequential and cumulative pattern, with each new step requiring but a small shift in process, which in each case provides a remarkable increase in adaptive facility. To repeat, a rather small, at times even relatively minor, shift in the complexity of process may allow a remarkable increase in the adaptation facility provided.

A brief summary of our discussion in chapter 2 helps clarify the meaning of progressive steps, suggests which one appeared first in the evolutionary context, and sets the stage for the derivation and analysis of the others.

It is recalled that the first motile life element was presumed to demonstrate certain characteristics. It had to be able to move away from harmful energy sources and to move toward energy source material. It also had to be able to respond differentially to various types and levels of energy. Differential response to pertinent energies, then, was fundamental in the sense that without it, the element could not exist. We may thus consider differential responsiveness to pertinent energies as the foundation upon which all of the progressive steps of evolution have been based. Indeed, this characteristic is so basic that all the progressive steps in evolution can be viewed simply as the emergence of more facile modes of differential response. The amoeba is a differential responder and so is the human being. Yet in the human being we find an organism that is capable of highly refined and efficient methods of information processing wherein a number of progressive steps are integrated to collectively provide facile resolution of imbalance states.

PROGRESSIVE STEPS IN EVOLUTION

(Step 1). Single Dimension Differential Behavior

This would involve an organism restricted in response potential to a single energy dimension. If such an organism did exist, what could we say about it? In terms of the inverse square law discussed in the previous chapter, we can infer that it would be a homeostatic negative feedback mechanism. That is, strength of response would be inversely related to distance from the energy source and would be more rapid to higher levels than to lower levels. For example, if molecular movement (heat and cold) happened to be the energy in question, such an organism would move away faster from a heat source when closer to it than it would at a greater distance. If the speed of movement of such an organism were plotted, a negatively accelerated curve would be predicted.

A variant of single-dimension differential behavior is immediately suggested. Since a single energy dimension may encompass a wide range of intensities, slight intensities might induce approach, while greater intensities might induce avoidance. Euglena demonstrates this point; it moves toward a dim light source (light is essential for photosynthesis), but away from intense light, which is potentially damaging (Hall, 1964). In each instance, a sort of *behavioral fulcrum* between the two types of response, away from and toward, can be predicted and determined experimentally.

It should be noted that the stasis at the fulcrum is a simple, albeit primitive, example of equilibrium, as the term is used in the postulate of process. The quiescence of an organism at that point demonstrates that balance among all relevant systems has been achieved. According to the postulate, any deviation from this point will initiate movement back toward it, and, according to the fundamental corollary, the greater the deviation, the more rapid such movement will be.

(Step 2). Dual-Dimension Differential Behavior

Since the processes of metabolism would require that even the simplest inferable examples of life would need to use energy, moving toward behavior would be predicted for any organism that could move away. Such behavior would also be an expression of the inverse square law and the homeostatic negative-feedback corollary just mentioned. That is, as depletion increased, the movement in question would increase in speed.

The simple away from and toward behavior just described can serve as a basis for the definition and clarification of certain concepts. In such an organism, the equilibrium range may be defined as the condition obtained until the response threshold has been reached. This equilibrium or tolerance range, as previously discussed in chapter 2, would include the extent of shift toward imbalance from the point of satiation or energy application until behavior was first initiated. Minimal disequilibrium may be defined as the degree of imbalance required for such compensatory behavior to be initiated. If the state of imbalance kept increasing, a point would finally be reached where the organism would suffer damage or perhaps death. We can define this as the point of trauma disequilibrium.

The postulate of process and the fundamental corollary are both explicable within the framework of simple dual-dimensional behavior. Movement would not occur until the point of minimal disequilibrium had been reached. According to the postulate, such movement would tend toward equilibrium, because the induced responses would occasion movement either away from the hazardous energy source or toward energy source material. The corollary would be demonstrated by the speed, latency of response that would be a reflection of the organism's distance from the energy source, or both.

It should be emphasized that an energy change, whether imposed from the outside or arising from depletion from within, rarely occasions only minimal disequilibrium. More typically such a change intrudes deeper in the disequilibrium range. The reason for this is obvious; no organism is either perfectly sensitive or invariably reactive.

(Step 3). Multi-Dimension Differential Response

The previous two examples of differential response probably never existed. I included them for purposes of logical development and as a basis for defining concepts. Very likely, when life first emerged, it was capable of responding to a number of different dimensions, both externally applied and those representing internal imbalances.

Even the simplest single-celled contemporary organism is a multi-dimensional responder. Indeed this is true for all cells of all organisms, regardless of where they are found. All such behavior, whether it involves actual movement or is restricted to changes in the cell membrane, is instigated by imbalances arising either from environmental changes or internal metabolic activity. For instance, if any one of the fundamental ingredients of a cell moves past the tolerance range, compensatory action brings about a return to equilibrium. At times, this compensation may involve movement of the cell itself as is the case with sperm, slime mold amoeba, and leukocytes. More often, it involves alterations in the cell membrane so that transport either in or out can take place. An amoeba can respond differentially to various intensities of light, saline solutions, various types of acids, and heat and cold, as well as to ingestible versus noningestible particles in its environment (Allen, 1962). Indeed, chemotaxis, cell motion directed by external chemical gradients, is a phenomenon of widespread occurrence and significance, as documented in chapter 2.

Differential behavior instigated by direct energy applications and metabolic imbalances provided a measure of adaptive facility. By putting distance between themselves and hazard, such organisms' away from behavior provided a functional, though limited, margin of safety; reflecting the inverse square law. Also, movement toward energy source material, or compensatory changes in cell membranes, provided the basis for the reestablishment of vital balances. Yet differential behavior limited to direct energy application is restricted and inflexible. Appropriate reaction cannot occur unless particular energies or metabolic requirements are present.

Since there is often a narrow gap in space and time between slight and potentially dangerous levels of energy, the margin of safety would be minimal. Since the chemical emanations from energy sources decrease rapidly with distance, the food-finding capacity of such an organism would be extremely restricted. Absence of appropriate energy or the lack of emanations from food material would result in immobility.

This leads us to the next two progressive steps in differential behavior. Before these steps are considered, however, certain basic points should be reemphasized. Continuity marks the development of progressive steps in evolution. Each new step represents only a slight alteration of a more primitive attribute and, in turn, provides the foundation for the next step in the adaptation ladder. The use of cues follows this pattern. In its most rudimentary form, cue use was a minor extension of the energy-bound behavior just described. In fact, the point at which cues were first used is very difficult to determine. It could be argued, for instance, that this capacity is implicit in all behavior, even in the most primitive unidimensional examples, as exemplified in response gradients to nontraumatic intensities of a given energy dimension. Is the amoeba's retreat from a chemorepellent an example of the use of cues? Certainly such behavior implies the capacity to respond to slight intensities, in themselves not damaging, resulting in the avoidance of energy levels that could be harmful. Thus, it might be argued that even the most primitive unidimensional behavior involved the use of cues, because both approach and avoidance gradients are implicit.

What, then, is the most elementary example of cue use in differential response? In my view, the use of cues (or cross-dimensional differential response) involves an appropriate avoidance or approach response to some energy other than the one representing hazard or food material. When this first happened, a remarkable increase in adaptive facility occurred. With the use of cues, avoidance behavior became possible, and approach behavior became much more flexible. That is, responses could be made that allowed the organism to avoid some dangerous circumstance or to approach energy source material even though neither one was present in the immediate environment.

Two major types of cue use are represented in contemporary organisms, and both have probably had a long evolutionary history. In the first type to be considered, genetic linkages are involved, while in the second, learning accounts for the relationship.

(Step 4). Inherited Associations

In all of the examples that are mentioned in this section, a genetic linkage exists between a given cue and a response that has adaptive significance for the organism. Since genetic linkage is present, a long-term relationship between cue and critical circumstance is implied, over sufficient time for selection pressures to ensure that the genetic blueprint coding the association can develop and become stable in the species.

Several kinds of inherited associations are examined in depth in chapter 6, but I list a few here briefly. The pecking of the herring gull chick on the red spot of the bill of the adult has already been mentioned. In this example, the cue (the red spot) initiates the behavior of pecking. The pecking, in turn, instigates regurgitation, and, through this double linkage of inherited associations, the chick is provided with food (Tinbergen, 1948).

The hawk-shape response of the chick is another example. Chicks that have been hatched in an incubator inside a building cannot have experienced anything resembling a hawk, but they will try to evade a hawk shape (the cue) the first time it is encountered (Tinbergen, 1948). How did the genetic linkage evolve? We can surmise that the flight response was initially made to any sudden movement close to chickens, and, by degrees, chickens that were more reactive to the specific cue (the hawk shape) tended to survive, until gradually the cue-flight association became genetically encoded.

All reflexes, instincts, innate releasing mechanisms, and fixed action patterns (Lorenz, 1970) are examples of inherited associations. Many language signals, whether in terms of calls made or movements performed, can likewise be so classified. Mating rituals and a great variety of sexual signals, and many other behaviors that are discussed in chapter 6, are also examples of genetic linkage between cues and responses appropriate to critical survival situations.

The emergence of cue use through inherited associations provided much greater facility in response than the direct energy mechanisms previously described. Both approach and avoidance behavior can be activated prior to the critical circumstance. Unfortunately, in such inherited associations, organisms are bound by the cues that initiate and sustain the behavior. If the cue appears, the genetic linkage ensures that the behavior will take place. This is very functional as long as everything is constant and orderly. Yet if the environment changes sufficiently, that is, if something impedes the cue-response sequence, serious damage or destruction can result. Indeed, if unusual circumstances are encountered, the automatic and stereotyped behavior triggered by the cue might well lead to the mass extinction of a species. For example, if the islands to which the arctic tern migrate each year were to fall below sea level, the tern would die. The pheromone-following behavior of the processional caterpillar is normally adaptive, but it also may ensure the destruction of large numbers, if by chance the leaders cross their own path and thereby set up a circular pattern of behavior that would prevail until starvation took place (Teale, 1949).

Thus far, we have dealt with cues only as they function in adaptation to the external environment. However, as organisms increased in size and complexity, they became differentiated into organ systems with varying characteristics and functions. Because a condition of relative balance must be maintained both within and between such systems, several inherited cue-activation mechanisms have developed to ensure that appropriate action within systems and interaction between systems (the internal environment) can take place.

Such dynamic interdependence is universal in all organ systems, from the single cell to the human being. Yet the extent of such activity depends upon the relative balance in all other systems, and when compensatory reactions occur, their effects reverberate to all functionally adjacent systems. For example, if an individual goes without food for a sufficient time, the repercussions are specific to that particular imbalance. After a time, however, adjacent systems become involved. If the imbalance continues or increases, the disequilibrium spreads to more and more systems until finally, unless compensatory changes occur, death results.

In this section, I have suggested that most internal activity and much overt behavior are based on genetic linkages, and that an organism of great complexity and some adaptability could have evolved and continued to exist without going beyond such inherited associations. Such inherited associations are reflections of long-term concurrences that were present during the evolution of a given species, concurrences that typically lasted for thousands, perhaps millions, of generations. Yet, as previously mentioned, organisms limited to inherited associations would be greatly restricted in the flexibility of their responses. If the environment were altered sufficiently, this mode of adaptation might well become the vehicle for the destruction of the species.

Because many of the regularities in the environment are of much shorter duration, in certain cases lasting for only a few moments during the life of an organism, the ability to capitalize on such short-term concurrences would be a major step in increasing adaptive facility.

(Step 5). Learned Cue-Response Associations

Learning involves the capacity for rapid changes within the organism, changes that reflect the short-term concurrences in the organism's own particular environment. When a cue consistently occurs prior to the appearance of either a hazardous circumstance or a usable energy source, the organism becomes modified so that this cue can occasion adaptive behavior in the future. When the cue appears, appropriate toward or away from behavior can take place.

Learning thus constitutes a remarkable progressive step in evolution. By responding to the cue, hazardous circumstance can be avoided, or food can be found with much greater facility. The organism is no longer stimulus-bound. It is released from the straitjacket of inherited mechanisms by this remarkable capacity for plasticity, for adaptive reactions to a multitude of different kinds of circumstances. The capacity for learning, in short, provided one of the greatest adaptive leaps that has taken place during the shaping process (i.e., selecting in and selecting out) that has been in operation since life began.

Although the various kinds of learning are considered in detail in chapter 6, a brief summary at this point provides an integrating overview. Learned associations, as just discussed, allowed appropriate away from and toward behavior relative to critical equilibration situations. In classical conditioning, as described by Pavlov (1927) and buttressed by hundreds of investigators since his time, we have examples of learned cue deneutralization. That is, the cue picks up the capacity to activate or prepare the organism for other equilibratory behavior. The bell (cue) that precedes the meat powder acquires the capacity to activate the preparatory behavior of salivation, thus allowing both mastication and digestion to proceed more smoothly. To the child who has been burned by a stove, the sight of the stove thereafter becomes a cue that activates avoidance behavior.

In operant or instrumental conditioning, as discussed and researched by Thorndike (1898) and Skinner (1938) and many of their adherents, the association that is learned is between a particular cue and a behavior appropriate to the maintenance of balance within the process systems of the organism. In Thorndike's problem box, the sight of the string (the cue) became associated with the pulling of the string that opened a door that gave access to food. In the Skinner box, the sight of the bar (the cue) became associated with the pressing of the bar, which led to the presentation of a food pellet.

In both classical and instrumental conditioning, the short-term concurrences previously mentioned are present, emphasizing the plasticity of the learning process. A few hours of training is all that is typically required to demonstrate classical conditioning, and sometimes a period of only a few minutes is required to establish an association between the bar-press response and the food.

(Step 6). World Internalization

Another type of learning, observation learning, has been neglected until fairly recently. Though it was implicit in the experiments of Ebbinghaus

(1885) and Underwood (1949), observation learning was not directly approached as such until the research by Bandura et al. (1961, 1963). This prolonged neglect is curious, since observation accounts for most of the learning that takes place in human beings, and probably in many other animals as well. I briefly describe it here, make a more general statement about it at the end of this chapter, and give it specific and detailed treatment in chapter 6.

Observation learning involves a capacity to encode replicas of sensory input. All that is necessary for such encoding to take place is for the individual to attend to (apperceive) some aspect of his or her experience. Thus from the time a person awakens until he or she goes to sleep, experiences are encoded and become available for recollection. Indeed, many of the experiences that occur during a person's lifetime coalesce to provide an enduring cumulative record of what has happened. It is this record, and our ability to shunt through the various categories of this record, that provides the most refined capacity for differential behavior that evolution has produced. I discuss other contributions that observation learning provides later in this chapter, but first I must consider a number of corollary adaptation mechanisms. It was necessary for these corollary mechanisms to emerge (although their initial expressions may have been primitive and rudimentary) before the more basic progressive steps in evolution could evolve and achieve their high levels of adaptive utility.

COROLLARY BUTTRESSING MECHANISMS

In our development up to this point, we have described an organism with remarkable plasticity. It has the ability to capitalize upon short-term concurrences present during even short periods of its own lifetime, and it has the capacity to encode a functional duplicate of its day-by-day experience. Yet the capacities for learned associations and for world internalization did not happen all at once. They emerged gradually as organisms became more complex and as the corollary mechanisms, discussed in this section, also emerged to provide the basis for their operation.

Although these mechanisms are corollary, they are vitally important. Indeed, certain of the attributes that human beings emphasize as most characteristic of our species are examples of these mechanisms. But at this point they should be considered as buttressing foundations that made the appearance of the various progressive steps in evolution possible. In later chapters, I suggest how they coalesce to provide the fundamentals of human nature and those aspects of ourselves that we find of greatest value.

Sensory Input

As organisms became more complex and were differentiated into various organ systems, each with separate functions, some kind of information technique had to develop in order for the status of these many systems to be monitored and to provide the basis for the homeostatic activity that is essential for all examples of life. It is my view that sensation, in all of its many varieties, constitutes the primary solution to the input problem in complex organisms, human beings most particularly. Vital systems within the body are continually falling out of balance. On the outside of the body, energy flux is a continuing process. Where energy changes have adaptive significance, input must be provided so that appropriate action can take place. Many such changes, whether on the outside of the body or on the inside, if they have been of critical importance in the survival of the species during its evolutionary history, are represented by appropriate sensory input.

Although sensation is the major solution to the input problem in complex organisms, it is not the only one. As discussed later, many homeostatic systems, particularly those that involve repetitive and continuing activities, are automatic compensatory mechanisms that can operate even when the organism is asleep or unconscious.

The various types and functions of sensations are considered in detail in chapter 4. A brief overview here describes the various categories of such input and states their general functions.

Primary Activation Input. Where imbalances relate to the immediate survival of the organism, appropriate sensory inputs have emerged that not only indicate the degree of imbalance but also drive the organism to take appropriate compensatory action. Historically these have been called primary needs. For my development, I call them primary activation inputs. For instance, if the fluid content of the body drops below optimal levels, the sensory input thirst in all its various intensities both provides information about the degree of imbalance and instigates behavior appropriate to its resolution.

Secondary Activation Input. In both avoidance and approach behavior (i.e., the use of cues) an interesting problem in motivation is introduced. Since the behavior occurs before the critical situation is encountered, the motivation for it cannot be derived from that situation. The cue that instigates the behavior must produce the motivation. Thus a type of secondary activation technique is required so that response to the cue can be made. The term emotion has traditionally been used to cover this class of secondary activation input. When the child touches the stove and is burned, it will avoid the stove thereafter. The sight of the stove constitutes the cue that, in turn, produces the secondary activating input of fear, which drives the child to avoid the stove in the future. All of the various secondary activating inputs such as anger, fear, and guilt, discussed at length in chapter 4, make a similar contribution. A cue produces the secondary activation input that, in turn, instigates behavior appropriate to an adaptation circumstance.

Cue Function Input. A third kind of sensation arises from the various sense modalities of the body. These information inputs are derived from the presence of certain energies that were of survival significance during the evolution of a species. For instance, everything that moves in the environment, whether predator or prey, causes pressure fronts in a medium. The emergence of a receptor system sensitive to such pressure fronts would be an almost foregone conclusion. The auditory systems of many species have evolved to provide this critical survival information. The same is true for all sense modalities. Each represents energies that were of importance to the survival of a species during its evolutionary history.

A critical point to keep in mind about such information inputs, and one that will keep them separate from the activation inputs just discussed, is that, in the normal ranges, they do not directly represent conditions of imbalance. Rather they represent energies that have either preceded or been consistently related to such crucial circumstances.

These and other kinds of sensation are discussed in detail in chapter 4, but from our brief consideration here it can be seen that all such input, taken together, makes a vital contribution to the information processing and reaction capacities of many organisms. Sensation not only provides the information essential for monitoring the various process systems of the body and the spur that activates behavior appropriate to imbalances, but also provides the basic components involved in learning and memory.

The term input, as I use it, stands not only for sensations from the sense modalities and those representing activation, both primary and secondary, but also for any input that may be introduced into the field of awareness. Thus the term also encompasses any memory that may come back into awareness.

Simplification Techniques

As evolution proceeded, the variety and refinements of sensory mechanisms increased until many organisms, humans in particular, were capable of having thousands of different inputs. This multiplicity was functional, in that it provided the basis for more subtle nuances of adaptive behavior, but at the same time it introduced a critical problem, the problem of information overload. The evolutionary solution to this burgeoning complexity was the emergence of simplification mechanisms that reduced complexity, with only a minimal loss of information. These simplification techniques, all of which make vital contributions to adjustment, are central to the development of psychology and to the very essence of human nature. I deal with each one briefly here and at greater length in later chapters.

Input Fusion. Most of the environmental situations that have provided the selection pressures responsible for the evolution of organisms are signaled by changes in a number of different energy dimensions. Also, in many cases, a particular dimension provides a number of different possibilities for sensory differentiation. When there are receptors appropriate to such energy shifts, a remarkable amount of information about each situation is available. The multiplicity of cues within modalities and the overlap among modalities helps insure that vital discriminations can occur, but it adds to the problem of complexity. Fusion permits a partial solution to this problem, with a minimum reduction of critical information.

Fusion is discussed at length in chapter 5, but the conditions that produce it may be stated here: Inputs that occur together, or simultaneously, will tend to become fused into unity. Fusion frees the organism from having to respond to each bit of input that signifies the presence of a critical situation by permitting it to respond to the situation as a whole. A tiger, for instance, is actually a composite of many inputs, including color, movement, sound, odor, and subtle nuances of texture and shape. After fusion has taken place, the tiger can be responded to as a whole, rather than to each of the particulars that are involved, and it is the total tiger that is dangerous.

Thus, fusion provides simplification with minimal loss of functional information. Responding to the whole rather than to the many different parts has allowed an increase in rapidity of response, and this increase in response facility has constituted the selective circumstance that gradually produced the capacity for fusion in human beings, and probably in many other animals.

The universality of fusion is emphasized by the fact that most of the experiences of human beings consist of phenomenological wholes. We typically respond to books, trees, houses, people, and so on, rather than to the multiple inputs that signify their presence. Following conventional usage, we shall call such fused representations percepts.

Another example of fusion produces is what I call motocepts. Every movement of an organism produces a variety of inputs from the kinesthetic receptors in muscles, tendons, and joints. When particular acts are repeated, the conditions necessary for fusion are provided. Such fusion undergirds the development of motor skills, such as walking, talking, or typing.

The survival contribution of this type of fusion is considerable. It allows behavior to take place much more rapidly, and it frees the apperceptual process, as discussed in detail later, from continuing involvement in the behavior.

Repeated presentations of similar percepts and motocepts result in still another type of fusion that is of equal or even greater importance. When similar phenomena are repeatedly experienced, they become integrated into a general memory representation. For example, after experiencing many houses, which may differ in certain respects although they are similar in many others, a summary representation becomes encoded in memory. This type of fusion not only reduces the number of items that must be encoded, it also reduces the complexity and increases the rapidity of later recall. Although this topic is treated at length later (chap. 6), such fused memory representations or concepts are foundational to thought and fundamental to the effective use of language.

We now encounter another phenomenological unity, also derived from input fusion. It has caused a remarkable amount of dissension among psychologists, but it, in my view, must be included in any general theory if human beings are to be adequately explained. For continuity with the other fusion phenomena, I shall call it the autocept. First, let us consider how it develops and then suggest certain of its functions.

Even as the percept house is derived from the recurrent presentations of many sensory components (particular colors, shades, textures, shapes, etc.), and even as the successive presentation of similar houses results in a concept that summarizes this diversity, the autocept likewise is derived from simultaneous and repeatedly imposed inputs, in particular those inputs that are most immediate, persistent, and critical to the welfare of the person. Thus, the experience of self (the autocept), as a unified phenomenon, emerges according to the same principles accounting for the development of all fusion phenomena.

The autocept makes a vital adaptive contribution. The organism experiences itself as a unity that provides the basis for more rapid and appropriate behaviors. Trauma, for example, affects not only that part of the body that is damaged, but also the entire organism, facilitating total and immediate response.

The autocept also provides a basis for continuity and identity. How does a person know that he or she is the same individual that he or she was yesterday, or for that matter, a moment ago? Simply because certain inputs that have been recurrently imposed since childhood, such as those derived from the rhythms of breathing and heartbeat, provide a familiar background that can be continually monitored to provide self-awareness. These familiar visceral inputs are joined by enduring memories of family members, home, pets, prized possessions, and so on, as these in varying degree have become a part of what a person is. Inputs from continuing processes of the body, and from basic memory encodes, engender self-awareness in much the same manner as other inputs engender an awareness of thirst or hunger.

The most important function of the autocept is to provide the basis for volitional behavior. It is argued in the following pages that although man does not have free will in the philosophical and theological sense of that term, he is capable of volitional action. Since the autocept is a summary and accumulation of the most basic and recurrent experiences in an individual's life, the behavior arising from it will not only be lawful, it will be the most functional expression of what an individual has become.

Constancies. Although the intrinsic dilemma (see chap. 2) is an ever-present problem, it particularly complicates matters at this point. Input fusion, as discussed in the last section, and the constancies to be considered now, are never separate in terms of their development. Input fusion provides the basis for the emergence of phenomenological wholes, while the constancies help provide the stable characteristics of these wholes.

The shape, size, and brightness of each object vary with distance, angle of view, and the amount of light being reflected from it. Consequently, even though the sensory elements are effectively fused into phenomenological wholes (as a function of the fusion processes previously discussed), the organism is still confronted with many versions of such phenomenological wholes, as each object is encountered from hundreds of different perspectives.

Since a given item in the environment has essentially the same survival implications regardless of perspective, a simplification mechanism reducing such complication would have remarkable adaptive significance. The constancies (shape, size, and brightness) provide the basis for this vital simplification. Because of them, the organism need not cope with the multitude of different shapes, sizes, and brightnesses projected upon the retina but, instead, only with the sameness provided by the constancy mechanisms. A window remains the same in terms of its adaptive significance, regardless of perspective, and so does a tiger.

It may be objected that the constancies provide simplification at the expense of information (little retinal tigers are harmless because of their distance, while big ones are dangerous because of their proximity, i.e., the size constancy factor blunts this basis for discrimination). Some information may be lost, but the adaptive contribution (simplification) provided more than compensates for this. As suggested in chapter 5, other cues provide precise information about distance, so the distance discrimination loss due to size constancy is not critical.

Input Focalization. In spite of the remarkable amount of simplification provided by the overlapping processes of fusion and constancy, an intolerable (from an adaptive stand-point) amount of complexity still remains. Hundreds of phenomenological wholes (percepts) are being encountered during each moment, and thousands of fused-memory encodes (concepts) are also potentially available. If two or more inputs are imposed simultaneously, to which does the organism respond? Certainly an organism cannot respond to a number of different inputs simultaneously. Were such to happen behavior would be chaotic and undirected. Thus, there must be some kind of selection—or, as I term it, focalization—which will ensure that out of all the inputs available (whether from sensation or from memory), the particular one that is most pertinent for survival at that moment will occasion appropriate response.

The simplest solution to the focalization problem was probably occasioned by the fact that, as a unity, an organism can move in only one direction at a time. Thus, the direction of movement was determined by a simple resolution of forces, as in the slime mold amoeba discussed in chapter 2. For example, when two chemorepellents are imposed at different places on the surface of an amoeba, it will follow the path that minimizes their combined effect. Yet after the emergence of sensation and memory in higher level organisms, the number of information components available made such simple force resolution inadequate. It was no longer sufficient to respond according to simple additive and subtractive outcomes. It became necessary to respond to the specific input that, out of the multiplicity available, was most crucial to survival at a given time.

Thus, some process similar to attention, as this concept is presently used in psychology, had to appear if the organism were to respond effectively to the myriad inputs that are always potentially available. For this type of resolution, there had to be a central area, called the field of awareness in today's terminology, that was continually accessible to the inputs from both sensation and memory, and that provided the arena for the focalization under discussion.

The emergence of this mechanism of singularity (or focalization) was a critical evolutionary development. Without singularity, reactions would be spastic and multidirectional and would resemble a seizure more than adaptive behavior. Integrated behavior requires that the organism respond to only one item of information at a time. Effective behavior requires that the response be to that item of information that is most crucial to the welfare of the organism at a given moment.

The mechanism whereby inputs, from either memory or sensation, come into focus, I call apperception, following Wundt. The term attention might seem more appropriate except for the fact that this term has typically been restricted to focalization on sensory components alone. I use the term apperception to emphasize that both sensory components and memory components are involved, and that the process of focalization may shift rapidly back and forth between these two major sources of input. As discussed later (see chap. 7), the ongoing process of apperception is automatic and (within the limitations imposed by structure, both as innately determined and as shaped by experience) is a homeostatic, negative-feedback mechanism. That is, of all the dynamically interacting input components potentially available from memory and sensation, those (in terms of the species and personal history of each individual) that are most likely to facilitate the restoration of equilibrium are those that are most likely to be brought into focus at each given moment.

Even as the most primitive examples of life are viewed as homeostatic in nature, so are the most complex. As such, apperception fits within the framework of the postulate of process and the homeostatic negative-feedback corollary. It can easily be imagined how an amoeba, caught between two fields of force, will take the path that allows the greatest degree of equilibrium to be maintained, but it is difficult to see how the complex process of human apperception is an extension of this same principle. This view is developed and defended in chapter 7.

Thus far, in describing the progressive steps in evolution (guided by the postulate of inference), we have derived an organism with remarkable adaptive capacities. The use of cues facilitated by inherited mechanisms made effective responses to long-term environmental regularities possible, while the emergence of learning and memory permitted more subtle adaptations to short-term, even relatively transitory, concurrences obtaining during an organism's own lifetime. Observation learning allowed the emergence of an internalized world that, as a function of memory, both accumulates and summarizes the events that occur during an organism's life. In our examination of corollary mechanisms, we observed how response to more and varied types of cues presupposed the development of refined sensory and retention equipment, which in turn necessitated the appearance of simplification mechanisms that effectively reduced the burgeoning complexity to tolerable levels.

We are now ready to consider the next progressive step in our development. It rests upon the other steps; its appearance required very little change in the processes already described, but it allowed a quantum leap in the capacity for refined information processing and for increased facility of differential response. It is that characteristic, which although it may not be unique in human beings, represents one of our highest adaptive achievements.

(Step 7). Internal Locomotion: Thinking

In our discussion of step 6, it was suggested that observation learning provided the basis for the internalization of a replica of the individual's environment. The development of this internalized substitute world laid the foundation for the appearance of the most refined mechanism for facilitating differential behavior that evolution has produced. With the emergence of the substitute world, the stage was set. Once such a world was present, locomotion within this world could almost have been predicted.

Such internal movement, though requiring but a small increment in process, represents another giant stride in adaptive facility. This remarkable progressive step, which we call thinking, permitted the organism to solve problems of great survival import without actually confronting the danger. In the safety of the cave or the trees, our ancestors who had this capacity, even though in relatively primitive form, could make plans for the coming hunt, the next home camp, the next place to find salt, etc.—plans that were based upon the individual's total collective experience as represented in the internalized world of memory encodes. Thus, internal locomotion, or thinking, allowed foresight. It, in essence, provided a readout of all the past experience relevant to a similar problem that might be encountered in the future.

Thinking can be defined as the continuous process of apperception relative to inputs from memory. Sensing, on the other hand, is the continuous process of apperception relative to inputs from the various receptor systems of the body. The processes of thinking and sensing are functionally and perpetually interrelated. Both sensory input and memory encodes are available for focalization, and apperception typically shifts between them in rapid fashion, with the shifts depending upon a number of variables (see chap. 7).

It should be emphasized that internal locomotion (thinking) demonstrates the postulate of process that all activity, whether overt or internal, tends toward equilibrium. Although hundreds of coalescing variables may be involved at each moment, movement from one encode to the next is fully as lawful as the tropistic behavior of the moth or the most equilibratory path of an amoeba.

Although there is continuing apperceptual flux between inputs from sensation and memory, there is a qualitative difference (as Hume suggested years ago) between the internal and the external worlds. From an evolutionary point of view, organisms that could not make this fundamental discrimination could not have survived. Danger exists primarily in the external world, and the organism that tried to solve external problems by internal locomotion alone would soon be selected out in the evolutionary struggle. The ability to discriminate between the memory image and the sensory impression has been and yet remains essential to survival.

(Step 8). Language

Once again we observe how each progressive step is based on those lower on the adaptation ladder and how each provides a marked increase in adaptation facility without extensive change in process. Once locomotion within an internalized world (thinking) emerged, some technique for sharing and accumulating the various internalized worlds of each individual would mark a major further step in adaptation facility. Language provided this. It provided the ability for collective foresight, for group planning, and for group cooperation. Human beings, more than any other animal, are group animals. We survived as a group and we evolved as a group, and it was language that provided a major basis for the circumstances of our evolution. Language not only brought the experiences of all members of the group into functional coalescence, but also accumulated the experiences of past members of the group to further increase the amount of information available for coping with each survival circumstance. In short, language was the fundamental instrument in the development and transmission of culture.

When spoken language developed is, of course, impossible to determine. Sounds leave no fossils. Yet it must have emerged slowly, at first, perhaps, in grunts and hoots and signals. Then as the underpinnings of brain and vocal chords became more refined, language increased in corresponding complexity, until great precision in expression and communication was achieved.

In my development I have limited the progressive steps in evolution to eight. Why not extend it? It could be argued, for instance, that symbolic representations of language, as in writing, marked a major addition to the adaptive repertory, and that electronic methods of communication and problem solution as in radios and computers greatly increased adaptive potential, particularly in relation to issues that are now important.

These, as well as other innovations, certainly increase man's ability to resolve contemporary problems. I believe, however, that such innovations are better classified as exaptations arising from the cumulative influences of culture, rather than as biologically based evolutionary derivatives.

RELATED PROBLEMS AND CRITICISMS

From the preceding discussion of the progressive steps in evolution, it may seem that a given attribute contributes only positively to differential responses facility. This is not the case. Regardless of the contribution of a given progressive step, its very inclusion in the process repertory of organisms brings with it certain negative repercussions. No attribute can be incorporated without an increase in complexity, and complexity itself inevitably increases the possibility of malfunction in any system. The question becomes, does the increase in adaptive facility resulting from the inclusion of an attribute contribute more than the complexity detracts? If so, it is a functional attribute and this net, rather than absolute, functional value is the condition of its incorporation.

It may also seem that each progressive step is somehow separate and distinct from the rest, both in their emergence and their present use. This also is not the case. Many if not most of these steps developed in conjunction, with perpetual dynamic interdependence prevailing to determine relative function. Indeed, organisms that demonstrate the highest steps in the ladder also, without exception, still utilize those lower down, in varying degree. In the action of the human immune system, the most primitive examples of gradient behavior are widely represented. Cue utilization through inherited mechanisms is retained in the repertory of most animals and in the activity of the human autonomic nervous system. World internalization is a direct and necessary precursor for thought.

The adaptive contribution of a progressive step (or any attribute) cannot be determined in isolation. An organism is a totality comprised of many organs and organ systems working in dynamic synchrony. The functional value of any attribute (or progressive step) depends not only upon its own contribution, but also upon its synthesis with other segments of the system. A student of mine once asked, "If the length of a frog's leg is a functional attribute, why didn't the legs of frogs keep growing longer and longer?" The answer is obvious. Every attribute brings increased complexity and must be integrated into the total system. Legs can grow only as long as added length contributes more to survival than increased complexity and integrative difficulty detract. This judgement must depend upon how a particular length of leg is integrated into the total frog. Years ago, Leonardo Da Vinci remarked that "man would not invent anything more economical or more direct than nature, for in nature's inventions, nothing is wanting and nothing superfluous." Nature does provide the most efficient adjustment with the least complexity possible, but the contribution of each attribute can never be understood alone. It must always be considered as a part of an integrated whole.

Another question that may have troubled the reader involves the method used in the development of the argument of this chapter. He or she might well say, "Come on now, you can't really insist that your conclusions are completely derived from a consideration of what must have happened in the course of evolution. Surely a good part of the logic was influenced by prior knowledge of the very attributes that have been so laboriously inferred." The accusation is valid, but the objection is not. Certainly a being freshly arrived from Andromeda could not simply examine our world and infer the inevitable emergence of the human being, let alone the progressive steps that have been described. Yet I am not from Andromeda, and neither is the reader. We are both products of the selective pressures that have been present on this earth during our evolution. We represent within our processes and structures (indeed in the very knowing capacities that we are now turning upon the questions of our nature) the shaping conditions that determined our evolution and this nature. Since the reader, as well as the writer, is a product of the selection pressures being described, and since he exemplifies within himself the claimed principles and attributes, she or he provides an informal check derived from a continual questioning: "Is this really how I work? Is this really a valid description of that totality that I call 'myself'?"

A much less complex problem involves the sequence taken as we moved from one progressive step to the next in our development. The journey from primitive life elements to complex organisms may seem to have followed iron rails fixed beforehand, with the human being's appearance predetermined once life first appeared. I certainly do not share this point of view. It seems to me that humans are the outcome of a remarkable number of adventitious events. If the evolutionary clock were turned backward two or three billion years, it is quite unlikely that humans would evolve at all. There have been too many tangents in evolution, too many fortuitous circumstances to permit the assumption of inevitability. Yet, as improbable as the event may seem in retrospect, humans did appear, and their appearance was lawful.

It is not my task to demonstrate the circumstances that might have made the emergence of human beings inevitable. I simply try to explain the nature and function of the attributes that both define them and permit their existence to continue. In the following pages, I defend the view that humans are both the lawful outcomes of natural processes and the present manifestation of these laws, and I do so without accepting the restrictive notion of predestination or the circumlocutions of teleology.

CHAPTER SUMMARY

In the second chapter, a number of general principles were inferred. It was suggested that the first moving life elements were made of vulnerable protoplasm and that they resided in a context of fluctuating energies, some degrees of which might harm or destroy them. Thus, techniques for regaining and maintaining the balance essential for life to continue were necessary. The entire course of evolution can be understood in terms of the emergence and refinement of such techniques, gambits that allowed the balance necessary for life to be maintained within tolerable limits until replication could take place. Three fundamental techniques were discussed, namely, reproduction, armor, and differential response.

In this chapter, we examined a number of progressive steps in evolution. It was suggested that in each succeeding step a small change in process typically resulted in an enormous facilitation of differential response.

Differential behavior (step one) began when a primitive life element, in order to maintain the balance necessary for existence, responded differently to energy shifts on a single dimension. The next step (step two) was reached when two energy dimensions became the occasion for response. Very likely, such simple examples are purely hypothetical. The first moving life elements were all probably multienergy responders (step three). Even at this stage, the adaptive contribution of such differential behavior was considerable. Because of the inverse square law, the movement away from potentially harmful energies or toward energy source material provided a considerable margin of safety. A much greater increase in response facility came when cue use emerged, that is, when a noncritical energy could occasion either away from or toward responses relative to situations that were of immediate survival importance. Two different types of cue use were discussed: the first (step 4) involved inherited associations between cue and behavior. Inherited associations provided appropriate adaptive responses to long-term regularities that were present during the evolution, or at least a part of the life history, of a species. The major drawback of inherited associations was their lack of flexibility. If the environment changed sufficiently, their mode of operation could occasion the destruction of a species. Next, learned associations (step 5) were considered. These allowed appropriate behavior to short-term, even transitory, concurrences existing during the organism's own life span.

The most refined type of learning that has appeared in evolution, namely, observation learning (step 6), was then examined, and the notion of a surrogate internalized world was postulated, a world that included a summary and consolidation of the basic experiences in an individual's life. With the emergence of such internal replicas, the individual could bring all pertinent aspects of his or her past experience to bear on the solution of present problems.

At this juncture, corollary mechanisms were discussed, so-called because their development was both foundational to and corollary to the progressive steps in response facility being examined. These corollary mechanisms included: (a) sensation as the evolutionary solution to the input problem in complex organisms, and (b) simplification mechanisms that prevented the organism from being inundated by the sheer amount of information that sensation provided. These simplification mechanisms included fusion and constancy, which, when taken together, ensure the development of such phenomenological wholes as percepts, concepts, motocepts, and the autocept. The process of apperception as the fundamental focalizing mechanism ended our discussion of corollary mechanisms. Next, the notion of internal locomotion (step 7) was considered. It was emphasized that each progressive step rests on those that came before, and that, in each case, a small change in process allowed a remarkable increase in adaptation facility. Thinking (i.e., locomotion within the internalized surrogate world already provided by observation learning), is another example of such a remarkable adaptive achievement. With this ability, individuals could remain in the safety of their home camps or caves and solve problems crucial to their survival. That is, planning and foresight could take place.

The final and most recent progressive step came with the development and use of language (step 8). Language provided the basis for bringing together the experiences of each living member of the group and it also, and this is of vital importance, allowed the accumulation of the experiences of past members to further increase the amount of information available for coping with each survival circumstance.

A word of caution at this point. By progressive steps, I do not mean to imply that organisms that have achieved the higher rungs in evolution are necessarily better or, for that matter, even better adapted. Every species is, in a given slice of time, effectively adapted to its environmental niche; its sheer presence is assurance of that fact. Yet organisms that achieved the higher steps in evolution were more capable of responding effectively to more complex kinds of circumstances. The progressive steps discussed imply greater facility in both information-processing and compensatory reactions. The human being, as the most obvious example of all of these steps combined in various proportions, is the most adaptable organism that evolution has produced.

These are, in my view, the major steps in increasing adaption facility that have taken place in the evolution of complex organisms. Let me emphasize again that all living things are functional analogues, reflections of the selection pressures that defined the particular ecological niches in which they evolved. This is also true of human beings. Gravity is reflected in our muscles and bones, a segment of the electromagnetic continuum by the color red, molecules sifting through the air by the odors we sense. The inverse square law that obtains in the physical world is demonstrated in the negative-feedback curve that depicts our behavior, and the Euclidian three-dimensional deployment in the external environment by the surrogate cognitive duplicate through which we move and interact.

Our cognitive world is a functional not a literal analogue; colors, sounds, and odors exist only in our heads, but they are functional translations of energy shifts in the external environment. The three-dimensional character of our cognitive world is not a literal reflection of the external world; selection pressures (see chap. 5) insured that certain vital features are emphasized, while less important ones are diminished or missing altogether.

This analysis suggests that the only reality is our subjective experience, that we can never compare your red with my red or your rat with my rat, etc. This is true, but this does not mean that each of us is isolated in solipsistic loneliness. As members of the same species with a shared evolutionary history, we may assume that our information-processing machinery works in similar ways. Insofar as our cognitive world is a functional duplicate, we can know about and interact within the external world, and, insofar as we are functionally alike, we can communicate. Our job as psychologists is to determine the nature of an experience, the circumstance that produces it and the role it plays in information processing and reaction.

Thus, human beings, and all living things, are functional duplicates of the forces that have shaped them down through evolutionary time. Here, as demonstrated in the coming chapters, lies the key to understanding our nature. When we examine any aspect of what we are, whether physiological, behavioral, or cognitive, we shall always ask and try to answer the question: what is or was its function in survival?

In the following chapter, I try to indicate how the basic postulates and corollary derived in chapter 2 provide the base of a conceptual structure integrating and lending meaning to much of the subject matter of psychology. In this effort, the present chapter serves as a convenient overview, a kind of skeletal model, which guides us as we proceed. I hope the reader can move comfortably along the path that I have indicated. The reader will and should make his or her own critical forays into the issues being discussed, and I encourage him or her to continually ask the question, "Is this really the way I work?"

4
Sensory Input

The human hand is a marvel of engineering. It took millions of generations of natural selection to produce this functional appendage in its present form. There is certainly nothing remarkable in such a statement, but let us extend it. Not only do we have such functional attributes as hands, feet, and legs; we are also capable of experiencing such inputs as pain, red, sweet, cold, fear, and so on. The capacity to have such input is also the outcome of millions of generations of evolutionary shaping. Every organism, particularly as we move up the phylogenetic scale, is bombarded with myriads of such inputs that must be processed before effective behavior can occur.

In any information-processing system there has to be an area, central point, functional integrator—call it what you will—in which the information pertinent to the operation of the mechanism in question is represented so that effective action can take place. In the human being, following Wundt, I call this area *the field of awareness*. At this moment, for instance, the field of awareness of the reader includes such sensations as odors, pressures, sights and sounds, inputs from physiological imbalances such as hunger, thirst, or pain, as well as those more subtle inputs typically called emotions such as anger or fear. During the waking state any of these inputs may intrude into the field of awareness to occasion action. Yet this is but half of the story; inputs from memory may also enter the field of awareness when the circumstances are appropriate.

Here again we face the intrinsic dilemma and the carryover of traditional terms of classification. This chapter considers only those components of experience that are immediately and irresistibly introduced into the field of awareness from systems both on the inside and outside of the body. Thus, it includes sensory inputs from the sense organs, activation systems, and those characterized as affect. Encodes (memories), which may also be available as inputs into the field of awareness, are treated in the two following chapters.

The terms I use are the outcome of a three-way compromise between: (a) the communication advantage of traditional usage, (b) an effort to indicate the function of a given process, and (c) the availability and appropriateness of computer concepts.

PROBLEMS OF EXPERIENCE

History

Experience as a topic for investigation is certainly not new to psychology. Indeed, it comprised much of the subject matter of structuralism, the first school in our discipline. These first psychologists spent many years analyzing experience into its various categories. They finally wound up with three basic divisions: memory images, feelings, and sensations. They then proceeded to classify sensation in terms of various modalities such as vision, audition, and olfaction, which were then broken down into various dimensions, such as brightness and hue or pitch and loudness. In their search for the smallest brick in the structure of sensation, they finally wound up with just-noticeable differences along a particular dimension.

It seems curious that the structuralists were never concerned with how experience functions in the survival of species, nor how it helps the individual adapt to its own environment. Perhaps we can excuse this oversight, for structuralism was dying even as Darwin's theory was gradually finding acceptance. Yet what about functionalism, our second school in psychology? Its orientation was based on Darwin's theory, and James, one of this school's originators, had declared early and often that the mind would never have come into being if it did not have adaptive function (James, 1892). Why, then, did his followers not turn to the categories established by the structuralists and seek to evaluate their adaptive contributions? The foundation had already been laid. They believed that the mind was the subject matter of psychology, and they insisted on an evolutionary perspective. What happened?

Boring (1929/1950) suggested that the explosion produced by behaviorism effectively fragmented functionalism, but this explanation does not go far enough. Functionalism was fragmented from the beginning. The real obstacle that kept the Functionalists from centering on the adaptive contribution of the categories of experience seems to have been their preference for descriptive generalities. James became so involved with his stream of consciousness that he never effectively moved beyond it. He talked about consciousness as one might describe an automobile, pointing out that it runs continually on roads, stops and starts frequently, and belongs to somebody. Even in his description of thought, where he probably achieved his greatest specificity, we gain little insight into the variables affecting the stream of thought or the particular role that thought plays in adaptive behavior.

The functionalists of Chicago also seem to have been limited to global generalities. Carr suggested that mental activities are those processes responsible for the "acquisition, fixation, retention, organization, and evaluation of experiences, and their subsequent utilization in the guidance of conduct" (Carr, 1925, p. 1). There is little to quarrel with in this statement, but where do we go from here? It is one thing to point out that these processes are important for the understanding of behavior, but it is quite another to evaluate their specific contribution to behavior.

The Chicago functionalists might well have dealt with the manner in which each sensory attribute contributes to more effective behavior, but their consideration of this topic was skimpy at best. Carr did posit increased adaptive potential as a function of the type of sensory modality found in different organisms. He thought that one of the basic criteria for evaluating an organism's relative place along the phylogenetic scale was its capacity to respond effectively to objects at greater and greater distances. Animals higher on the phylogenetic scale increasingly use distance receptors like those in the human eye and ear, while there is a corresponding reduction in the use of contact receptors (Carr, 1925). Yet Carr has little more to say about the adaptive significance of specific sensory mechanisms.

However functionalism did provide a benign environment for the development of the many branches of psychology that were emerging in the first part of the 20th century. At his redoubt at Cornell, Titchener, the last great structuralist, kept insisting that the only psychology was the study of consciousness by trained adults (Titchener, 1909). Since animals, babies, and mentally ill people cannot indulge in introspection, Titchener cast them out of psychology. The functionalists, on the other hand, allowed them in, along with the rapidly growing testing movement and the data from the physiologists.

Perhaps the eclecticism of the early functionalists was more a reflection of the demands of the time than a statement of an intrinsic broad-mindedness. However it was, the scene was set for the emergence of a hard-nosed scientific objectivism, and Watson provided this approach. Psychology was suddenly self-consciously scientific, and anything that smacked of mentalism or subjectivism was unacceptable. Thus, beginning with Watson's (1913) review article and continuing through the 1950s, behaviorism, in one form or another, became the mainstream of our discipline. Considerations of experience or consciousness, either in terms of its structure or of its function, became eclipsed for a time.

Of course, many psychologists never relinquished experience as a major basis for understanding the workings of the human being. The Gestalt psychologists, the followers of Freud, the existentialists, and the humanists all emphasized the importance of experience as a basis for understanding human nature and human conduct.

Since 1971 because of the emergence of interest in information processing and the corollary resurgence of concern with cognition (see chap. 1), experience has again become of central importance in our efforts to understand ourselves. Feeding into this rapidly growing trend are the data and research from ethology and evolutionary theory (also see chap. 1). Symptomatic of this interest in evolutionary thinking as a frame of reference for psychology is an article by Skinner (1981) that analyzes behavior in terms of its consequences for adaptation, and the very recent wholehearted endorsement of an evolutionary orientation as demonstrated in books by Barkow, Cosmides, and Tooby (1992), and Pinker (1994). Since 1996 over 800 articles have been listed under the designation theory of evolution in the psychological abstracts.

Psychologists, previously and at the present time, have been concerned with three evolutionary outcomes: (a) hardwired inherited behaviors such as reflexes and instincts, (b) aspects of the information- processing machinery, as in Broadbent's studies on the latency of attention shift, and quite recently (c) inherited predispositions sometimes called modules, which often lie hidden beneath the tangles of cultural overlay. This latter involvement marks a resurgence of interest rather than a new direction. We have long been concerned with precisely this issue in studies on intelligence, the source of personality traits, the specifics of emotional response, the development of language, and the like under such descriptive dichotomies as nature versus nurture, heredity versus environment and learned versus innate. The major new emphasis has been an avowed commitment to an evolutionary perspective and the development of ingenious experimental designs.

In this and the following chapters, an effort will be made to continue the tradition of the Chicago functionalists and to extend their unrealized program of construing the many facets of experience and behavior in terms of their adaptive contribution. In this chapter, I consider the various categories of sensation and try to infer the manner in which they provide information about conditions (both internal and external) critical to survival and, equally important, how they instigate behaviors appropriate for survival. Yet before I begin, I must consider one of the oldest and muddiest issues in psychology, an issue of awesome complexity: the so-called mind-body problem. Anyone who accepts the view that experience influences behavior must grapple with it.

Very recently, demonstrating both the longevity and worrying persistence of the problem, two books recapitulate some of the ancient arguments and achieve some dubious rapproachment. In the first, Kim (1998), after long argument, wound up with an identity position, the view that there is only one event. He posited a reductionism of the psychological to the physical for many processes. But not for all. It turns out that he has what may be termed a partial identity hypothesis. He is quite willing to accept that a number of mental activities such as intelligence and decision making are simply brain activity but curiously, at least to me, such feelings as well-being and jealousy and the big one, conscious awareness, are not reducible to brain states.

McGinn (1999) started out in a quandary. According to him, conscious states cannot be identical to brain states, and yet they cannot be different either. He finally concluded that the human intellect simply is not up to the job of resolving the dilemma and then suggested that consciousness perhaps sprang into being and was propelled into reality by the same remarkable conditions that produced the Big Bang. Not only did McGinn drive consciousness into the implausible and the unknowable, but also he so relegated the issue of free will and the nature of the self.

I am not in agreement with these interpretations. I offer them because they are very recent and both represent serious efforts to treat this difficult issue. It is my own view that experience (mind, consciousness, awareness, etc.) is a lawful outcome of the selection pressures encountered by organisms during evolution, and that it can be understood from an adaptive frame of reference. The same conclusion is drawn concerning the nature of the self and freewill, both to be treated in the following chapter. But let us now consider the mind-body problem.

Mind Versus Body

What is sensation? How does physiological excitation or activation result in psychological experience? Let me give a simple example that may clarify the issue. Assume that a particular wave length of the electromagnetic continuum strikes the retina of the eye. The consequent action on the receptors sends an impulse down the optic nerve, across the optic chiasma, and finally to the occipital lobe of the brain. Then, somehow, it happens. Suddenly, the physical process appears to be transformed, and the individual has the psychological experience of red. The historic dualism between mind and body comes into focus here. The psychological attribute experience has inspired the formulation of such traditional constructs as spirit, mind, consciousness, and soul. Yet its precise nature remains an unsolved question. Experience has been ignored, evaded, and defined away by many psychologists, but just when this tenacious ghost seems to have been laid to rest, it revives with new life and greater ambiguities, as demonstrated by Kim (1999) and McGinn (1999) in their recent books.

The reason experience must be dealt with by psychologists is simple, yet profound. It is in experience that humans, and probably many animals, find their reality. Sensation is the outcome of processes in our neurological machinery as these are activated by energies in the environment. Thus, sensation is the way the knowing processes of our body transform the external world so that we both have knowledge of it and can respond appropriately to it. It is the most important evolutionary solution to the input problem in complex organisms. Sensation, and indeed all experience, evolved because it allowed organisms to adapt more effectively to their environment. This assumption leads to a very difficult, yet very basic question. Do components of experience have some function beyond the physical processes from which they are derived or with which they may be identified?

I assume that all aspects of experience are simply the manner in which physio-chemical-electrical processes in the nervous system are manifested, and that they have no existence apart from these physical processes. So when I speak about a particular experience, though I may use such terms as sweet, sour, or red, I am including in these concepts the total physio-chemical-electrical event that accounts for the experience in question.

The previously mentioned position is not very different from the identity hypothesis held by many thinkers and originally stated by Aristotle, but it does place the emphasis on the experiential side of the total process, and assumes that this experiential side is functional in adaptive behavior, and that this functionality was instrumental in the evolutionary emergence of such components in the first place.

Views other than the one proposed invariably require the existence of an impalpable, unextended (i.e., not occupying space or time) mindstuff. Man becomes an open or split system, and those areas in which the mindstuff works are typically considered beyond scientific investigation. Yet there is more than simple pragmatism to justify a functionalistic identity hypothesis. It is the simplest hypothesis available, it is not incompatible with any knowledge that we have, it is consistent with the available facts, as sketchy as they are, and it is amenable to further development. For science, these advantages are considerable, particularly when the unsatisfactory implications of the alternative positions are evaluated.

It might reasonably be concluded from this analysis that I am an epiphenomenalist, and that experience is considered as a nonfunctional artifact of physical changes. However my view is precisely the opposite. Even as light is the natural and functional outcome of the various physical processes in a flashlight, so components of experience are considered to be natural and functional outcomes of physical, chemical, and electrical transitions.

Yet if we accept the hypothesis that mind and body are one, why do we need to assume that experience has any function whatsoever? Why could not the organism function just as well with only the physical aspect of the process? Does not experience simply add a nonfunctional, unnecessary complication? I do not think so. Can the function of the filament of a bulb be considered apart from the light that it produces? The light and the filament cannot be separated because they are essential aspects of only one process. It is the perspective that creates the dualism. I add parenthetically that improvements in flashlights have been directed toward producing more effective light, and that many a filament was tried and discarded along the way (see Wagoner & Goodson, 1976, for a more extended treatment of this issue).

From a research standpoint, we are forced to draw an interesting conclusion: it is the experiential aspect of the process that is available for evaluation. The physical aspect is hidden from us. The color red is given to us immediately and so are all the other categories of experience, but we know nothing about the physiological process that occurs with the red, or, for that matter, any of the other physical aspects of experience. Thus, the components of experience are the only data available on such processes and must serve as the basis for our evaluation of the nature and function of these processes. It is as if we were seeking to learn how flashlights work, with the light produced being the only data available for study.

As knowledge and techniques expand, researchers may well isolate the particular physical processes that are intrinsic to the various categories of experience, but that time has not yet come. Even when this advance occurs, there still may appear to be two processes rather than one. When measuring instruments are used, the process may be described in terms of recording units, but in studies by the organism in which the process takes place, it will still be described in terms of components of experience. There seems to be no solution to this vexing problem. It will be clarified, if not completely resolved when techniques are developed that allow the projection of a subject's moment-by-moment experiences upon a screen, but that achievement will probably not occur for some time.

I thus assume that there is but one event, composed of the experience and the physical process, with our only knowledge of the event being that which is given directly in experience. When we use terms like pain, pleasure, and red, all of which have traditionally referred to categories of experience, the total process is intended, and not just its subjective aspect. Why do I use such terms, fraught as they are with ambiguity? I must use them; they constitute most of the language available in the area. Yet I also lean heavily on both computer and information theory when the concepts from these disciplines seem applicable.

THE CATEGORIES OF SENSORY INPUT AND THEIR FUNCTION

In the statement of the Postulate of Inference, it was suggested that any attribute that has been present in a species for an enduring period can be understood in terms of its adaptive contribution. Gradually, during the millions of generations of life's existence, the various input systems emerged. The general function of all sensation is to provide information and activation relative to problems of survival encountered during the day-by-day existence of each individual. Yet there are many different categories, each of which makes a unique contribution to particular adaptation problems. In this chapter, I analyze these various categories and try to infer something about the contribution that they make. However first, let us consider a number of general questions relating to the emergence and evaluation of sensory mechanisms.

Could an equally effective organism have evolved without sensation? As a matter of fact, it is easy to conceive of an entity that could respond appropriately to many different energies without this attribute. Indeed, every complex contemporary organism is a composite of many types of input–output mechanisms. A casual examination of the human being, for instance, reveals several homeostatic systems that involve sensation only secondarily, if at all. First, and probably the most primitive, are the processes intrinsic in cell functioning, where imbalances within the cell occasion compensatory changes in the membrane. Other homeostatic activities include various reflex reactions and such finely modulated processes as those demonstrated in healing, body defense against microorganisms, and the digestion of food. The autonomic nervous system controls many continuing activities without the intrusion of sensation. All such nonsensory mechanisms, as well as many others not mentioned or unknown, very likely appeared early in living organisms in various combinations, and many still remain as a vital part of the adaptive repertory of each individual.

Sensation is required for the monitoring and reaction to more variable and generally more complex adaptation circumstances. But where homeostatic activities are rhythmic or when relatively invariant, automatic balance maintaining or regaining mechanisms are typically involved. Indeed, the autonomic nervous system evolved to provide the monitoring and regulation of such activities, freeing sensory systems for more variable information processing and reaction.

Somewhere along the line, as organisms became more complex, simple homeostatic devices were no longer sufficient. A method of representing the multitude of energy changes taking place (both inside and outside the organism) became necessary to provide the information processing essential for balance maintenance. Though many different techniques may have appeared over the course of evolutionary trial and error, sensation provided the most enduring and effective solution to the input-activation problem. It is the solution that is predominant in human beings, and its remarkable facility for information processing and appropriate reaction is a direct reflection of its richness and variety.

What can be said about the beginnings of sensation? Nothing with any degree of certainty, but we can assume that its emergence was gradual, and that it took millions of generations of natural selection to achieve the highly refined input systems found in organisms today. The first hints of sensation may have been present in the earliest and most primitive examples of moving life. An electrical potential is intrinsic to every living cell, and this potential was likely a precursor that, as a function of evolutionary honing, finally resulted in the polarization–depolarization sequence of contemporary neurons (Rosenzwieg & Leiman, 1982). Sensation may have had its origin in similar rudimentary processes.

It seems likely that the emergence of sensation and its basic divisions reflected the two kinds of survival requirements imposed upon every example of life. Since all life regardless of simplicity requires energy, primitive creatures must have been able to ingest suitable material. Such ingestion was somehow activated by growing imbalances within the organism, so that compensatory activity could occur and essential balances could be restored. In their most rudimentary forms, such balance-trending activities were probably similar to the osmotic shifts that occur in contemporary cells, where changes in the membrane allow appropriate material to be transported through so that optimal balances can be maintained. As such, they represented the initial phases in the development of the highly refined activation systems of complex organisms. The amoeba represents such a rudimentary system. After ingesting large quantities of usable material, it remains quiescent for long periods. As time passes, however, it becomes increasingly active and increasingly prone to ingest further material. Even at this level, however, the organism is selective. The incorporation of carbon or glass into the plasmagel does not affect behavior, whereas assimilation of digestible material markedly reduces reactivity (Jennings, 1906).

The first motile life elements must also have been able to respond differentially to energy changes arising from the external environment. Initially, these responses were probably at best only diffuse reactions to the impact of particular levels of energy. Yet as evolution progressed, zones of sensitivity without doubt developed, allowing organisms to respond to ever more subtle energy shifts. As a contemporary prototype of such crude initial differentiation, consider the flagellate protozoan, Euglena. On the anterior portion of this one-celled animal, there is a small reddish spot. Sudden shifts in the intensity of light falling on this spot occasion abrupt changes in the animal's movement (Carlile, 1975).

The receptor sites on the surface of such eukaroytic cells as slime mold amoeba and leukocytes (discussed in chap. 2) may also be considered as primitive prototypes of sense modalities, particularly those of taste and smell. Different classes of such sites are reactive to different chemicals, so that each such cell can aptly be called a multisensory system. It seems likely that all sensory systems evolved from such rudimentary beginnings. Initially, differential response was no doubt crude, but as organisms became more complex, the specialization and refinement of sensory systems that partly accounted for this complexity allowed greater and greater precision for readout and reaction to more subtle energy shifts. At approximately the same time, and perhaps contingent upon such burgeoning complexity, receptor systems allowing reaction to more subtle and varied changes within the organism emerged.

It is important to state again the notion that every living organism is a homeostatic, negative-feedback system, and that the many varieties and shades of sensation are reflections of the inverse square law, which prevails relative to energies in the environment.

It should also be emphasized that the term sensation is used in a broader sense than is typical in contemporary psychological texts. Under the term sensation I include not only such traditional sensations as light and sound, but also the many shades of input representing affect and those representing the various activation systems of the body. The inclusive character of the term is demonstrated in the following statement:

> Each type of sensation represents energy changes (whether internal or external) that have been of enduring importance in balance maintenance during the evolution of a species.

In the human being, for instance, there are thousands of receptors, some widely distributed, others coalesced into sense organs, both on the inside and outside of the body, which are continually providing inputs that make the behavior essential for survival possible. The input from such mechanisms may be divided into several relatively distinct categories according to their function.

Primary Activation Input

As organisms became more complex, those conditions that were hazardous—either critical internal imbalances or external energy applications—came to be represented by appropriate kinds of input that have traditionally been called primary needs. The remarkable character of such input is that it not only signals that vital systems are approaching critical imbalances, it, as introduced, constitutes the spur for appropriate action. Organisms react not to a hot iron, but to pain; not to loss of body fluid, but to thirst; not to hours of deprivation, but to hunger. In all these instances and in others that are mentioned later, the input has a double function. It signals that a state of imbalance is present, and it also drives the organism to do something about it.

From a functional point of view, such primary activation inputs can be divided into two basic classes. Those arising from internal imbalances having to do with the ongoing physiological operations of the body, such as thirst, hunger, and the need to urinate, and those arising from hazardous external energy applications, such as pain, heat, and cold.

All such inputs exist only as the sensations of the individual, and typically have little similarity to the precipitating circumstances. For instance, the inputs of heat and cold are instigated by molecular motion. If the molecules in a medium, whether air or water, are moving too rapidly the activating input called hot is introduced. If the molecules are moving too slowly, the activating input called cold is introduced. So in a real sense, such primary activating inputs are psychological transformations of body imbalances. Transformations that provide the basis for appropriate compensatory behavior, but transformations that may be very different from the instigating circumstance.

A number of steps may, and often do, intervene between the circumstance of imbalance and the introduction of activating input. For instance, when energy deficit in the body reaches a certain level, a number of subtle changes take place in the chemical proportions in the blood. Apparently, there is a monitoring mechanism in the brain that reacts to this disproportion, even as receptors in the nose are differentially responsive to shifts in the chemical composition of the air (Oomura, 1976).

Other changes in the body may provide overlapping indications that energy imbalances are being encountered. Stomach contractions, olfactory cues, and eating habits may all contribute additional bases for reaction. But the major functional outcome of all these coalescing indicators is the sensation of hunger, and it is the hunger that both signals the imbalance and drives the individual to do something about it. The same statement applies to the other primary activation inputs. Though a number of indicators may overlap to represent the imbalance circumstance, the input that both signals the imbalance and occasions compensatory action is the sensation, whether it be cold, hot, pain, or another primary activation input.

Although the most important adaptive function of primary activation input is to signal imbalance and occasion compensatory reaction, such inputs may also serve as cues that provide the basis for behavior appropriate to the specific imbalance circumstance. For instance, increasing hunger is often the cue that it is time to return home for dinner. This example suggests a characteristic of primary activation input that is fundamental and necessary for appropriate behavior to occur. The input representing each type of imbalance must be qualitatively different from the others. If this were not the case, it would be impossible for the individual to orient appropriately to each imbalance circumstance. This leads us to postulate a doctrine of specific need qualities reminiscent of Muller's (1843) doctrine of specific nerve energies. Stated formally, this doctrine reads:

> Each type of activating input must be qualitatively distinct from all the others so that behavior appropriate to each imbalance state can take place.

Such specificity is of prime importance when learning (see chap. 5) enters the picture to provide the basis for response. If the activation inputs were not qualitatively different—that is, not distinguishable from one another—the organism would not be able to learn behaviors appropriate to food versus water sources. How can we describe such differences in quality? We cannot. They are simply given as such in sensations. Thus hunger and pain are just as different and as distinguishable as red and sweet.

A characteristic that may contribute to such qualitative differences is what I shall call prepotence. Certain activation inputs are simply more intolerable than others. In general, those that are more critical to survival generate more prepotent inputs; pain is more intolerable than thirst, and thirst more intolerable than hunger. These intermechanism differences in prepotence are predictable from an evolutionary frame of reference. Body trauma is more immediately critical for survival than either fluid or energy requirements.

The intensity of activation input is even more important for adaptive behavior. This intramechanism continuum exists for all primary activation input and is a direct reflection of the homeostatic negative-feedback principle and the inverse square law. Where a primary activation mechanism can be isolated, a relationship can be found between the degree of imbalance and the intensity of the input. For example, pain varies from the barely noticeable to the excruciating, and so do heat and cold.

Although many experiments have investigated the relationship between the stimulation of trauma receptors and shifts in experience (registered as pain, heat, or cold), few have described the relationship between the extent of either deficit or surfeit and corresponding changes in sensation. This lack of research reflects partly the difficulty of isolating such mechanisms and partly psychologists' continuing emphasis on overt behavior. In such research the behaviorists have provided clear evidence supporting a homeostatic point of view: that there is a consistent relationship between both food and water deprivation and compensatory activity (Keller & Schoenfeld, 1950; Mackintosh, 1983). The data from many different studies support the homeostatic negative-feedback corollary of the present theory (i.e., the longer the period of deprivation, the more rapid the compensatory activity).

Before ending this section, I must mention another class of input that is very troublesome to classify, one in which homeostasis is abetted by inaction and quiescence. Although the relevant transmitting mechanisms have not yet been identified, the sensory derivatives have long been recognized as fatigue and nausea.

Fatigue occurs when a given muscle or muscle complex falls into physiological imbalance during prolonged or excessive activity. When this happens, either oxygen or sugar, or perhaps both, is depleted, waste products build up, and fatigue is the sensory result. Such input makes a vital adaptive contribution. Cessation of activity allows balance to be regained in the affected muscles. Blood flow provides both sugar and oxygen and removes the waste products that have built up as a function of the exertion. Since blood flow is relatively constant, such compensatory changes occur automatically and continuously as time passes. The rate of movement toward equilibrium, according to the homeostatic negative-feedback corollary, will describe a negatively accelerating curve. The survival function of fatigue is considerable. Without such input, a given muscle or muscle complex might be used to the point of tissue damage.

Nausea also produces quiescence, as anyone who has been seasick can readily testify. It is also correlated with vomiting when it becomes extreme. Yet as soon as this quite specific and sometimes violent activity is over, the individual typically wants to lie down and remain immobile. Vomiting also makes an adaptive contribution; it may occasion the removal of noxious materials or allow a return to equilibrium in adjacent systems that have initiated the input reflexively. Hart (1988) suggested that the behavioral changes that occur in illness are evolved techniques for coping with viral and bacterial infections.

In summary, primary activation input both signals the presence of imbalance and provides the impetus for compensatory behavior. It is in relation to such input that the homeostatic negative-feedback corollary inferred in chapter 2 has its clearest and most readily demonstrable application. The greater the imbalance in physiological systems, the greater the tendency toward compensatory action; the thirstier an animal is, the more rapidly it will drink, the greater the oxygen deprivation, the more rapidly it will gulp air. Even the output of urine is a reflection of the extent of bladder distention, being high in volume at the outset, diminishing gradually, and finally stopping as balance within the system is reestablished. A negatively accelerated curve, which reflects the homeostatic negative-feedback corollary, will be found whenever accurate and appropriate measurements are taken of any example of balance-trending behavior.

It has been suggested that all primary activation inputs vary in quality, prepotence, and intensity, and that each of these dimensions is important in adaptive behavior. Without differences in quality, the organism could not respond appropriately to circumstances pertinent to a particular imbalance; without variation in prepotence, the latency and speed of response would be the same for imbalances in vital systems as for those less critical; and without fluctuations in intensity, the organism would respond as rapidly to inconsequential imbalances as to those more immediately important to survival.

Information Input

Many receptor systems of the body introduce input that, at least in ordinary ranges, is essentially neutral in activation significance unless inherited or learned associations are involved. Inherited linkages are not uncommon. Wave lengths within the visible range of the electromagnetic continuum provide the most important source of information about the individual's world, but they also bring about reflexive changes in the pupil of the eye. The color red has little significance for human beings until learning takes place, but it is a vital wired-in signal to the herring gull chick that indicates that dinner is ready to be served, and it activates the pecking response.

The major function of information input is, as the term implies, to provide the organism with information about its environment and its position within that environment. In human beings and in most complex organisms, such input includes the experience of light, sound, taste, odor, and a variety of sensations derived from stimulation of the skin. It also includes input from receptors in the muscles, tendons, and joints typically subsumed under the term kinesthetic sense.

Information input makes a twofold contribution to adaptive behavior, at least in the human being. Although much of it is essentially neutral at the outset, as a function of learning such input may become deneutralized (i.e., take on activation significance) so that both away from and toward behavior can appropriately take place. Such input also provides the basis for the development of an internalized world that serves as the arena for thinking (internal locomotion) as described in chapter 3.

All such information input represents energy changes that have been important to the survival of the organism during the evolution of its species. For instance, during the entire process of evolution, the earth has encountered the vast range of the electromagnetic continuum, which extends from waves (sometimes described as quanta) that are only one-billionth of a millimeter in length, to those responsible for the broadcast band of radio, which may be miles from crest to crest.

For many organisms, this is particularly true of mammals, the most important information input is derived from eyes, ears, and noses. Energy from the electromagnetic continuum is, at least during daylight, perpetually emanating from the sun. Certain components of this energy (between 400 and 700 millimicrons) have remarkable reflection characteristics so that every item in the environment betrays its presence with a reflected image. Responding appropriately to this vital information has provided the selective conditions that has resulted in the evolution of vision.

Everything that moves in the environment causes pressure fronts to fan out in all directions and this universal condition signifying both predator and prey has resulted in the appearance and refinement of hearing.

All living things are surrounded by a halo of molecules and sensitivity to the cues provided has resulted in the emergence and refinement of olfaction.

Vision, hearing, and olfaction provide the major windows into the world in which most species evolved and as such are widely distributed throughout many species. They are clear examples of domain-general attributes. Attributes that demonstrate the influence of selective circumstance shared by many diverse creatures as evolution took place. They also alert us to the probability that where shared selective circumstances have been encountered, regardless of diverse appearances, such domain-general outcomes should be expected.

Taste may well be the most ancient of all receptor systems. We find its prototype in such microorganisms as protozoa (Hall, 1964) and bacteria (Hazelbauer, 1988) in their responses to chemical gradients and the wide representation of chemoreceptors in many species such as snails, slugs, and worms (Jennings, 1906). It has kinship with olfaction in that both respond to molecules in a medium, the one in air and the other in liquids, and in the overlap in the taste and smell of food, as anyone who has tried to eat with his or her nose stopped up can verify. We recall again the observation of Carr (see chap. 1) that we may infer the relative recency of sensory systems by their responses to greater and greater distances.

The inputs from the receptors in the muscles, tendons, and joints are commonly subsumed under the label kinesthetic sense. Each and every movement induced by the striated muscles of the body, insofar as the skeletal structure is affected, stimulates such receptors, and sensory feedback is introduced into the field of awareness. Although on most occasions such input is hardly discernible, it does provide a basis for monitoring the various activities in progress. This function is important in all overt behavior and particularly vital when new motor skills are being acquired. Then, as each movement takes place, the resulting sensory feedback provides the input that ultimately becomes fused into motocepts, which in turn provide the basis for highly refined motor skills. When actions have become habitual (as we discuss at length in chap. 8) as in typing and walking, such kinesthetic input becomes attenuated (i.e., less of it is apperceived) and is, as long as the behavior proceeds smoothly, relatively less important to adjustment. Input from the receptors in the semicircular canals of the middle ear serves much the same function except that the position and relative movement of the entire body is involved. The input from such receptors allows continuous monitoring of gross movement and position.

Both the inputs from receptors in the semicircular canals and those from the muscles, tendons, and joints provide information vital to the ongoing behavior of the organism. Without this information, behavior would be erratic and disorganized. Moment-by-moment information essential for compensatory adjustments would not be present, and the refinement and precision of motor skills would be impossible.

In our discussion of primary activation input, it was suggested that each type is qualitatively different from every other type. Differences in quality are even more apparent in information input, and these differences have been of great interest historically, as stated in Muller's doctrine of specific nerve energies (Muller, 1843). They also have vital implications for adjustment. If there were no discernible differences among the inputs from different sensory systems, differential responses would be impossible. A nonfunctional increase in complexity would result, and the condition would inhibit, rather than abet, survival.

Once again, it should be emphasized that sensory input need not be, and indeed usually is not, similar to the energies that produce the experience. Visual sensation is the external environment transformed by our knowing equipment. Red does not exist in the external world, it exists only in our heads. So it is with all the other colors and all the shades and textures and shapes of our visual world. Indeed, so it is with the inputs from each and every sensory system of the body.

Although there is no literal similarity between the energy that activates a sensory system and the sensation that results, there is a vital correspondence between the two. The hundreds of experiments in psychophysics give testimony to this relationship. Thus, increments in stimulation, regardless of which energy is involved, are invariably found to be correlated with increments in sensation. Although this relationship is logarithmic, it is profoundly functional in that it provides the basis for differential behavior. Over the millions of generations of organic evolution, more intense energies have always signified the lessening distance between predator and prey (the inverse square law), whether such energies involve changes in the visible spectrum, the displacement of pressure fronts, or the density of molecules in a medium. As the distance between organisms decreases arithmetically, the density of molecules, the size of the reflecting image, and the amplitude of pressure fronts increase geometrically, and the odors, visual images, and sounds resulting increase in corresponding intensity.

According to the Weber-Fechner function, sensory systems are most sensitive at the lowest ranges of stimulus intensity; that is, a subject can notice a smaller shift in energy. This would make sense from an evolutionary frame of reference. From an adaptive standpoint it would be much more important to react to the first signs of either predator or prey than to any signs thereafter. Thus, regardless of the energy dimension and the corresponding shifts in sensation that result, the logarithmic function holds, at least in the central ranges. The reader might well object at this point and suggest that since proximity results both in greater stimulation (see above) and in greater chances of either eating or being eaten, the sensory sensitivity should be greatest at the higher levels of stimulation. Sensation, of course, does become more intense as stimulation increases; it is just that at the higher levels of stimulation the sensory system is relatively less sensitive, suggesting again the greater survival importance of initial predator or prey detection.

Information input has another characteristic that makes a vital contribution to adaptive facility: its apparent locus varies. Visual components appear to originate in the eyes, and taste in the mouth. More importantly, tactile input appears to be localized on particular areas of the skin. This characteristic is probably as critical to adaptive behavior as the differences in quality and intensity previously mentioned. If it were absent, organisms would be unable to make precise responses to threats to particular parts of the body. When a cockroach crawls up a person's leg, that person knows with some precision exactly where it is.

Although all sensation, whatever its source, has this locus characteristic, it varies in precision from one transmitting mechanism to another. The input from skin receptors is unusually specific in its locus, whereas that from visceral mechanisms, and those responsible for hunger and thirst and other activation inputs, are typically more diffuse.

A related problem, and one that has long puzzled thinkers, involves the following question: although all sensation is occasioned by processes within the central nervous system, how is it that the input appears to be located within the receptors or—as is the case with visual and auditory input—at varying distances from the receptors? The words on this page appear to be located a number of inches from the reader, although the processes representing them are certainly within him or her. Are we born with this capacity for externalizing and loci, or is it learned? One conclusion seems certain. We are born with the capacities for differences in sensory quality and intensity. Whether the ability to localize and externalize is innate, learned, or both, is still open to debate. (See chap. 5 for an extended discussion of this problem.) Yet even if future research reveals that learning is predominant, it will still be necessary to assume that the capacity for such cues and for their organization is innate. As emphasized by Lotze (1852), innate differences in quality, which he called local signs, will undoubtedly be found to play an important part in the learning of locus specificity. Without them, discrimination among inputs from different loci would be impossible, and learning could not occur.

Finally, information input also varies markedly in discreteness and variety. From a functional point of view, this is readily understandable. There is, by contrast, only one condition that ordinarily occasions thirst (i.e., loss of fluid) and hunger (i.e., lack of food). There are, therefore, only three requirements that such input must fulfill. Each input must vary along the dimensions of intensity, quality, and prepotence. Yet there are literally thousands of pertinent environmental conditions; hence the many kinds of information input from the various sense modalities of the body, particularly the visual and auditory.

The discreteness and variety of information input powerfully abet discrimination, and are essential to learning (as we see in chap. 6). Without variety and discreteness, the myriad situations that are important to the survival of the individual could not be differentiated and specifically encoded.

In summary, information input provides the almost infinite number of sensory variations that make the highly refined and subtle differential behaviors in higher organisms possible. It also provides unique information about each survival circumstance so that appropriate and functional encoding can take place. In short, information input allows a moment-by-moment monitoring of how things are progressing in the external environment, contributes vital information about the organism's relative position, velocity, and progress as it moves within the environment, and provides the basis for the encoding of salient features of the environment.

Information input varies not only in quality and apparent locus, but also in intensity. Generally speaking, there is an inverse relationship between the proximity of an energy source and the amount of energy impacting (the inverse square law) on the organism. As distance from the source increases, pressure fronts decrease in amplitude, reflecting surfaces decrease in size, and molecules in media become diffused. That such changes would be represented by corresponding reductions in sensory input is predictable from a survival standpoint. There is usually a vital relationship between the intensity of an input and the proximity of the situation (whether hazardous or beneficial) that it signals. Sensory systems are most sensitive at the lowest amplitudes (the Weber-Fechner function) reflecting the survival importance of initial discernment of either predator or prey.

In the next section (secondary activation input) one of the adaptive dilemmas introduced by cue utilization becomes resolved. Such input provides cues with motivation, for both avoidance and approach behaviors.

Secondary Activation Input

In early research and thinking on secondary activation input (emotions), we find one of the clearest expressions of both an adaptive orientation and an effort to integrate all three levels of psychological data, the physiological, experiential, and behavioral, into a synthetic view. The Cannon-Bard theory (Cannon, 1929) postulated that the major function of emotion was to prepare the body for emergency; anger and fear had the adaptive function of inducing fight or flight. In this process Cannon and Bard postulated a particular sequence of events: (a) the individual encounters a critical circumstance, and (b) information is shunted to the thalamus in the brain, which both activates the sympathetic nervous system with resulting visceral upheaval, and the cerebral cortex, where the actual experience of the emotion takes place. This theory, with minor additions and refinements, remains relatively intact. Further, it stands as a clear example of a functional orientation and the synthetic view of the subject matter of psychology outlined in the first chapter.

Much of the contemporary research and controversy of evolutionary psychologists has to do with the issue of whether there is an inborn linkage between particular secondary activation inputs and specific reactions and deals with such questions as "do men and women experience jealousy to different circumstances?," "are we born with a fear of snakes, high places, etc., or is learning completely responsible for such linkages?"

A major difficulty in studying secondary activation inputs derives from their multiple functions and their occasional occurrence in combination. As Nesse (1994) suggested, fear may not only occasion flight and avoidance, but in extreme cases may cause fixation or inappropriate behavior. Or emotions may occur in combination as when jealousy, envy, fear, and anger unite in response to the loss of a mate to a competitor. When these factors are further complicated with the intrinsic and overlapping problems of cultural overlay, the difficulty of determining the origin and function of particular emotions becomes evident.

Although the differences are not as discrete as those for cue and primary activation, secondary activation inputs vary in terms of quality; Muller's doctrine of specific nerve energies may be expanded to include them. Otherwise appropriate response to the situations that they represent could not occur, that is, anger is different from fear, and each activates behavior that is appropriate to the adaptive circumstance that has produced them.

The Activation of Cues. In chapter 3, it was suggested that cue use provided a major increase in adaptive facility but that it also introduced an interesting and vital problem. Since the organism responds to the cue rather than to the critical situation itself, from whence comes the motivation for response? In avoidance behavior, for instance, the response occurs prior to the critical situation; thus, primary activation input cannot be present at the time the response occurs. The gradual emergence of secondary activation systems that would provide the motivation for action when only the cue was presented became the evolutionary solution to this somewhat paradoxical problem. The sight of the stove produces the secondary activation input of fear, and the fear instigates appropriate avoidance behavior. The sight of food brings about both preparatory and motivational changes that occasion appropriate approach and eating behaviors.

Yet such secondary activation input is not required for all cue-instigated behavior. This seems to be the case with reflexes in human beings, many of which occur automatically with activation input being only peripherally involved, often after the behavior has taken place. For example, there seems to be no motivational involvement in such behaviors as the grasping and patellar reflex, and the withdrawal reflex that takes place when an individual touches a hot stove, is elicited prior to the pain, which occurs after the fact.

Many other kinds of inherited associations between cue and response may likewise be free of activating inputs, since they are manifestations of genetic linkages between cue and behavior. The motivation required for such behavior may arise from the ongoing metabolic processes of the body—such as those controlled by the autonomic nervous system—which have few, if any, experiential repercussions. Indeed, many automatic processes, representing rhythmic oscillations in equilibrium maintenance, are energized by extraexperiential mechanisms. Certainly the same conclusion may be correct for many of the different kinds of cue-response associations found in animals lower down on the phylogenetic scale. This does not mean that a secondary activation system cannot play a part in inherited behaviors. Where general flight responses are involved, the cue may occasion secondary activation by way of a direct genetic linkage. For example, the shape of the hawk probably automatically produces fear in the chicken, and this fear activates flight behavior (Tinbergen, 1948).

Yet in most cases of learned cue-response associations, a secondary activation system is necessary to energize the behavior. This adds greatly to adaptive flexibility. For instance, the human baby may be born without (this is still an open experimental question) specific fears. This may have certain drawbacks such as failing to respond appropriately to insects, snakes, and high places, but since human beings encounter thousands of different potentially harmful situations, the establishment of appropriate learned associations between cue and activation input markedly increases the flexibility of the individual's adaptive repertory.

The many studies performed under the classical conditioning paradigm provide a wide range of data on the various facets of learned cue-response associations. Where a particular cue has been paired with an important circumstance, whether meat powder or electric shock, the cue soon picks up the capacity to instigate appropriate preparatory and approach or avoidance behavior (see chap. 6).

KINDS OF SECONDARY ACTIVATION INPUT (AFFECT)

A wide variety of different inputs provide the secondary activation (commonly called affect) necessary for the approach or avoidance behavior that occurs when a given cue is presented. Most such input has been subsumed under the term emotion, an area notorious for its ambiguity, some of it under an even more indefinite label, feelings, a third category called pleasure, and a fourth, which is common in the experience of all of us but has no name in the literature.

The subtle shades of affect (emotions, feelings, and pleasures) are almost too numerous to mention. Both their vagueness and importance are emphasized by the multitude of different adjectives that have been invented to represent them in our language. All the categories are vague, but some—like envy, greed, awe, and suspicion—are so ephemeral that they are almost impossible to describe.

Emotion

Of the basic kinds of emotion fear is probably the most important. As the many experiments by behaviorists have demonstrated (see Keller & Schoenfeld, 1950), it activates the individual so that harmful situations can be avoided. Without fear, the adaptive role that cues play in the behavior of the organism would be limited.

Guilt also makes a vital adaptive contribution, at least in human beings, and probably in all mammals that must cooperate in order to survive. It is the input that forces individuals to abide by the rules of the group, rules that are (or were at one time) essential for group and, thus, individual survival. If an individual breaks group rules, guilt drives him to seek expiation, and to reaffirm the rules that make group action and group living functional.

The adaptive contribution of anger is easy to infer. As the many experiments on frustration have demonstrated, when a barrier is placed between an organism and the resolution of a need (the reestablishment of optimal balance in a crucial body system), anger is a direct and very functional outcome. The activation that results from such input is turned against the barrier (at least when the barriers are tangible) with increased chances for its being surmounted. Thus, anger provides a fundamental adaptive input; it spurs the organism to greater efforts in overcoming obstacles and adversaries.

Love, of course, comes in many varieties. Yet in general, such input provides the motivation to create and to sustain situations that are important to the welfare of the young, both during the time that the female is pregnant and during the period when the infant is unable to care for itself. Love (along with guilt) welds us together as a functioning group, as a bonded couple, and as solicitous and caring parents.

Jealousy has been the subject matter of many studies during the past few years. It is a powerful input that initiates behavior designed to protect vital relationships between people, relationships that, in most cases, are critical to sexual reproduction or the welfare of children. Thus, jealousy helps safeguard the conditions that both occasioned its emergence and insured its inclusion into the genetic material of our species.

As with other secondary activation input, the capacity for jealousy is innate, though the circumstances that occasion its introduction may vary. Is such variation due to innate predispositions or is learning (culture) to some degree involved? This issue, perhaps fueled by the present-day concern with gender differences, has inspired a remarkable amount of research by contemporary evolutionary psychologists. Buss (1989a), in a series of studies, has apparently demonstrated that there is an innate, evolutionarily derived gender difference in reaction to sexual infidelity: Males react with jealousy to sexual infidelity because it introduces paternity uncertainty, while females react to such infidelity because it indicates a weakening of the pair bond essential to their welfare and the welfare of their offspring. Such differences in reaction are apparently robust across cultures (Buunk, 1996).

While not questioning the experimental findings, Harris and Cristenfeld (1996) suggested that the differences found may be related to culturally derived perspectives rather than to innate propensities; a woman thinks a man can have sex without love, while a man thinks women do not have sex without love.

This issue is still inspiring much research. It stands as an exemplar of contemporary evolutionary thinking influenced perhaps by the sexual orientation of the researchers, the present-day topicality of gender studies and the pervasive difficulty of extracting and clarifying innate, evolutionarily derived propensities apart from the tangles of cultural overlay.

DeSteno and Salovey (1996) emphasized this difficulty and include a caveat with which I completely agree, that evolutionary conclusions should not be resorted to before rigorous attempts to rule out plausible cultural explanations have been made. In passing, I half seriously suggest that evolutionarily derived predispositions may not only be responsible for gender differences in response to sexual infidelity but also for the different interpretations given by the researchers.

Feelings

Feelings are highly diffuse sensations, the origin and function of which are difficult to pin down. Actually, the psychological status of feelings has not changed appreciably since the structuralists argued about their classification (i.e., as to whether there were one or several different dimensions involved; Boring, 1929/1950). I have more to say about the classification of feelings later, but would like to posit here that they may have a function not fulfilled by less amorphous input: that of providing information on the general welfare of the organism. This monitoring function, though not as discrete as pleasure or kinesthetic feedback, is still important. To feel good or bad is not without adjustment significance. The activity level of the individual is affected. When such input registers that the general status of the organism is less than optimal, activity level is reduced, and more rest is assured. When such input signifies that the general status of the organism is at a high level of efficiency, more rapid and more effective activity commonly results.

Pleasure

This category of affect brings with it a long history of ambiguity. The hedonists, beginning with Epicurus and Lucretius and extending down to LaMettrie, have insisted that the search for pleasure is either the major motivating impulse in man or that it should be. We find the pleasure principle central to Freud's theory, and, as reinterpreted and clarified by the behaviorists, we see it reemerging again in the concepts of reinforcement and satisfaction.

Pleasure is certainly a kind of sensory input, but the search for the relevant transmitting mechanisms has been fruitless, at least until quite recently. It seems to be located at certain places, such as the genitals during intercourse and the mouth during eating, but neither specific receptors nor particular free nerve endings are apparently responsible. The same stimulation that occasions pleasure when the organism is deprived may produce discomfort when it is sated. Pleasure appears to be correlated with the activity that moves systems back towards balance, as in eating, drinking, and sexual intercourse, but there is no discernible pleasure input when pain is alleviated.

Findings pioneered by Olds (1960) have shed some light both on the general locus of such input and on the manner in which it influences behavior. When a microelectrode is implanted in a certain region of a rat's brain stem, behavior can be influenced by small voltages of electrical current. Olds, for instance, was able to shape a rat's behavior on a moment-by-moment basis by introducing a short pulse of shock immediately after a particular movement. Furthermore, he found that rats will learn a multiple-unit maze and sometimes cross an activated shock grid in order to stimulate themselves in the brain. This research suggests that there are particular areas of the brain that function as pleasure centers, and that organisms will actively pursue pleasure for its own sake.

Do these findings demonstrate that certain activities are not occasioned by disequilibrium, running counter to the postulate of process as stated in chapter 2? Careful consideration of this problem suggests that this is not the case. It is the loss or reduction or absence of pleasure, not its presence, that activates the organism. A child may remember that pleasure was derived from eating an ice cream cone, and this may instigate activity designed to get ice cream. Yet it is remembered pleasure, rather than pleasure being experienced, that initiates the behavior. Also, the realization that pleasure has been reduced or is no longer present may be sufficiently disequilibrating to motivate the individual to persevere in such activities as eating or masturbation.

The Functions of Pleasure

What is pleasure's contribution to adaptation? It seems to function neither as activation nor as information input, at least not primarily, but it does make important contributions that facilitate effective behavior.

Monitoring. All animals comprise a number of systems that are in perpetual flux between imbalances and compensatory action. This is particularly true of human beings. In some of the systems, balance reestablishment can be achieved immediately, as when a painful stimulus is removed, while others require prolonged activity. In those systems where compensatory activity is of some duration, pleasure provides continuous feedback on the progress of balance reestablishment, and its cessation is a cue that balance has been regained. The monitoring function of pleasure, coupled with the activation occasioned by its reduction (as previously discussed), helps sustain compensatory activities at reasonable levels.

As monitoring is much more important for homeostatic systems that require prolonged activity, a more intense pleasure component can be predicted in such cases. Thus sexual behavior should be and is more pleasurable than defecation, eating more pleasurable than drinking.

Our discussion leads us again to a principle similar to Muller's doctrine of specific nerve energies. In order for pleasure to function as a monitor, the input from a given homeostatic system must be distinguishable from that arising from other systems. If this were not the case, the pleasure input from the different systems would result in complication and confusion and would hinder rather than abet survival.

Discrimination. It appears that particular loci in the brain stem are responsible for different kinds of pleasure input. Deficit within a given system apparently results in the activation of centers appropriate to that deficit. If there is no deficit there is no pleasure. Olds (1960) dramatically confirmed this. He found an area in the brain of the rat that was peculiar to eating and another that initiated sexual activity. He demonstrated that an animal will stimulate itself in the appropriate area only when the corresponding drive is present. Deficit is thus critical in determining whether or not stimulation of the brain will produce pleasure.

If several such pleasure centers do exist, and if they are peculiarly responsive to various imbalances within the body, we may ultimately solve one of the most baffling problems of motivation: how both animals and human infants select adequate diets in a cafeteria situation. A rat maintained on a calcium-free diet for an extended period and then allowed to choose between one food with a low calcium content and another with high calcium content will choose the latter. How does the rat make this discrimination? It may be that the deficit sensitizes an appropriate center in the brain so that the food with the high calcium content will produce more pleasure than will the other. Pleasure would then provide both the basis for discrimination and the motivation for choice. Olds and Peretz (1960) demonstrated that a castrated rat would not stimulate its sex-reward area unless the hormone androgen had been injected into its blood stream. If the presence of a hormone can sensitize a sex center, it seems not unreasonable to assume that a lack of calcium may sensitize an appropriate hunger center.

This discussion suggests then that pleasure may contribute to adaptive behavior—extending the monitoring function already described—by providing the basis for certain fundamental discriminations. It is true that human beings soon lose their ability to select an adequate diet in a cafeteria situation, but this probably does not happen until learning interferes with an inherent pleasure discrimination mechanism.

Emphasis. In chapter 6, I suggest that learning (i.e., the establishment of memory encodes) is dependent upon the apperception of pertinent sensory input. It is also argued that pleasure makes a vital contribution to the encoding process by emphasizing both the compensatory activity and the situation in which the balance reestablishment occurs. Although specific experiences of pleasure may not be remembered per se, the situation in which the pleasure input occurred is more readily recalled because of the emphasis provided.

From the viewpoint of adaptation, pleasure thus fulfills several functions. A remembered pleasure may add a motivational increment that helps initiate and sustain compensatory activity. It also provides feedback for the monitoring of compensatory activity, furnishes the basis for discrimination among specific hungers, and abets learning by emphasizing situations in which balance reestablishment takes place.

The reader may feel that I have overlooked the fundamental contribution of pleasure: that it provides the seasoning that makes life worth living, and that an organism totally devoid of pleasure would end its life if it were capable of comprehending the significance and means of suicide. Certainly, many people feel that joy and happiness constitute the experiences that give value to human life.

If joy and happiness represent the highest goals and greatest fulfillment for human beings, is it reasonable to place such hurting experiences as pain and thirst and hunger at the center of human evolution and present human nature? It is true that we appreciate most those aspects of our life marked by pleasure. No utopia has ever been built around pain and hunger, and our projections for an afterlife often include a heaven where eternal joy prevails, and another realm where anguish is our perpetual lot. Yet we are not concerned here with value judgments about either a present or a future life—we are seeking to discover the attributes that were critical in human evolution and that even now are crucial to our every act. We could continue to exist without joy and pleasure, and perhaps without pain and hunger, but not for long.

Flare Input

Although this fourth category of secondary activation input (affect) makes an important contribution to behavior, I could find no record of it in the literature. Despite its universality in human beings, and perhaps in all animals, it is difficult to describe, probably because its manifestations are so transitory. An example may help. Let us assume that a person's family has gone out for the day, leaving him or her alone in the house. The person goes down into the basement and suddenly sees a man in the shadows. Within a second he or she will experience what I call flare input. It appears to originate in the viscera, but it also seems to be derived from nerve endings that are widely distributed throughout the body.

Although psychologists have apparently either overlooked or ignored this phenomenon, it may be related to the psychogalvanic reflex (PGR). Both occur within less than a second after the stimulus has been applied. Furthermore, the PGR can be measured in any part of the body, suggesting that it, too, is a generalized response. From a behavioral point of view, the startle response may be correlated with flare input. It, too, is a highly generalized reaction, suggesting the activation of widely dispersed nerve endings. Flare input and its corollary physical manifestations may be the atavistic remains of a generalized emergency reaction as found in the hair-raising reflex demonstrated by many animals, including humans. When an intensely frightening situation is first encountered, the generality and suddenness of the response may serve, or at one time served, to delay or circumvent attack.

Discussion. Affect (emotions, feelings, and pleasures) makes a vital contribution to certain of the most finely tuned capacities that humans demonstrate in information processing and reaction. It provides an important aspect of the context that gives color and diversity to meaning. It abets both encoding and recall. It enters into the associative matrix that undergirds the refinement of both thought and language. Under certain conditions it may carry the major burden of communication. This is particularly true in music and poetry, and to a lesser degree in prose and ordinary conversation.

Affect not only becomes intrinsic to many encodes, percepts, and words, but also is represented in the posturings and facial expressions of many animals, particularly human beings. Consider the subtle posturings and movements of Marcel Marceau, or the finely modulated affective nuances in the face and eyes of Meryl Streep. These artists go much deeper than an artificial depiction of a mood; their projections resonate in all of us. They strike at chords, whether intrinsic in our nature or reflections of our culture, that have universal affective implications. And consider such cultural and religious symbols (sometimes called collective representations) as Old Glory, the cross of Christ, or the Star of David. When presented to the appropriate reference group they speak a common language, evoking affective responses that provide the basis for group action, whether constructive or bizarre.

The manner in which the different kinds of affect enter into the stages of information processing is discussed in detail in appropriate sections of the following chapters. Yet at this point in our development I would like to again quote Darwin, indicating, as is so often the case, that he not only had remarkable insight into the processes that he studied but an uncanny prescience of their later treatment. After an in-depth analysis of the many facets of emotional expression both within and across species, he stated:

> The power of communication between the members of the same tribe by means of language has been of paramount importance in the development of man; and the force of language is much aided by the expressive movements of face and body.... We have also seen that expression in itself, or the language of the emotions, as it has sometimes been called, is certainly of importance for the welfare of mankind. (Darwin, 1897, pp. 154, 166)

Can we have an affective experience without feedback from visceral or body changes? This issue has remain unresolved since it was much discussed by James and then by Cannon during the early part of the last century. I do not restate their arguments; they are well known. But I would like to posit that certain affects (the stronger emotions such as fear and anger) achieve at least a part of their prepotency from body changes. In more amorphous affects, such as awe and disgust, an added component from body changes may be less important, perhaps nonexistent.

One can remember a circumstance in which he or she was angry or fearful and such memories are sometimes haloed with affect but in terms of intensity and reaction involvement, such affect is but a pale shadow of the real thing. Invariably the knowledge that these are just memories, in good measure, reduces the affect. Yet in dreams, which often have the immediacy and realism of actual situations, affect can at times be extremely powerful. That this can happen is strong evidence that memories, when they are perceived as real, are capable of profound affective repercussions, and the common experience of residual visceral upheaval upon waking from certain dreams suggests its importance, at least in the more intense emotions. It should be noted that most dreams progress with little discernable affective involvement. This is perhaps because of the raised sensory threshold (to be discussed in chap. 7) that keeps input from visceral changes from being experienced unless they are extremely powerful. On the other hand the subtler affects, such as disgust, joy, or sadness are often reported in dreams (Freud, 1900/1980), suggesting that the source of such affects may be purely cortical.

The shades of affect are as numerous and as subtle as the multitude of taste or color experiences humans are capable of having. Taste experiences, it will be recalled, are derived from a mixture of only four different kinds of receptors: sweet, salty, bitter, and sour. The almost infinite variety of colors is the outcome of only three primaries: red, green, and blue. Are there affect primaries as well, with all the other affects, regardless of how pervasive or ephemeral, simply the result of these primaries experienced in varied proportions? Wundt seemed to think so. For him there were three dimensions of feeling: relaxation–strain, grief–happiness, and pain–pleasure. Much more recently, Plutschik (1980) thought so too. To him there are eight different affect primaries: joy, acceptance, fear, surprise, sadness, disgust, anger, and anticipation. The other affects such as awe, disappointment, contempt, etc. are experiences derived from these primaries in various mixtures. Although Plutschik's work is impressive, the derivation of affect primaries is hedged with great, if not insurmountable, difficulties. They must be derived from analyzing the experience of subjects, per se. There is no knowledge of the underlying neural substrate as there is in taste, and no accessible stimulus manipulations as there are in vision. We must deal with affect experiences, as ephemeral as they are, with manipulations that are relative (because of each individual's unique history) in their effects.

The inclusion of affect in the total association matrix of mentation has produced a remarkable number of studies, many of them inconclusive and contradictory during the past twenty years. For a survey of such research and an introduction into the complexities of design required to cope with the relevant variables, one should read the review article by Blaney (1986), a chapter by Isen (1984), three important papers by Bower (1981, 1987) and his colleagues (1981), and in particular the analysis by Nesse (1990).

IS THERE A GENERALIZED ACTIVATION INPUT?

In our discussion of activation inputs it was suggested that there were two major types of input: primary activation input, which results from vital imbalances in the body, and secondary activation input, which provides the impetus for avoidance and approach behavior, and the myriad nuances of affect involved in social interchange. Even if we were to stop here, a fairly sophisticated view of motivation would be provided, one that describes the organism's behavior as sustained by drives that are in perpetual flux as imbalances are resolved and new ones develop.

Yet this picture, as realistic as it may seem, is probably oversimplified. It makes no allowance for movement without specific input, and it implies that movement will cease as soon as such input is no longer imposed. However organisms do not always move more slowly as specific activation inputs decline; they do not run down as soon as a particular balance is reestablished, as if they were toys with unwinding springs. Rather, they continue (at least during the waking state) to move in the environment and to react to it. They appear to be motivated by a generalized input that remains at a relatively high level, even after specific drives have been alleviated.

A generalized activation input, such as curiosity or alertness, could make an important adaptive contribution. It would not only maintain the individual in a state of heightened vigilance, but bring him or her into contact with broader segments of his or her environment. The learning that takes place during exploratory forays (i.e., world internalization) provides a cumulative record that can be drawn upon when the individual encounters survival problems in new situations. Despite this description of a generalized drive, and my interpretation of its function, there is no hard evidence for its existence. The primary activation inputs (i.e., primary needs) are so universal, and the shadings of secondary activation inputs (i.e., affect) are so diverse and persistent, that the notion of a nonspecific activator may be an inference from ignorance. Yet perhaps there is a *Big D* as Bolles has designated it (Bolles, 1975). The adaptive significance of such input, outlined in the previous discussion, suggests its likelihood, but future research must provide the answer to this question.

TRANSFORMATIONS

A key to understanding the contribution that cognition makes in the evolutionary emergence of information-processing capacities is the term transformation. Energy changes representing vital imbalances become transformed into the primary activating inputs that instigate and sustain the compensatory action required for balance reestablishment. Energy changes taking place on the outside of the body become transformed into the sensory components arising from the sense modalities. The adaptive requirements for activation implicit in cue use and the requirements for facile discrimination, precise monitoring, and emphasis become transformed into the subtle nuances of affect. The concurrencies encountered during a species' development, and during the individual's life, become transformed into inherited and learned associations. Perceptual structuring and observation learning transform the external world into a functional internalized replica. Thus, sensation not only is the evolutionary solution to the input problem in complex organisms, but also constitutes the manner in which critical situations of survival have been transformed so that appropriate behavior can take place.

PROCESS INTERDEPENDENCIES

Although the continuing interdependency between the organism and the agency that has molded it (i.e., the environment) is reflected in all attributes, it is most clearly indicated in the sensory systems of species. The reality for any species is always restricted to those ranges of particular energy dimensions to which differential response was important during that species' evolution. When a given energy dimension was not pertinent to survival, no transmitting mechanisms evolved. When a given dimension, or segment thereof, was critical to survival, that dimension is represented by highly refined transmitting mechanisms. Thus, the vulture has a better eye than the human being, the hound a better nose, and the bat a more discriminating ear.

Since replication is the ultimate event in evolution, any variation (see postulate of evolution, chap. 1) that increases the chances of reproduction, will be selected in. In certain cases this has resulted in a remarkable refinement of both sensory and signaling systems. The female silkworm moth, at the time of sexual receptivity, emits a large, highly volatile molecule that can be borne on the wind for great distances. The male silkworm moth has in turn evolved a receptor that is responsive only to the particular molecule in question, a receptor into which the molecule fits as a key into a lock. This sexual mutuality is of such refinement that the presence of a female can be discerned from miles away, and by following increases in the concentration gradient, the male ultimately finds her and copulation takes place (Gould, 1982).

Each species of firefly has a specific flash code that the female uses to signal her presence and to advertise her sexual receptivity. Where a number of species live in close proximity, these simultaneously emitted unique codes would seemingly result in imponderable complexity, but not, apparently, for fireflies. The receptor systems of the males of each species are attuned to the flash code of appropriate females so that confusion and nonproductive mating rarely takes place. It should be noted, however, as discussed by Alcock (1984), that predatory females of other species on occasion break the code of a prey species and lure the males thereof, not to connubial consummation, but to lunch.

Finding food is another critical and continuing requirement for organisms, and where the food source is restricted the honing processes of natural selection can be quite specific. The eye of the frog is a pertinent example. Investigators at Massachusetts Institute of Technology (Lettvin, Mauturans, McCulloch, & Pitts, 1959) studied the reaction of the frog's eye to different stimulus patterns by inserting microelectrodes into the optic nerve and measuring the impulses sent toward the brain over individual nerve fibers. The frog's eye apparently responds to four different and very appropriate kinds of information. Of most interest to us is that the small objects that produced maximum neural response correspond almost exactly to the dimension of flies and other appropriate insects when they are within striking distance of the frog's tongue. This finding suggests that the frog's eye is peculiarly responsive to the presence of insects in its environment, and allows us to surmise that this bug-perceiving mechanism has been honed and refined to its present state of specificity during the millions of generations during which frogs have been catching insects for survival.

Sensitivity to trauma is another factor critical to survival. This requirement is also reflected in the distribution of appropriate receptors on the surface of the body, as well as within certain organs and organ systems. For instance, we may infer that surface receptors near highly vulnerable or functionally crucial organs will be particularly responsive to energy shifts signifying potential hazard. We would thus predict greater tactual sensitivity near the eyes or on the neck than on the thigh or back. Heavily armored body organs should have fewer such receptors. It is not accidental that the human brain is largely insensitive to stimulation. Not only is it protected by the skull, but, also during the evolution of our species, circumstances sufficiently extreme to occasion trauma to the brain usually resulted in death. There can be little or no natural selection working to produce sensitivity in an organ that is so crucial that any and all damage has lethal repercussions.

Coevolutionary mutuality has been another circumstance that has affected the emergence and refinement of sensory systems, particularly where a predator-prey relationship has existed for a prolonged period. An organism that feeds on individuals whose vocal utterances occur primarily within a given range will, as a result of coevolution, develop an auditory sensitivity peculiar to the frequency and amplitude of these vibrations. The receptors of the organisms preyed upon will be particularly sensitive to the energy changes that signify the approach of the predator. The hawk and the owl both have ears that are remarkably effective in locating small sound sources, whereas small birds, their traditional prey, have evolved the capacity for emitting warning notes above the optimal frequency for phase-difference cues (Marler, 1955). The bat has developed both an effective sonar and the agility sufficient to prey on certain species of moth. The moth, in turn, has evolved an acute sensitivity to the bat's sound probes (Roeder, 1965). Small creatures such as field mice have developed acute hearing for the detection of predators, whereas owls that prey on them have developed soft fringes on their wings that dampen the noise of flight (Thorpe & Griffin, 1962).

These are but a few of the many examples of coevolution that could be given. One species evolves an exceptional eye, another develops camouflage techniques to puzzle it. One becomes hypersensitive to movement, another becomes capable of disengaging its tail, which remains wriggling to confound the predator. So the eternal struggle and the interdependent evolution of predator and prey goes on, with each new refinement in hunting being countered by an evasion, more or less effective. Curiously, it is the relative imperfection in both arenas that allows continuation of both species, for the greater the interdependency from coevolution, the greater the dependency of each species upon the other for survival.

Within the almost limitless editions of living creatures, there are many examples of the interdependence of process or structure that represent overlapping survival histories. It need hardly be mentioned that evolutionary interdependence can never be considered apart from the general environment that always serves as a more pervasive evolutionary influence than any particular ecological companion.

The importance of the general environment as a major arbiter of evolutionary shaping is emphasized by the precision and completeness with which fundamental energies are reflected in the sensory apparatus of most organisms. It does not seem unusual that the physical pressure front characteristics of frequency, amplitude, and complexity are faithfully transformed into the experiences of pitch, loudness, and timbre, or that the physical characteristics of the visual range of the electromagnetic continuum—again frequency, amplitude, and complexity—are transformed into hue, brightness, and saturation. It seems astonishing only when we think about it for a bit, when we contemplate the evolutionary honing required to leach out all of the potential information within these dimensions and render it into the service of adaptive behavior.

Also consider the evolutionary honing involved in the emergence of the various categories of sensation, a shaping process that has resulted in inputs uniquely designed to perform particular and necessary functions. There are literally thousands of different environmental conditions pertinent to the information requirements of the individual—thus the multiplicity of cue-inputs from such modalities as vision, audition, olfaction, etc. In contrast, since there is only one condition that occasions thirst (i.e., loss of fluid) and hunger (i.e., lack of food), there are only three requirements that such input must fulfill. Each must vary along the dimensions of intensity, quality, and prepotence. If the resolving power of the visual system were as limited as those responsible for thirst and hunger, we would see only variations in shades of gray. I am reminded again of Da Vinci's statement that "in nature nothing is wanting and nothing superfluous."

One might question here whether nature always works to get the most for the least. Consider again the electromagnetic continuum. On this dimension, wave lengths vary from one-billionth of a millimeter to 25 miles, yet eyes, the most universal of all transmitting mechanisms, are responsive only to a very narrow band between 400 and 800 millicrons. What a waste of information! Why not have x-ray eyes or orbs responsive to the total range? Two conditions seem to have limited eyes, whether those of newts or of Isaac Newton: the reflection potential of the visible band and the injury potential of wave lengths not within it. The eye can provide information only to the extent that it responds to reflected energy, and the wavelengths below violet tend to penetrate rather than reflect, while those above red tend to go around an object or to produce a diffuse reflection. This latter point can perhaps best be appreciated by looking at an infrared picture, where only vague contours are depicted and where the details are largely lost. When we consider the previously mentioned limitations, the restriction of response to the narrow band does not seem strange; rather, it is simply another example of the evolutionary imperative suggested by the postulate of inference.

CHAPTER SUMMARY

From a functional point of view, sensory input may be divided into three categories. One, primary activation input, which both signals that a vital system is in a state of imbalance and drives the organism to indulge in compensatory behavior. Although such input typically increases responsiveness, there are two notable exceptions, fatigue and nausea, where balance is reestablished through quiescence. Two, secondary activation input (or affect) permits the deneutralization of cues and provides the basis for the emergency reactions of the organism. In its more subtle expressions affect provides the basis for the multitude of reactions implicit in communication and social behavior. And three, information input though initially neutral may develop the capacity, through learning, to instigate secondary activation so the organism can avoid and approach situations pertinent to its welfare. This input also has a monitoring function, which provides the basis for moment-by-moment compensatory adjustment in direction, momentum, orientation, and position. Most important, such input provides the basis for encoding salient features of the individual's environment and, as such, is the source of an internalized world which, at least in human beings, is the functional arena for such higher mental processes as thinking, planning, and foresight.

Sensation, that major ingredient of mind, is the evolutionary solution to the input problem in complex organisms. It is neither some kind of transcendent intrusion nor some kind of epiphenomenal artifact; rather, it is a natural outcome of certain physio-chemical-electrical processes in the body that are functional primarily because of the experiential aspect. It should be emphasized again that the myriad shades and varieties of sensation, although they represent either internal changes or external energies, are not similar to these conditions. Experience provides a record and a replica of the individual's world, but it is a world transformed and truncated, both to reflect the individual's evolutionary history and to provide the basis for coping with present adjustment problems.

5 Perceptual Structuring

In chapter 4, the view that sensation is the major solution to the input problem in complex organisms was defended. On the inside and on the surface of the human body, there are thousands of specialized nerve endings. Some are scattered widely, others are clustered at particular loci, but all react to particular kinds and intensities of energy change. Thus, during each waking moment, the individual is bombarded by a multitude of information bits, representing both energies in display around him and internal systems that are in perpetual oscillation between imbalance and balance reestablishment.

The sheer magnitude and variety of inputs being imposed, although replete with functional information, would rapidly inundate the individual if simplification mechanisms were not available to reduce this complexity to tolerable levels. Perceptual structuring makes a remarkable contribution to this simplification, and it does so at a minimal functional cost. If the reader glances out the window, or at different parts of the room, he or she will observe that every item in the field of his or her awareness is neatly arranged within the context of three-dimensional space. Perceptual structuring accounts for this orderly representation; it takes the multiplicity of our raw sensations and arranges it so that a functional replica of the external environment is recreated. It should be emphasized that this is a functional, not an actual, duplicate. There may be similarities between the external environment and the internalized representation, but literal replication does not take place.

Even as our sensations constitute a transformation of the energies of the environment, our perceptual world also constitutes a transformation, one in which functionally insignificant aspects are diminished or not included and where crucial ones are given extra accent and clarity. In an adaptive sense, our perceptual world is a better than adequate representation of the external environment. It excludes nonsalient aspects and gives emphasis to features that have been of enduring importance in survival. Our perception of visual distance, for example, is of great evolutionary significance, so much so that a number of overlapping indicators have evolved to represent it. The binocular cues such as accommodation, convergence, and retinal disparity coalesce with such monocular cues as overlap, linear perspective, aerial perspective, and motion parallax to provide a series of back-up systems for this vital discrimination. A person blind in one eye loses the binocular cues just mentioned, but the monocular cues remaining still provide adequate, though less precise, information about how near or far things are from him or her.

The many overlapping cues that give flexibility and precision to our perception of visual distance are but one example of the manner in which redundancy has been selected into information-processing systems to provide a functional portrayal of critical features of the environment. Yet redundancy is not limited to the perception of visual distance. In general, as we shall see, the more important a particular environmental feature has been in the evolution of a species, the greater the tendency for that feature to be represented by overlapping indicators.

CLARIFICATION OF TERMS

Problems of perceptual organization have created long-standing confusion and ambiguity. A part of this confusion arises from an absence of clear-cut distinctions among sensation, perception, and attention. I try to reduce this ambiguity—or at least provide a clear context for my own discussion. Sensation, as described in chapter 4, includes all the raw experiential inputs from the various transmitting mechanisms of the body. Perception, on the other hand, refers to the particular unity, form, and organization (i.e., structuring) that such raw sensory inputs automatically assume because of either learning or innate processes. Attention (the term apperception, as I have previously indicated, seems preferable) refers to the process whereby inputs (whether sensory or memory) come briefly into focus.

A study of sensation, then, must deal with raw experiential inputs per se. An analysis of perception must concentrate upon the manner in which such sensory inputs become organized and structured. An evaluation of apperception must focus on the variables that affect the shifting from one component (whether sensory or memory) to the next. This last problem, which in my view constitutes the paramount task for psychology, is discussed at length in chapter 7. In this chapter, we examine the problem of input structuring that has classically been subsumed under the label perception.

A simple example will clarify the problem. An adult human being perceives a three-dimensional world populated by such phenomena as trees, chairs, and houses, which are always changing in shape, size, brightness, and proximity. This world and the items that populate it appear to have remarkable consistency and simplicity, despite such continuous fluctuation. A particular phenomenon is perceived as the same even though it rarely, if ever, activates receptors in precisely the same manner on different occasions. James (1892) surmised that the experience of the newborn baby consists of an undifferentiated panorama of input from various transmitting mechanisms, a "booming, buzzing confusion." We now know that the experience of the neonate is not as disorganized as James supposed (see Banks & Shannon, 1993; Fantz & Fagan, 1975; Steiner, 1979) but the organization that is present at birth is apparently rudimentary. By the time the child is four or five months of age the perception of phenomena within three-dimensional space apparently begins, so that when items shift in position within this space, they assume an orderly progression (McGraw, 1987).

Certain basic questions emerge from this brief consideration. What are the relative influences of heredity and learning in the development of this three-dimensional space, populated with phenomena? How are discrete sensory inputs synthesized into whole experiences? How do we perceive movement? From what source does our notion of time come? These and other issues are considered in the following pages.

THE GENESIS OF PERCEPTUAL STRUCTURING

There are two obvious positions that can be held on the genesis of perceptual structuring: the nativist and the empiricist. The most extreme nativist would insist that such organization results entirely from inborn processes, while the most extreme empiricist would insist that all such patterning is derived from cues given in experience. For obvious reasons, there have been few defenders of either extreme position. Even the most radical empiricist must admit (though some have failed to do so publicly) the existence of an innate process that arranges the pertinent cues in the necessary ways, and the most zealous nativist must agree that innate structuring cannot be demonstrated in the absence of sensory input. The issue remains very much alive, with one group insisting that the structure of perception is primarily determined by innate factors and the other insisting that it is due to the organism's transactions with its environment.

History

The early empiricists such as Hobbes and Locke believed that the mind of the newborn child was a tabula rasa and that all knowledge was derived from experience. Curiously, the first hint that experience could be affected by inherent factors came from the empiricist Mill (1829). Although he thought that each sensory component was somehow still represented in the whole, he spoke of an amalgamation of such elements into a composite different from its elements. It is true that he was interested in such relatively simple examples of composites as noise and hue, which result when different sounds or light wave frequencies are presented simultaneously, but he did contribute the notion that unique experiential wholes could be derived from fused components.

James Mill's concept of fusion was the direct antecedent of his son's (John Stuart Mill, 1848) mental chemistry and Wundt's creative synthesis. It was no longer assumed that the parts were still contained within the whole, but the principle that new experiences can emerge from the admixture of sensory elements remained pretty much the same. Thus, even among the ranks of the empiricists an inherent process was implicit, for how else could one explain the emergence of these new experienced wholes? Also in the concepts of fusion, mental chemistry, and creative synthesis, we find the first hints of experimental phenomenology: The view that the ways in which sensory inputs are combined into wholes can be determined through experimentation.

There were other thinkers, on the left wing of empiricism, who never acknowledged the possibility of any kind of nativistic contribution. Hume, Bonnet, and Condillac (see Boring, 1929/1950; Mach, 1897/1959), all insisted that the stuff of knowledge was restricted solely to experience, and that all knowledge could ultimately be traced back to this source.

Kant (1781) was the first avowed nativist in perceptual structuring. It is true that Descartes (1892) had previously emphasized the nativistic origin of geometrical axioms, and perhaps thus anticipated evolutionarily derived influences in perceptual organization; however, in both the categories of understanding and the forms of sensibility, Kant insisted that innate ways of knowing provide the organization that experience will take once it is introduced. Kant, of course, was not an extreme nativist. He did assume that experience was necessary before innate organizing processes could be demonstrated.

The first psychologist to fall within the nativistic tradition of perceptual organization was Hering (see Boring, 1929/1950), who held the view that a specific type of receptor in the eye accounted directly for the experience of distance, and that the influence of this receptor would be imposed the very first time experience took place.

More recently, the tradition of nativism in perceptual organization has been defended by the Gestalt psychologists. Expanding on the work of Ehrenfels (1890; who, in his theory of form qualities, insisted that the whole was more than the sum of its parts), the Gestalt psychologists developed the view that innate processes within the brain—the brain field—provide the basis for the emergence of our perceptual world. The coercing effects of the brain field, as exemplified in closure, figure-ground, continuation, and so forth, automatically provide the form and structure of experience. It should also be noted that the Gestalt psychologists, with their remarkable experiments on the Phi phenomenon, permanently shattered the simplistic notion that perceptual structuring can take place purely on the basis of raw experience per se. There is no movement in the flicker of the two lights, yet, when an appropriate alternation frequency is reached, the perception of movement suddenly takes place.

Empiricists have been primarily interested in the source of human knowledge, but they have emphasized cues given in experience as the basis for the structuring of the perceptual field. In general, they hold the view that perceptual organization is a result of the individual's transaction (transactionalism is a recent term for this view) with his environment. Although Berkeley, in the service of his theology, tried to deny the existence of an external world except as constructed by the perceptions of God, he still falls within this tradition. In his *New Theory of Vision* (1709), he stated the view that has been basic to the transactionalist movement from Lotze (1852) and Helmholtz (1924) to modern adherents. The structure of perception is derived from cues given in experience. Most transactionalists assume an innate capacity for cue arrangements, but they all insist that perceptual organization has its genesis in experience and develops gradually as the child grows older.

Relative Adaptive Contribution. Which view of perceptual organization would provide the greatest adaptation facility? Certainly, the nativistic interpretation would win by definition if the time involved for its appearance were the criterion. But would instant nativistic structuring impose sufficient inflexibility to outweigh the advantages gained? All inherited characteristics—reflexes, instincts, and possibly perceptual organization—contribute to rapid and efficient reaction, but they inevitably exact a price in reduced plasticity. If the nativistic view of perceptual organization is correct, even the youngest organism would profit, but it must then bear a burden of inflexibility, perhaps for life. Is early adaptive potential worth the price? To many species it probably is. Simple organisms restricted to narrow environmental niches require relatively less flexibility for adjustment, and those already bound by innate behavioral mechanisms might gain more from inherited perceptual structuring than they would lose. They are already so stimulus bound and their adjustment problems so rudimentary that a little more inflexibility probably would not matter much.

There is a related point that is somewhat difficult to formulate: that innate mechanisms seem to come in constellations; any reduction in flexibility in one area probably will be accompanied by other inflexible mechanisms. If, for instance, the evolutionary solution to a particular survival problem has been the development of an instinctive pattern, it is very likely that other inherited mechanisms will also be found. Conversely, the more that learning provides the solution for particular survival problems, the greater is the probability that flexibility will prevail in other areas as well. In the human being, where inherited instincts have been reduced in favor of learning, such flexibility may also prevail in perceptual organization. Thus, integration requirements alone may determine whether or not organisms will have plastic perceptual structuring or be bound by wired-in mechanisms.

In certain species there are quite specific survival requirements that may impose inherited perceptual structuring, at least to some degree. The young of many species must begin to move through space immediately after they are born. For example, the baby wildebeest can be up and running only seven minutes after its birth (Wilson, 1975). Since the selection pressures of predation against this species have been so great for such an enduring period, it seems likely that the perception of depth, which is necessary for effective running on variable terrain, may be innate.

Thus many animals are probably born with the ability to see depth and to respond appropriately to objects at various distances from them. Since human beings and other anthropoids are cared for by their parents during the vulnerable early stages of their lives, the survival value of inherited perceptual structuring would be minimal and would likely be outweighed by resulting inflexibility. Furthermore, at least in human infants, motor facility is so restricted during early infancy that such fixed organization would not be functional anyway. The notion of process constellations should not be forgotten. The almost universal human use of learning in other areas of adaptation may also tip the scale in favor of learning in perceptual organization.

Not even the most avid empiricist can deny that certain inherited factors profoundly influence the knowing processes of a human being.

Most experience comes to us already divided into separate and distinct modes. Each transmitting mechanism introduces input with characteristic and unique qualities. Learning does not alter these qualities. No matter how often a person has seen red or been thirsty, hungry, in pain, or nauseated, the characteristic nature of the input remains unchanged. Also, as previously mentioned, it is necessary for those who hold an empiricistic view of perceptual organization to accept an innate capacity for such structuring, or it could not occur. It is even possible that the axioms of Euclidean geometry may be innately represented. In the struggle for survival, the shortest distance between predator and prey, between the hunted and a tree, has always been a straight line, and this basic survival information may have been selected into the genetic blueprints of species, as may the other geometrical axioms as well.

Helmholtz (1924) was probably the greatest contributor to our understanding of the development of perceptual structuring as derived from learning, yet his attacks upon the nativists seem superficial and even simplistic. He tried to undercut a nativisitic interpretation of the geometrical axioms by positing another kind of world, perhaps inside a sphere, where the Euclidean axioms would not hold. In such a context, he suggested, the shortest distance between two points would not be a straight line, parallel lines might meet, and so on. He seems to have forgotten that we live on the outside of a sphere, where the same logic should apply. More importantly, he failed to take evolution into account. Even if we were to hypothesize a galaxy of non-Euclidean worlds, we cannot assume that a creature similar to a human being would evolve in any of them. The organisms that would develop in such strange (to us) contexts would also be strange, and might manifest within their innate process systems the non-Euclidean character of their world. When pursued, they might run in curves or indulge in spiral flight. Helmholtz's other argument—that the geometrical axioms must be derived from experience because experience is required to demonstrate them—is even more disappointing; that a principle requires experience for its demonstration tells us nothing about its genesis. We might as well argue that spiders learn to spin their webs because researchers observe the spinning process with their eyes.

Regardless of how we explain the geometrical axioms, or how we try to relate them to perceptual organization, one conclusion seems evident: a capacity for perceptual structuring must be inherited. Whether we hold the view that perceptual organization is present at birth or must await the presentation of certain cues, this nativisitic conclusion holds. The real issue, then, is whether learning is required for perceptual organization to emerge, or whether such organization is automatically imposed the first time sensory inputs are introduced.

Although a definitive answer to these questions may be literally impossible, as suggested in the following section, there are a number of pertinent experiments that provide valuable information on the process underlying perceptual structuring and the readiness whereby it can be altered.

The Malleability of the Perceptual Process. It has been noted by a number of observers beginning with Stratton (1897), that the process underlying perceptual organization is so malleable that it can be altered by relatively short experiences with inverting lenses. Although there is some disagreement about the extent of perceptual inversion, most studies indicate that, after only three to five days of wearing inverting lenses, the subject automatically begins to see his or her topsy-turvy world more naturally.

Schafer and Murphey (1943) demonstrated that when cues are almost balanced, a relatively short training period may be sufficient to establish a perceptual predisposition. They used drawings in which two faces in figure-ground rivalry were presented to different groups of children. After establishing an approximately 50-50 reversibility ratio, they selectively rewarded and punished the children when the faces were separately presented. The faces were then recombined and presented tachistoscopically for very short intervals. Under these conditions, the children tended to perceive the rewarded, but not the punished, faces.

Perhaps the most dramatic evidence on perceptual malleability comes from our laboratory (Goodson, 1981). For many years the students in our introductory psychology course were required to trace star patterns while following their progress in a mirror that both inverted and reversed the situation. After only two hours of such training, between 10 and 20% of these students would automatically, to their own astonishment and that of the instructor, write the mirror image of their signatures and, indeed, in certain cases, long paragraphs that were legible only when the material was viewed in a mirror. These findings indicate that the process underlying perceptual organization is extremely malleable, and they further suggest that the issue of nativism versus empiricism in perceptual structuring may never be resolved. That is, by the time a baby is old enough to understand the instructions and perform the activities necessary to demonstrate a nativistic interpretation, it will already have had ample time for learning to have produced the structuring process. For instance, the many experiments on the visual cliff (Campos, Langer, & Krowitz, 1970; Gibson & Walk, 1960) which purport to demonstrate nativisitic factors in depth perception, are all performed with infants who had already had months of experience, which could easily account for the avoidance behavior.

ENVIRONMENTAL FACTORS IN PERCEPTUAL STRUCTURING

However one may interpret the process underlying perceptual structuring (either as inherent or as the outcome of cues given in experience), this process provides a functional replica of the external environment in which the individual must move. Perhaps some insight can be gained about the emergence of percepts (i.e., the perception of phenomena) by examining features of the environment that have been present throughout the evolution of most species.

A realistic position is implicit in evolutionary theory: the existence of both organisms and an environment in which they evolved is taken for granted. Three important inferences are possible concerning the nature of the external environment. First, it apparently contains many relatively distinct items that vary in distance from one another. Second, such items are often changing in position; some of them, rocks and trees for example, shift only in reaction to impacts from other items, while others, such as organisms and machines, react in such passive fashion, but are also capable of movement on their own. Third, each of such items has a number of stimulus characteristics. The external environment of every organism thus consists of fluctuating entities with multiple characteristics located at varying distances from one another. It is in such a context that species have evolved, and the honing effects of natural selection have ensured that both items in the environment and their relative movements will be functionally represented in perception. From this, we can understand why the capacity to perceive phenomena within three-dimensional space would emerge in complex organisms. Now the question becomes, which aspects of the environment are primarily responsible for the emergence of phenomena (i.e., percepts) in perception? There appear to be a number of such contributors.

Edgeness

One fundamental and universal characteristic of items in the environment is that they have edges. Items have edges that are represented in sensory input—particularly visual input—as blobs bordered by continuing lines. In the visual modality, such continuous linearity arises from shifts in the reflection characteristics between thing and background, or between thing and other thing. Since the receptors respond automatically to light-wave frequencies and amplitude variations, the perception of edgeness is probably affected very little by learning. Thus, a human baby's first visual experience likely consists of a two-dimensional mosaic, with the various parts separated from one another by edges resulting from varying reflection characteristics.

That edgeness is a fundamental characteristic of perception is supported by an almost universal finding in nature. Wherever we look, we see the use of camouflage techniques. It appears to be a survival imperative; if you wish to survive, break up the edges, or do something to make the edges less apparent. The chameleon changes color to meld into the background, as does the snow rabbit, the quail, and the arctic fox. Certain succulent plants have come to have the appearance of rocks or bird feces. In some cases, the edges are broken up by variegated colors, shadings, and contours, as in the carapaces of turtles, the stripes of zebras, the spots of leopards, and the designs on many snakes. The use of mimicry or camouflage in nature is emulated by man in the various ways he paints his ships and planes and clothing during war, and for the same reason. The ubiquity of edge-breaking emphasizes the importance of edgeness as a basic and foundational characteristic of perception.

Although there is always danger in such extrapolation, some pertinent work with animals has emphasized the fundamental role of edgeness in the visual transmitting mechanism. Lettvin and his associates (1959) discovered two types of receptors for edgeness in the leopard frog's eye: contrast receptors for stationary contours, and moving-edge detectors for moving items. In even more striking research, Hubel and Wiesel (1962) isolated cortical neurons in the cat that fire only when the retina is stimulated by edges at specific angles. That is, groups of retinal receptors must be stimulated along specific straight lines in order to excite these particular neurons in the visual cortex.

It seems very likely that the resolving mechanism for edges became more central as organisms became more complex, but, regardless of the manner in which this characteristic is fed into the system, it remains foundational to the perception of all things in the environment. That it depends upon gradients of receptor stimulation—whatever the specific nature or distribution of these receptors—seems beyond doubt. Without edges to stimulate these receptors differentially, the field of vision would be unpopulated—a bland, amorphous continuity. It is true that edges need not be sharp, but there must be some shift in the activation of appropriate receptors, however gradual, or the field would remain undifferentiated.

In this discussion, I have dealt primarily with vision, but the same logic holds for all transmitting mechanisms. Sounds also have edges, as do tastes and tactile input, and in every case the characteristics that produce the edges in perception are breaks or gradients in receptor activation.

Movement

As previously mentioned, items in the environment are always changing in position relative to one another. If an organism moves, it always does so relative to other moving items or against a more static background. Movement, as we shall see, is a characteristic of the environment that is as basic as edgeness, and it undoubtedly plays a crucial role both in the emergence of percepts and in the perceptual structuring of three-dimensional space.

The universality of movement has affected the adaptive repertoire of many organisms, which is perhaps as basic as the break up the edges dictum described in the last section. The rule here seems to be, "If you don't want anybody to notice you, don't move." This is true for both predator and prey. The lioness waits without movement by the waterhole; the cheetah approaches the herd of impala gradually, immobile for long moments, until the final and often fatal rush. The most vulnerable of life's creatures respond with immobility to any sign of danger or to the mother's signal. The quail, thus, remains unnoticed in the grass and the rabbit undetected in the bush. Life feeds upon itself, and living things move. This dual circumstance has provided the selection pressures that have not only resulted in the motion-cessation gambits mentioned previously, but also honed the knowing processes of most organisms to be hypersensitive to any indication that movement is taking place. Thus, in a static field, anything that moves stands out, and is immediately brought into apperceptual focus (as discussed at length in chap. 7).

As the experience of movement is derived from shifts in edgeness, its perception is likely to be as basic as the perception of the edge itself. I refer again to Lettvin's remarkable study and to the bug-perceiving mechanism discussed in chapter 4. Lettvin and his colleagues (1959) demonstrated that movement provides such a basic form of information for frogs that it is represented by specific receptors in the eye. The eye, rather than the brain, is the critical resolving organ for this information. Does the human eye also have receptors specific to movement? Probably not. Our perception of movement without doubt originates from shifting receptor stimulation, but actual resolution is likely accomplished by some central, and as yet unknown, mechanism.

The problem of how we perceive movement has had a fairly long history in psychology and is still of great interest to researchers in the field. The method by which we perceive movement at all is still open to conjecture. How is it, for instance, that we see continuous movement—a bird flying, for example? As an item moves across the visual field, different receptors are activated. Why, then, do we not see movement as a series of little jumps?

Wertheimer's (1945) experiments on apparent movement probably answered this question, and perhaps contributed much more. He found, as will be remembered, that when lights at different positions flash alternately at approximately 20 times a second, the subject sees the light moving from one position to another, but when the frequency of alternations is increased to approximately 40 flashes a second, the subject sees each light as stationary and without flicker. It appears that Wertheimer actually measured the duration of the neural reverberation occasioned by each light stimulus, and that movement emerged when the reverberation from one light overlapped the neural process from the other. Increasing the frequency demonstrated the same principle, except then the reverberatory overlap occurred relative to each light individually. Although there may be a gap between the units of energy imposed, there is no gap in the sensory input introduced. Each such reverberation must last approximately one-twentieth of a second, and when any imposed energy exceeds this frequency, the gap in the sensory input disappears.

Thus, it would seem that the same principle that accounts for the lower absolute threshold for pitch also accounts for the apparent movement demonstrated by Wertheimer. In the former instance, a continuous tone, rather than separate sounds, is heard when reverberations from succeeding stimuli overlap. If the same frequency is involved in each succeeding unit, a constant unvarying tone is heard, but if the frequency shifts, the tone moves up or down the scale. Thus, in both the perception of tone and in the perception of apparent movement, continuing experience is derived from discrete inputs occurring sufficiently close together to occasion reverberatory overlap.

Wertheimer's findings can be extended to explain the perception of all movement. Usually, when any object is changing in position relative to a background, the shift in light energy imposed occurs so rapidly that there is reverberatory overlap between the object as just seen and the object as now seen. When this happens, we have the experience of continuous movement across the visual field. The most obvious example of this is the motion picture. A series of stills, differing slightly from one another, is imposed at a frequency that allows appropriate reverberatory overlap. There is no movement in the physical moving picture. The movement is provided by processes in the individual. The shifting reflection characteristics of the stills provide the reverberatory overlap in the same way that an object moving across the visual field provides the necessary reverberatory overlap to result in our everyday movement perception.

This interpretation suggests another problem that every organism must resolve if it is to survive: How to differentiate between changes in receptors produced by its own movement and those occasioned by the movement of other items in the environment. There is a shift in the reflection characteristics of objects impinging on the retina, whether the eyes, the head, or the entire body is moved. Yet, somehow the effects of the subject's own movements are automatically subtracted and he or she makes a fair judgment about the rate and direction of motion of things external to him or her. Obviously, this ability is critical to adaptation. An organism that could not distinguish between the retinal changes produced by its own movements and those produced by other moving things would have very little chance of surviving.

What cues provide the basis for discounting the effects of one's own movements in such a vital, yet complicated situation? Certainly they must include some characteristic of the visual input that is present when the organism moves, but is not present or is altered when something in the environment moves. Careful analysis and some research in our own laboratory (Goodson, Ritter, & Thorpe, 1978) suggest that a critical cue for this vital subtraction may be movement parallax. When an organism moves forward, the items in its environment appear to shift backward in corresponding and consistent progression, but when something in the environment moves and the organism is stationary, this consistent backward progression is missing completely.

Yet what happens when both the organism and items in the environment are moving simultaneously? The organism can still subtract the effects of its own movement and make fair judgments of the velocity of external items. How? Once again, movement parallax, or a subsidiary cue implicit in movement parallax, seems primarily responsible. From the vantage point of a moving observer, there is a consistent increase in the velocity of items as a function of decreasing distance. This notion is easy to check. If one moves one's head from side to side it will be noted that near items move backward more rapidly than those further away. When any item fails to shift with a velocity consistent with other items at the same distance, that item is automatically perceived as moving. It stands out as does an erratic goose in a migratory formation.

Our own research (Goodson, Snider, & Swearingen, 1980) provided evidence supporting this notion. Moving subjects were allowed to observe a stimulus ball moving between two static planes of similar balls of varying size. When speeded up to synchronize with the nearer plane, the ball was seen to speed up and then become stationary. When slowed down to synchronize with the farther plane, the ball was seen to slow down before becoming stationary as synchronization took place. When moved out of synchrony with one plane into synchrony with the other it appeared, as in the goose example, to be moving from one formation to another.

The issue is likely more complex than I have indicated. Many cues may combine, as in depth perception, to provide the basis for effective and expedient solution to movement problems. No information is more critical to survival, and it is likely that any process so important would involve several complementary and overlapping indicators. Fortunately, the problem is open to research. It would be necessary to devise a context in which a background, various intervening stimuli, the subject's eyes, and his or her head can all be systematically manipulated to determine the contribution of different cues to the perception of different kinds of movement.

I have explored the problem of movement perception at some length, not only because of its obvious and vital relationship to adaptive behavior, but also because it is basic to the emergence of phenomena in perception. Regardless of the number of different inputs characterizing an item, all such characteristics move in synchrony as the item moves and, as such, are perceived as unified. Thus, uniform movement is one of the conditions underlying the emergence of percepts, but the one yet to be considered is just as important.

Multiplicity

Things in the environment not only have edges and movement, but also vary in terms of a multitude of different physical characteristics, such as shape, texture, shading, and various combinations of parts. A tiger, for instance, is composed of a multiplicity of physical characteristics, many of which, such as stripes, body, head, and tooth configuration, are unique to this particular species. This is true of all things in the environment. Multiplicity prevails, whether we are talking about houses, rocks, trees, or other human beings. Such multiplicity invariably increases the problem of adjustment by adding to the problem of complexity. And complexity without compensatory simplification can create the problem of information overload and contribute to its own demise in the evolutionary struggle. Several mechanisms have evolved to reduce the problem of multiplicity, and they make their contribution with only a minimal loss in adaptive information.

Fusion. As discussed in chapter 3, fusion occurs when simultaneous inputs are recurrently imposed. Fusion provides simplification without appreciable loss of functional information. The multiplicity that is characteristic of each thing in the environment becomes reduced to unity, and it is the thing as a unit to which organisms must respond. Thus, a tiger can be heard or smelled, and vision alone yields multiple cues to its presence. It is the total tiger that is dangerous, not its stripes, teeth, or claws alone. The same is true of all items in the environment. It is the thing as a whole, whether predator, prey, or other individual, that typically has survival significance, not just one particular indication of its presence.

It must be admitted that fusion results in a loss of information since individual cues have become melded into the whole. Yet when circumstances are appropriate and necessary, the integration effects of fusion can be overridden. Blind people can become sensitized to particular cues that, under normal circumstances, would have become lost in the total percept. Hunters, both in early history and in contemporary societies, may become particularly attuned to minor cues representing the presence of prey.

The importance of fusion is demonstrated by the fact that most of the inputs to which human beings respond consist of phenomenological wholes (percepts) such as books, trees, houses, and people, rather than the multiple inputs that signify their presence. In passing, it may be noted that phenomenology, as emphasized in writings of Brentano, Husserl, and certain of the contemporary existentialists, can find support and clarification in an evolutionary perspective.

As mentioned in chapter 3, fusion not only effectively simplifies the individual's environment, but also markedly reduces the complexity of encoding. It also reduces the time required for effective responses. An individual who had to apprehend each and every cue that represents the presence of danger, say a tiger, would long since have been selected out. This example suggests the selection pressures that resulted in the development of the capacity for fusion. Undue complication by itself is lethal, and so, at least in certain circumstances, are long reaction times.

Constancy

Consider the size of the image that a particular item, say a coin, makes upon the retina of the eye. When the coin is close to the eye, the image imposed is quite large, but as it moves away the image becomes smaller and smaller. Also, consider the shape that the coin projects upon the retina. When we are directly facing the coin, the shape is circular, but if the coin is turned in either direction, the shape on the retina shifts through a series of ellipses. The same variability characterizes all things in the external environment. Each particular thing is represented, not by one invariant retinal projection, but by hundreds that are perpetually in flux as distance and perspective vary.

Size Constancy. Size constancy insures that, regardless of distance and the resulting variation in the size of the retinal image, an item will be perceived as being the same. Thus, to use our coin example again, even though the retinal image becomes quite small as the coin is moved away from the eye, the coin is still seen as having the same size that it had when it was held close. People in the distance do not appear to be midgets, and cars in the distance do not appear to be toys.

Without doubt, learning participates in the emergence of size constancy, and general perspective appears to be a factor in its operation. People have had many experiences with things that are on the same level with them and that are moving toward them, but very little as seen from above or moving away. Thus, when they first fly in an airplane, they often note that cars and houses and animals below them appear to be remarkably diminished in size. Wagoner discovered that signs and other items such as mailboxes tend to shrink in the perception of an observer who is moving away from them at considerable speed, approximately 70 miles an hour. In a laboratory simulation situation (Wagoner, Goodson, & Nunez, 1980), subjects reported a significantly greater breakdown in size constancy when replicas of signs were moving away from them than when they were moving toward them.

Size constancy thus provides simplification of the individual's environment, and it does so without appreciable loss of information. The survival significance of an item is very much the same whether it is near to us or further from us. It is true that the cues necessary for discrimination are blunted to some degree. Large, near tigers are much more dangerous than small, distant ones, but the reduction in complexity provided by size constancy apparently outweighs this information loss.

Shape Constancy. In spite of the hundreds of different shapes a particular item in the environment imposes upon the retina as it varies in perspective, shape constancy insures that a constant shape will be perceived. Thus, although a coin projects many different ellipses upon the retina as it is turned, it is perceived as having a constant circular shape. The same is true for many items in an individual's environment. Doors and windows are perceived as rectangular even though the shape projected on the retina is trapezoidal except when they are directly faced.

Shape constancy also provides functional simplification. The individual need not respond to the thousands of different images that characterize each object as it shifts in perspective, but only to the constant one that appears in his or her perception.

I should not end this section without mentioning brightness constancy. It also simplifies the individual's effective environment. Brightness constancy refers to the fact that, regardless of the amount of light being reflected from a particular item, it is still seen as having a certain characteristic shade. Thus, coal appears black regardless of the amount of light being reflected from it.

Helson's adaptation level theory (1948) appeared to subsume the facts of constancy development very neatly, and it is in complete accord with the functional interpretation given here. Helson believed that from our many experiences of brightnesses, shapes, and sizes, a residual representing an average of all such inputs emerges, and that perceptual judgments are reflections of this average.

Summary and Comments

At least three basic characteristics of the external environment coalesce to provide the basis for the perception of phenomenological unities (i.e., percepts). Although each of these characteristics has been treated separately, they do not function in isolation. As suggested in the description of the intrinsic dilemma (chap. 2), any discussion of a dynamic, perpetually interacting set of processes imposes a chopped and segmented view. The influences of edgeness, movement, and multiplicity were not imposed separately during the evolution of the capacity for perceptual organization, nor are they manifested separately now.

Likewise, regardless of the mix of heredity and learning, fusion and constancy mechanisms have always been inseparable overlapping processes, and, in concert with the foundational characteristics of edgeness, multiplicity, and movement, they provide the basis for the perception of our phenomenological world.

The analysis in this section may seem similar to Gibson's (1966) theory that the senses function as perceptual systems, but my view is far different. In Gibson's theory such invariants as texture gradients, edges, and corners are aspects of the environment that, when encountered by the sense organs, directly without intervening learning, result in perceptual organization. In my view, such basics as edges, multiplicity, and movement simply comprise the sensory layer of perceptual structuring. Before the visual perceptual world can develop such monocular cues as areal perspective, overlap, movement parallax, and linear perspective and such binocular cues as convergence, accommodation, and binocular parallax must be repeatedly experienced. In Gibson's theory, the perceptual world is given ipso-facto the first time the sensory layer is presented; so that the sensory system, rather than central processing, automatically provides an analogue of the environment. Gibson called his theory perceptual ecology, emphasizing that what we perceive about the environment is a realistic and nonmediated (by central processing) display of those segments of the environment that were of enduring significance during our evolution and still remain fundamental in our everyday encounters.

No one, at least no one with an evolutionary orientation, would disagree that our perceived world is an analogue of the external environment, and that the selection pressures of evolution have insured that it is a functional, if not a literal, representation. The issue is whether the processes that produce our perceptual world are a centrally manifested product of learning or reside solely in the sensory system. In Gibson's theory the eyes appear to be holes in our head, with filters in them—filters that emphasize features of the environment that were important in our evolution and deemphasize or cancel others that were of little or no importance. Here, again, we can agree that the selection pressures of evolution have insured that certain features of the environment are emphasized and others deemphasized or excluded. The issue is whether such emphasis or deemphasis is achieved solely by the sensory system or is primarily the result of central processing. The works of Stratton (1897) and of our own laboratory (1981) demonstrated that remarkable restructuring can automatically occur when the cues that have been organized to provide our perceptual world are placed in opposition. Thus Stratton's topsy-turvy world began to be perceived as normal, and subjects in our laboratory with only two hours training in mirror drawing automatically wrote the mirror images of their signatures. This simply could not happen if the sensory system were solely responsible for perceptual organization. Does the bug-perceiving mechanism and edge detectors isolated in the frog and cat (by Hubel & Wiesel, 1962 and Lettvin et al., 1959) support Gibson's position? The long history of interdependence between survival and catching bugs has probably been sufficient for selection pressures to insure the emergence of this primitive but efficient mechanism in frogs. In cats, the edge detectors may give some support to the view that certain perceptual capacities may be resolved in the sensory system, but it does not demonstrate that complex perceptual organization can occur solely at the sensory level.

WORLD EXTERNALIZATION

In chapter 4, I defended the view that sensation in all of its variety was the evolutionary solution to the input problem in complex organisms. I also suggested that sensation provided a functional representation of energy changes that were of importance in the evolutionary history of complex organisms, particularly of the human being. It was further surmised that this sensory representation was a functional transformation rather than a literal portrayal, a transformation that included and emphasized features of the environment that were of enduring importance, while diminishing or excluding those of little or no importance.

It turns out that a double transformation is involved. Not only is the world internalized, but also, before adaptive behavior can occur, the perceptual processes just discussed insure that it is projected out again. Thus, the words on this paper appear to be some distance from the reader, although the processes resulting in the experience are certainly within the central nervous system. The same is true for many of our experiences. The processes take place within the body, yet we perceive phenomena (i.e., percepts) at various distances from us, neatly arranged in three-dimensional space.

The perception of phenomena at varying distances makes a profound adaptive contribution. All organisms reside in an environment of moving items located at varying distances, and their behavior, whether successful or not, must take place within this context of three-dimensionality. Thus, effective and accurate judgment of distance, with its vital implications for survival, has provided the selection pressures that have resulted in our capacity for world externalization.

Are we born with this capacity for world externalization, or must experience play a role in its development? As previously discussed, this remains an open question, but the evidence appears to support the view that learning is important. In my view, Helmholtz (1924) gave the clearest interpretation not only of the emergence of depth perception, but also of the entire process of perceptual structuring. He agreed, as all thinkers on this subject now agree, that the individual is born with the capacity for perceptual organization. Yet he further assumed, as did Berkeley and Lotze, that our externalized perceptual world is built up from cues given in experience.

Helmholtz considered most of the cues just enumerated as basic to depth perception, and then supposed that their continuing reoccurrence provided the basis for perceptual organization. He viewed this development as an inductive process, one in which the effects of individual cues accumulate to result finally in our three-dimensional perceptual world populated with phenomena. He also pointed out that once such perceptual structuring has taken place, it is automatically and irresistibly imposed. Thus, he would maintain that it is very difficult for us not to see things at varying distances, and that it is likewise almost impossible to see a window or a door as trapezoidal even though the retinal image usually is trapezoidal.

There is much evidence to support Helmholtz's view that perceptual structuring, once it has taken place, is automatically and irresistibly imposed. The many demonstrations by Ames and his colleagues (see Cantril, 1950; Ittelson & Cantril, 1954; Kilpatrick, 1952) showed clearly what can happen when cues that have been previously integrated into the perceptual structure are placed in opposition. Rotating trapezoids seem to oscillate, familiar objects appear distorted in size and shape, marbles seem to roll uphill, and totally scrambled patterns of stimuli appear meaningful and familiar. Perhaps the most dramatic demonstration is provided when objects without cue distortion are presented simultaneously with rotating trapezoids. Cubes appear to fly through space and cardboard tubes seem to bend and even break through the trapezoid to which they are attached.

Recently in a remarkable series of interrelated experiments reviewed by Aoki and Siekevitz (1988) it was demonstrated that experience during a particular and restricted developmental window (during the first four months after birth in cats) is vital for the cortical changes essential for perceptual organization. The experience of light triggers the adding of phosphate groups to a large protein molecule (termed MAP2) that accounts for the plasticity of the organizational process. Aoki and Siekevitz emphasized that the search for the manner in which experience alters brain physiology and experience is just beginning, and they added: "As we try to answer such questions, we hope to get closer to understanding how the external world comes to be mirrored in the microscopic structure of the brain" (1988, p. 60).

These demonstrations and experiments powerfully suggest that the human perceptual world is constructed from cues given in experience, though the assumption of an innate capacity for appropriate cue utilization seems required. They also emphasize the fragility of this perceptual structure. That is, when the cues that have been melded to provide this construction are placed in opposition, the familiar furniture of our world dissolves into a series of question marks.

PERCEPTUAL STRUCTURING AND MEMORY STRUCTURING

One of the more fascinating aspects of cognitive organization is not only that sensory inputs are arranged to provide an orderly, three-dimensional perceptual world, but also that this same organization is present in memory as well. Thus, when we indulge in flights of fancy or let imagination play over previously experienced situations, the memories assume the same order and structure that were present in perception. This fact may not be readily appreciated by some people—those poor in the ability for pictorial recall—but even in such supposed eidetic-free individuals, pictorial imagery is still potential, as expressed in their dreams. Indeed, the most striking evidence supporting the similarity between perceptual and memory structuring comes from dreams. Our dreams do not come to us as formless blobs within a two-dimensional mosaic; rather, they appear as realistic and full-bodied representations, including both phenomena and a three-dimensional context within which the action of the dream takes place. In dreams, the individual moves within a three-dimensional memory world that is so similar to the externalized world of his or her perception that, in retrospect, he or she sometimes has difficulty deciding whether what was experienced was just a dream or the result of some actual interaction with the environment.

This memory analogue of the perceptual world is remarkably adaptive. The plans, choices, and judgments made while locomoting through the duplicate world of memory are, thus, much more readily applicable to the external environment that, as previously discussed, is represented in the truncated, but functionally more adequate, perceptual world.

It may seem that unnecessary complication is introduced by the transformations implicit in the sensing, perceiving, and memory processes of the human being. First, energy changes, both within and without, are transformed into sensory components. Next, the organizing and structuring processes of perception take such input and transform it into a functional three-dimensional representation of the physical environment. Memory processes then encode both the salient items from sensory input and the form and structure of the perceptual world. There is complication in such a three-phased process, but the adaptive facility provided by these transformations far outweighs the negative effects of complication. Nature does work to get the most from the least, but where a great deal is provided in terms of adaptation flexibility, the least may be very complex.

In my treatment I have given primary emphasis to the manner in which visual sensory input is coerced by the innate organizing processes of perception, but the inputs from all other modalities are in varying degrees also affected. Certain fundamental cues—time, intensity, and phase—provide the basis for auditory localization so that the directions of sound sources can be determined. The recurrent inputs derived from repeated movements become fused into the phenomenological unity that I call the motocept so that motor skills can be perfected and operate smoothly and efficiently. Sensory inputs, unique to each point on the skin, become structured to provide the basis for tactual spatiality so that the person can precisely locate stimulation on any part of his or her body.

Two important equilibratory characteristics (reflecting the postulate of process) are demonstrated in perceptual organizing and structuring. First, there is an active coercion of sensory inputs so that a functional duplicate of the physical environment will take place. Second, once such organization and structuring has occurred, there is, as Helmholtz (1924) emphasized, a near irresistible tendency for it to be imposed thereafter.

In summary, perceptual structuring involves two functionally interdependent, though analytically distinguishable, outcomes. First is the emergence of phenomena, based on the intrinsic capacity of receptor systems to respond to edges and movement. The emergence of phenomena is abetted by fusion, which reduces the burdensome complication of multiplicity, and by the constancy mechanisms, which provide sameness despite perpetual changes in brightness, distance, and perspective.

Second, world externalization, (i.e., the development of three-dimensional visual space) is provided by many overlapping cues derived from the eyes working together, or from a single eye operating on its own. This externalized three-dimensional context provides the stage on which the ordinary flux and flow of phenomena takes place.

Whether the effects of the organizing processes of perception are primarily innate, primarily due to learning, or the result of a mix of both remains an open question. Certainly, however, regardless of the empirical bias that one may have, (a bias that, incidentally, I share) one must assume an innate capacity for the cues given in experience to be arranged in appropriate ways. These innate processes apparently take such cues and reconstruct a functional duplicate of the individual's physical environment. Parenthetically, the sheer redundancy of back-up cues contributing to the perception of depth indicates the importance that accurate distance perception has had in the evolution of our species. It also powerfully suggests that the perceptual world of the individual cannot be organized haphazardly. The manner in which innate processes organize experienced cues is present before cues are introduced and insures that, when they are, they will be organized in a particular way.

That is, perceptual organization is not idiosyncratic; the same cues that account for the perception of far and those that account for the perception of near function in the same manner for all of us. The following statement is a reflection of the postulate of inference given in chapter 1:

> The innate processes underlying perceptual structuring coerce sensory input so that a functional duplicate of the external environment is represented in the individual's experience.

TIME

The simple fact that it is impossible for any organism that is in perpetual interaction with an everchanging environment to maintain a steady state suggests that rhythm is intrinsic to life. Systems are perpetually falling out of balance and then returning to balance as appropriate compensatory action takes place. There are rhythms in every aspect of life: in the homeostatic activities of the cell, in the beating heart, in food ingestion, in the elimination of waste, in the estrus cycles of most female mammals, and in the menstrual cycle of the human female. These and countless other cycles are characteristic of all examples of life. Every organism is reminiscent of Ptolemy's solar system, in which a swarm of cyclical mechanisms constitute an interlocking system of wheels within wheels, each spinning at its own tempo, but each influenced by all the rest.

Rhythm also marks the fluctuations of energy and the movements of things in the natural environment. The earth turns and night becomes day; it tilts, and winter turns into summer; the moon waxes and wanes, and the tides rise and fall. These and hundreds of other independent and derivative cycles characterize the environment, and this has probably been the case for billions of years.

Organisms evolved in a context of such rhythms, and they reflect them in their processes. A fundamental synchrony exists between biological and environmental cycles. The daily rhythm of color shift in the fiddler crab continues even in perpetual darkness (Brown, 1958). The mussel placed in still water perseveres in its tidal rhythm of water propulsion (Rao, 1954). The palolo worm, true to its lunar impulses, breeds only once a year, usually on a single particular day (Clark & Hess, 1943). The golden mantled ground squirrel continues its annual hibernation, synchronized with those of its species in the wild, although kept at a constant temperature and in total darkness (Pengelley & Asmundson, 1974). There are many examples of biological clocks that reflect internal homeostatic cycles or recurrent changes in the environment, and each of them determines the tempo of some critical activity. The heart beats out its vital rhythm, and the grizzly bear seeks out an appropriate place for hibernation. All such rhythms are important, not only because they demonstrate the close cyclic harmony between environmental change and an organism's response, but also because they suggest an intriguing way of viewing the problem of time in human beings.

What is time? Does it exist as an objective dimension, or is it simply an expression, or an outcome of the previously discussed organizing processes of perception, particularly as these processes are manifested in the constancy mechanisms? I believe that this latter view is probably correct. In an organism's daily and other rhythms, a biological clock appears to be in operation. The pace of change within the organism may be learned, or it may be represented genetically, but time need not exist as an objective dimension. Motion, process, cyclical rhythm—whatever we may call it—characterizes both the environment and the individual, but time probably exists only in the latter as a subjective constancy mechanism. If this view is correct, time is nothing more than a summarization of our experience with motion. As such, although it is derived from motion and can have meaning only in relation to motion, it is not identical to motion any more than the sensation of red is identical to the wavelength of 680 millimicrons.

Yet human beings' experience of time is much less discrete and definitive than their sensation of color. It is a learned perceptual structure built up in much the same way as those representing size and shape constancy. Perceived cues, derived from the pace at which input fluctuates, become fused into a subjective framework for other experiences. In size constancy, our most common experience with an object becomes summarized into the representation that is cast upon it, regardless of the size of the retinal image. In time constancy, the same thing seems to happen. An average of experienced tempos becomes the reference frame used each time the individual thinks about time in the abstract or makes a judgment of how much time has passed.

In other animals, the cadence of events important for survival may be internalized in different ways. However this happens, certain recurring events become the instigators and sustainers of adaptive behavior. Time becomes synonymous with a rhythm of responses triggered and sustained by some cyclical stimulus. A rat placed on a fixed-interval schedule is able to respond with remarkable accuracy, demonstrating that it can tell almost exactly when a particular interval has passed. Somehow, the rat makes this discrimination in terms of the rhythms of its own body, so that appropriate responses to the cycle of the schedule can take place. In the human being, the internal rhythms of breathing, heartbeat, and other biological processes combine with such recurrent and systematic external changes as the alternation of night and day, the arrival of the postman, the movement of hands on a clock, and so on to become subsumed into the time constancy subjective frame, which then provides the basis for time judgments thereafter.

This time constancy frame of reference is automatically imposed, and provides a functional background relative to which all tempos and events are automatically evaluated. Thus, human beings not only have a succession of awarenesses, but also (as Kant insisted years ago) have an awareness of succession. The chronocept, as I call this time constancy phenomenon, provides an enduring basis for the judgments of tempo, and it is what we are referring to when we use the word time. Since the rhythms of human physiology and the cycles and tempos within the environment are very similar for all human beings, the time constancy phenomenon must be quite similar in all of us.

The previously stated view of time may seem similar to that given by Kant, who assumed that time was a subjective frame that gave order to the pace of human experience. There is some similarity, but where Kant believed that time was an inherited subjective intuition imposed upon experience, I believe that our perception of time is derived from repeated tempos given in experience. Whatever turns out to be the case, there can be no doubt about the adaptive contribution that an awareness of time brings to the individual. Since most critical events are cyclical, whether within the individual or in the environment, an ability to make accurate judgments about when events will be repeated is critical for effective planning and foresight, and it is our talent for these judgments that provides our remarkable adaptive flexibility. Also, it is the existence of "if A then B" concurrences in the environment that makes cue utilization, whether through inherited or learned associations (to be considered at length in chap. 6) appropriate in both avoidance and approach behavior.

THE AUTOCEPT

The concept of self has been around for a long time. It was touched upon by both Aristotle and Plato. It constitutes a central aspect of Freud's development and is included with varying degrees of emphasis in contemporary theories of personality and social psychology. It has been a pivotal notion in the psychologies of both the humanists and the existentialists. Other psychologists have either disregarded or relegated it out of the field. But regardless of its exclusion or neglect, the concept of self keeps reemerging with new coloring and ambiguity. Its vagueness is perhaps derived from the tenuous nature of the experience that represents it, but also because it has been central to so many considerations (in theology and philosophy, as well as psychology) that are in themselves amorphous and unclear.

In spite of these difficulties, I believe that an adequate theory of psychology cannot be developed without a consideration of how the self emerges and how it functions in behavior. The importance of the self is emphasized by the fact that it not only is identified as the most essential aspect of each person, but also is experienced as a vital motivational system that may initiate and sustain behavior.

In my view, the self, or what I have termed the autocept (to make it congruent with other fusion-constancy phenomena) comes into being in much the same manner as do the phenomena of our perceptual world. It is a fused phenomenological unity derived from the most consistent and recurrent inputs that have been imposed upon an individual, particularly those encountered during the formative years of life.

Hume, in a classic interpretation, concluded after much thought, that the self is nothing more than a succession of awarenesses, a point of view that inspired one of the most telling retorts in the history of thought. Kant observed that there is much more to the self than a succession of awarenesses; there is an awareness of succession, indicating that there is a stable aspect to selfness relative to which the flux of other experiences takes place. This enduring aspect continues across the gaps of sleep and the loss of self-awareness that sometimes takes place when a person is absorbed in some task or problem. Yet whence comes this enduring aspect of selfness that serves as a backdrop to the passing flux?

Autocept Emergence

Even before birth, the processes underlying learning and memory are without doubt operative, as are those that result in perceptual structuring. Also, during embryonic life, a fair amount of sensory input is being imposed, and much of this input is recurrent. The beating of the mother's heart, the rhythmic changes induced by her movements and her breathing provide consistent and recurrent sensory reverberations, as do the processes within the fetus' own body, such as its beating heart. Even while an embryo, the fetus touches parts of its own body and may indulge in sucking behavior, providing further consistent and recurrent input. The repetitive and cyclic character of such input fulfills the requirement of fusion previously discussed, and it is from such inputs that the basic core of the self or autocept is derived. If at this time the fetus had the ability to comprehend and speak and was asked the question "Who are you?," he or she would respond "I am Boom-Boom" or "I am Slosh-Slosh," indicating that the essential aspect of what it was at that time consisted of these predominant recurrent inputs.

The mother, in turn, feels the baby's presence in her body when it changes its position, or kicks. She feels her body growing, and she feels the increasing activity of the fetus. Indeed, its increasing and recurrent activity insures that, even prior to its birth, the baby has become an integral part of the mother's autocept. This fundamental inclusion is sometimes unhappily demonstrated in the postpartum depression and anxiety that occurs in many mothers when the structure of the autocept is altered by the loss of the familiar pressures and rhythms of the baby within her body.

After the birth of the child, the number and variety of consistent and recurrent experiences are dramatically increased. The rhythms of hunger, thirst, eating, elimination, and breathing become added to and fused into the basic core previously mentioned. When the exteroceptors have matured sufficiently, certain of the recurrent sights, sounds, and touches of the child's immediate environment also become integrated. The child's mother's face and touch, the changing of its diaper, the sounds of its mother's voice, all coalesce and make a contribution to the emerging self. A given autocept is thus at its core a composite of the experiences that have been most frequent and recurrent during an individual's early life: pain, pleasure, mother's face, elimination, eating, drinking, and so on. It is a complex of inter-dependent and overlapping memory representations that have become fused into phenomenological unity.

The optimal time for the emergence of the autocept (i.e., the incorporation of recurrent inputs within its structure) is during infancy and early childhood. Children have a remarkable penchant for recurrent input. Anyone who has raised children will recall their interest in hearing the same stories repeated again and again, seeing the same commercials over and over again on television, playing the same games, and singing the same songs. Anything that smacks of "Hickory, Dickory Dock" becomes a likely candidate for inclusion. Many parents will also recall their child's involvement with a particular toy or favorite blanket. In such cases, the item in question literally appears to become a part of the autocept, at least for a time. Where such an inclusion has taken place, it is quite typical for the child to insist that the item be placed with them when they go to bed at night. Thus, parents fondly remember their children, thumbs in mouths, hands clutching their beloved blanket, at peace with the world, and ready to go to sleep.

This discussion suggests that the structure of an autocept not only tends to be retained, but also manifests a persistent and perpetual tendency to achieve optimal balance; and, to the degree that this condition is realized, contentment and happiness are the experienced result. It also suggests why certain situations, individuals, and activities become so important, particularly when there are serious threats to autocept organization. When children are insecure, they often suck their thumbs, want to be fondled by their mothers, or desire the security of their own room.

Touching and fondling have long been critical in the evolution of our species. Their prototypes are found in the grooming behavior in many primates. Grooming appears to be the favorite pastime of most monkeys and lemurs, as well as of larger apes such as gorillas and chimpanzees. The importance of fondling in human beings is indicated by the fact that babies who do not receive adequate handling and body contact fail to develop normally, and occasionally will wither and die.

Fondling and other types of bodily contact are important to the inclusion of other people in the individual's autocept. An individual's mother becomes a part of him or her, and, later, so does the father, as he becomes more involved with the fondling and handling of the child. Indeed, all members of the family, insofar as they constitute recurrent inputs, likewise become fused into the individual's autocept. Grief over the death of a member of the family can almost literally be gauged in terms of the extent to which that member has been incorporated into an autocept. We grieve not only for the loss of the other person per se, but also for the loss of that aspect of ourselves that that individual had come to represent.

Autocept Boundary

If the autocept is a phenomenological unity as described, one might wonder what establishes its boundaries. Such boundaries seem to be present in all of us. There is that which is me and that which is not me. There are even certain aspects of what we are that can shift rapidly from self to nonself. One can collect saliva in his or her mouth, and, as long as it remains within the mouth, it is still a part of him or her. Yet if this saliva is spit into a glass, it immediately becomes excluded, so much so that the thought of drinking it becomes repugnant and unacceptable.

Reinclusion may also occur. I have had the experience of being a disembodied selfless entity when first emerging from sleep. Immediately, I indulge in a frantic search for self by touching various parts of my body, touching my teeth with my tongue, or recalling certain well-established memories such as my mother's or my wife's face. Gradually, as these experiences coalesce, my selfness returns again as a unitary phenomenon.

If there is a boundary to the self, what is its source? It appears that the overlapping of dual or, in certain cases, multiple input provides the sensory characteristic that determines the boundary. If one touches something exterior to oneself, there is only one input, that derived from the fingers; however, if one touches some part of one's body, there is input not only from one's fingers but also from the part of the body that is touched. If one watches the touching of some part of the body, triple overlapping input is imposed. Everyone who has had a child has observed the remarkable amount of time it spends watching its own hands moving, putting things into its mouth, and feeling various parts of its body. These are all, I believe, boundary-establishing behaviors, which lead, by the time the child is only a few months old, to a fairly clear distinction between self and nonself. As the child grows older and has more experiences of the difference between single source and multiple source inputs, the boundary between self and nonself becomes more precisely established.

McGraw held a position on the circumstances that account for the emerging self that is very similar to my own. He believed that self is not distinguished from nonself in the neonate.

> For example, the pacifiers on which infants suck no less than the thumbs on which they suck are sensed as a part of self. But through experience, the thumb becomes 'self' and the pacifier 'other.' This occurs because there is a quite fundamental difference between the way the two are experienced. Specifically *each sucking movement is matched by a thumb sensation ... but not by a pacifier sensation.* (McGraw, 1987, p. 722; italics added)

Experiments on Autocept Emergence

Many recurrent inputs from the various sensory systems of the body become integrated into the emerging self. These inputs, such as those from points on the skin, pain, hunger, thirst, anger, and fear remain relatively unaltered as they are repeated again and again. As Ackerman et al. (1998) suggested, anger, when it reoccurs, is the same anger, and such other inputs as shame, hunger, fear, etc., are essentially the same regardless of how many times they occur. Mascolo and Griffin (1990) pointed out that, even though there may be some change in the quality of such input during development, there is similarity in such experiences. I would agree that the circumstances that may induce such input may shift as the child matures but would insist that the quality of the input, pain, hunger, fear, and the like does not change, and as it reoccurs becomes a stable component of the self. I once asked a three-year-old girl, "How do you know that you are you?" She answered, "Because when I pinch me it hurts." And she thoughtfully pinched herself.

From such a hypothetical core the self system expands as more and more recurrent input becomes integrated. As a corollary, the infant soon demonstrates an increasing capacity to control its own movement and to instigate and monitor goal-centered behavior. At eight weeks infants can learn to control a toy suspended above them by shifting pressure on a pillow, and they appear to garner satisfaction from such control (Watson & Ramey, 1971). Sullivan and Lewis (1993) interrupted a similar contingency in four- to six-month-old infants and concluded that the resulting anger reactions were due to the infants' perceived loss of control. At seven to eight months, Watson and Fishler (1977) found that infants will try to grab objects above their heads when such are seen only in a mirror. At 24 months (Ferrari & Sternberg, 1998), children associate their names and their belongings, shoes, toys, etc., with themselves, and at 35 months most are able to correctly identify their gender. As more and more recurrent conditions are imposed the self system becomes better organized and more inclusive, self-generated (volitional) behavior appears, and intentional acts become more common.

McGraw (1987) suggested that the most rigorous test of self-recognition is the mark-on-the-face test in which subjects can see the mark, usually a paint smudge, only by looking in a mirror. If, upon observing the mark, the subjects then react to that spot, usually by rubbing on their own face, it is assumed that an identification between mirror image and self has taken place. Using this test Brooks-Gunn & Lewis (1984) reported that 20% of 15-month-old human babies wiped their noses when they observed in a mirror that red rouge has been placed on them, and that such rubbing behavior was typical by 20 months. Gallup (1970) found that chimpanzees at first greeted their images in full-length mirrors as unfamiliar companions. Yet after 10 days they used the mirrors to groom parts of their bodies they could not otherwise see. When marks were placed on their faces while anesthetized, the majority of the chimpanzees reacted to the mark when seen in a mirror by rubbing that spot on their own faces. Although only primates have demonstrated self-recognition in the mark-on-the-face test, we cannot assume that other animals are devoid of autocept development. Mirror recognition is a very complex operation, and this complexity, rather than the absence of an experience of self, may be the factor accounting for the negative results.

Autocept Functions

In chapter 3, it was suggested that the autocept has several adaptive functions. It is a simplifying mechanism that effectively integrates the most vivid and recurrent experiences of an individual's life. Each individual becomes the kind of person that is appropriate for adapting to varying kinds of environments, whether demanding or benign. The unity aspect of the self also makes a vital contribution to adaptive behavior. Since the individual experiences himself or herself as a whole, a threat to any aspect of what he or she is (whether a part of the body or some belief or value) involves the total person, thus facilitating integrated and directional response.

Automaticity

An important aspect of activity arising from the autocept is the ease and automaticity that prevails. When I decide to do something and then do it, the action occurs in most cases without apparent effort. I have decided to touch my nose, and now I do it. The I or autocept retains its unity and identity and the action unfolds immediately and with little discernible effort.

Automaticity (i.e., the degree that a particular process can occur without awareness) is a direct outcome of the number of times inputs have reoccurred either simultaneously or in sequence, as happens in the fusion that results in the emergence of phenomena and in the development of associations between inputs or between inputs and responses. Both such eventualities take place in the emergence of the autocept and in the behaviors that are associated with it. The inputs that have been fused into unity and the acts that have been associated in a particular individual are so numerous and intricately represented that isolation of particulars is essentially impossible. Both the autocept and the associated behaviors, as I discuss at length in chapter 8, are subsumed within a remarkable *Composite Action System* (CAS) resulting in behavior that is automatic, effortless, and efficient in its contribution to the information-processing capacities of all humans and probably most animals.

Such automaticity is implicit in many acts from the simplest reflexes, both learned and unlearned, to complex behaviors. Every day or so I drive through a series of stoplights, turns, and varying traffic to a store where I shop. Often I make the trip with little awareness of the fact that I have correctly performed many appropriate activities along the route. As a function of long practice with the salient features of the trip, it takes place automatically, with only occasional apperception monitoring.

Complex behaviors such as typing, piano playing or speaking can thus, after practice, proceed efficiently with only occasional monitoring. Many experiments (see Logan, 1991) have demonstrated the manner in which such automaticity develops. The typical design involves having subjects perform two different tasks simultaneously. When one can be performed without interference with the other, automaticity has been achieved (Hasher & Zacks, 1979; Shiffrin & Schneider, 1977). Since the fused components representing the autocept have been repeatedly associated with many such integrated behavior sequences, the motor threshold (see chap. 8) between the autocept and such behavior is lowered to such a marked degree that self-directed behavior can proceed easily and automatically.

Continuity

During every moment there is a continuing self-awareness that remains in the background of the variable flux of other inputs, except for lapses during sleep or when one is totally involved with particular activities. There is nothing exceptional in this. The autocept is a fused phenomenological unity consisting, in part, of the recurrent inputs from the physiological activities of the body, conjoined with constellations of lasting memories and familiar scenes. These central aspects of self continue during our waking moments even as a chronic pain, hunger, or fear sometimes hovers in the background of more transient inputs.

The autocept thus provides for continuity. Perpetual feedback from the viscera, touches on various parts of the body, and enduring memory residuals provide a basis for identity from one moment to the next—a bridge across the gaps of sleep and through the flux of other inputs.

Volition

Volition has been a persistently troublesome issue in efforts to explain the behavior of human beings. At least some activity seems to be directed by an agency that is independent of, or even opposed to, the basic requirements of the organism. For example, a person apparently can disregard other imperative needs and spend his or her time thinking about an inconsequential subject. Thus, although I might be very hungry or thirsty, or even in pain, "I" might, if "I" choose to do so, spend a number of minutes thinking about sedimentary rocks, a topic that interests "me" not at all. This apparent ability to direct behavior willfully, and the experience of freedom that goes along with such activity, have been sources of great satisfaction to those who believe in freewill and a challenge to those who view human behavior as lawful. Any comprehensive theory of psychology must deal with this problem. A failure to do so is a fundamental failure, for volitional activity, at least in my view, is a basic characteristic of human nature.

It will be recalled that all perceptual structuring, including the autocept, is assumed to have two fundamental equilibratory characteristics. First, there is a strong tendency for such organization to take place, and second, once such structuring has occurred, there is a strong tendency for its integrity to be maintained. The individual is a product of what has happened to him or her, whether ordinary or bizzare, and will fight to maintain the integrity of what she or he is.

Not only are certain early recurring experiences fused into the autocept, but the principles and precepts that represent the rules and regulations of society also tend to become included, as do those values and prejudices that are consistently and recurrently imposed. If something threatens the integrity of the autocept, powerful secondary activating input will be introduced, and behavior directed toward reducing such input will take place.

What is the specific source of such activation input? Apparently it is derived from the various processes that are responsible for emotional experiences. There can be a number of different kinds of threats to the integrity of the autocept, from physical attacks upon the person to attacks upon one's cherished beliefs and values. Thus, in certain cases, anger may be the most appropriate activation input, while in others, fear, guilt, or one of the other social motives, such as jealousy, might instigate the most appropriate compensatory response.

Whatever the source of threat and the particular type of activation input, the resulting behavior is construed by the individual as arising from the autocept and appears to him or her to be the result of his or her own free choice. However, the behavior is not free in any absolute sense; it is a direct outcome of the particular recurrent experiences that have been fused into the autocept in question. The person may have an experience of freedom in such choices, but it is an illusory freedom because fusion often precludes the identification of particular causes. The feeling of freedom that pervades volitional behavior, at least in part, is the outcome of this submersion of the specific into the fused totality.

As mentioned in chapter 4, behavior is also sometimes assumed to be volitional when it has only indirect reference to the autocept. Certain activities have become so habitual that no specific input appears to instigate them. Thus, the habitual smoker may automatically light up one cigarette after another, and then declare, when such behavior is called to his or her attention, that he or she is making the choices in such cases. Other activities may be motivated by generalized drives that are so diffuse that the inputs characterizing them are difficult to isolate, and, here again, the individual may mistakenly attribute them to the autocept.

The behavior that results when something threatens the integrity of the autocept is directly dependent upon the degree of fusion that has occurred. This point is supported by the general finding that little children cannot sustain concentration on a topic for extended periods unless the activity is directly related to one or more of the obvious need systems. As age advances, and as more and more recurrent inputs are integrated into the autocept, the ability to concentrate upon the uninteresting becomes fairly refined until, by the time the child reaches college, many professors and textbook writers can capitalize on this ability with varying degrees of success.

In summary, volitional behavior is viewed as a fundamental aspect of the human being, but the occurrence of such behavior cannot be used to support the view that man has freewill. A long time ago, Hobbes pointed out that "Man can do what he wills, but he cannot will to will what he wills." This statement, I believe, is essentially correct. Humans have the capacity to make choices, but the choices that are made are those that are congruent with the unity and structure of the autocept as it, in turn, has been derived from the most critical and recurrent experiences in a person's life. As such, the autocept becomes a motivational system that instigates and sustains behavior that serves to preserve the integrity of what each individual has become.

A person can do what she wants when what she wants to do is congruent with what she has become. To that degree he or she has achieved something that many societies, at least in the western world, define as the most desirable outcome of a political system, freedom (a state that is central to the yearnings of people almost everywhere and one that is said to be requisite for the condition of human happiness). Coercion, or indeed the imposition of any circumstance that runs counter to the structure of the individual's autocept, can result in just the opposite condition; misery and unhappiness prevails.

Here we find the roots of the conflicts and schisms that have confronted people down through time. When the autocepts of two or more people are at variance because of the imposition of antithetical principles and precepts, the barriers between them can be almost insurmountable. The differences in religion, custom, color, sex, ethnic identification, national traditions, and so forth can insure that groups will have divergent autocepts, and different definitions of freedom. Thus the conditions required for the freedom of one group may interfere with the freedom of another, often with unfortunate repercussions, as has happened recently in many parts of Africa, the Middle East, Ireland, Bosnia, and Kosovo.

There have been many causes of war, of course, but those conflicts based on variant autocepts are the most obdurate and intractable because they find their roots in the core of selfness that unifies and defines each people and constitutes the wellspring of their freedom.

A given autocept is a reflection of the conditions that have been most persistent and exclusive during an individual's development. If children have been repeatedly starved or punished, they will find their basic identity in the derivative inputs of such treatment and, if questioned, might well respond that he or she is pain or hunger. It has often been remarked that a child becomes a reflection of the way he or she is treated, and, according to the argument being expressed here, this would certainly appear to be the case. If a child is given much love, he or she will develop into a loving person; if treated cruelly, he or she will develop into a cruel person.

As more and more of the inputs characterizing the principles and practices of a given culture are imposed, these become incorporated into the autocept. This incorporation determines the kind of person, whether rigid and intractable, amenable and flexible, loving and benign, or hostile and hating, that the individual will become. In essence we tend to become a fusion of the most consistent and recurrent inputs that have been imposed during our lifetime. The language reflects this very well. When asked who he or she is the individual may well respond, "I am a woman," or "I am Catholic," or "I am an American," or "I am a terrorist."

All of this has vital adaptive implications. The kind of individual that develops may well be the type that will suit the particular requirements of the conditions in which he or she lives. Certainly a loving and generous individual would not have survived long among the Terra Del Fuegians described by Darwin, and a mean and selfish person would soon be an outcast among the gentle Arapesh described by Mead.

The Experience of Freedom. If one lies down on a bed relaxed, he or she can notice a continuing background input from each part of the body. We have no name for it. It is a low, barely perceptible, emanation that I call neural hum. It may be related to the psychogalvanic reflex that can be recorded from all parts of the body, or to the sound waves that can be heard with an electronic stethoscope (Wagoner, 1999). These sounds emanate from any and all areas of the body and can be heard if one has an acute ear by listening to one's own hand or someone else's head. This all may seem extremely fanciful, if not downright funny, but I assure the reader that such sounds do exist. With an electronic stethoscope they can be picked up easily on all parts of the body. Please bear with me, there is more to come.

Now try a small experiment. Sit down in a chair and prepare to get up. But make this preparation without overtly moving. It will be found that the implicit changes corollary with the intention to move provide continuing sensory feedback that appears to be derived from the parts of the body involved in the intended movement. Now get up from the chair. Notice the input from the muscles and other parts of the body (i.e., inputs from the kinesthetic receptors) that are derived from this movement. Now imagine that all such input, both from kinesthetic receptors and the neural hum previously mentioned, were suddenly removed. If this were to happen, the repercussions would be considerable. The capacity for volitional acts and the corollary feeling of freedom might very well be taken away.

All of this, of course, is extremely hypothetical. Yet there is some evidence, tenuous though it may be, that the experience of volition is, at least in part, derived from the previously described neural hum, conjoined with the inputs from implicit and overt movements.

The first such evidence comes from individuals with neurological damage. Where such has happened there may be at least two consequences with implications for our discussion. The removal of sensory inputs ordinarily derived from the affected part of the body, particularly those arising from movement, causes a disruption in the integrity of the autocept. Indeed, in paraplegia and quadriplegia both a loss of self and the loss of volitional control are commonly reported (Bucherhof, 1998; Dias-de-Carvalho et al., 1998; Miller, 1993). In quadriplegia, the loss of self is devastating and the experience of volition is essentially terminated.

One of my colleagues, E. Ypma (personal communication, 1999), fell from a station platform into the path of an oncoming train. His legs were severed midway between the knees and the hip. I asked him a series of questions about the alteration of selfness that resulted. First I asked if he could feel the previously mentioned neural hum in various parts of his upper body. He immediately agreed that he could. He called it "a continuing state which entered awareness as his attention shifted to the particular part in question." I then asked if he had such an experience when his attention shifted to his legs. He said, "No, there is nothing." I then asked if he had the feeling that he could move the upper parts of his body. He said that he did. That there was "the continuing presence of the potential for the freedom of movement." Then I asked him about his legs. He said that this feeling of the freedom for movement was not there. "It isn't there, it just isn't." Another colleague, B. Ivy (personal communication, 1999) perhaps put it best. "It would be like trying to move a part of your body that is out beyond the tips of your fingers." I tried to do this myself and found that the feeling of freedom for such a movement simply is not there.

Another source of information about both the experience of self and volition comes from the application of drugs that may cause paralysis in parts of the body. Recently I was given a spinal anesthetic for a hernia operation, which allowed me to make a number of observations. Most remarkably, there was a separation of the lower part of my body from my self. I felt my leg but it was not my leg. No dual input. I even felt my penis but it was not mine. It was as though it belonged to someone else. There was no input from areas that previously had been integrated into my autocept. Furthermore, there was no neural hum, no feedback from implicit or actual movements.

Nothing. Even more dramatic, the feeling that I had volitional control was gone. I no longer had the experience of freedom of movement in the lower part of my body. I no longer had the feeling that I could move my legs when I wanted to.

The systematic control of sensory inputs with the use of drugs is an important avenue for the study of the autocept and its function, and the wide use of spinal anesthesia provides a readily available source of subjects that should be utilized.

A much more common source of pertinent information, one that is available for everyone, comes from dreams. When a person goes to sleep both the sensory and motor thresholds are raised (see chap. 8 for an extended treatment of dreams), so that a more intense input than ordinary is required to enter the field of awareness and to occasion action.

The raising of the sensory and motor thresholds has the adaptive function of allowing rest and recuperation, but the inputs that are essential to the experience of volition are cut out. The person is still present in the dream, but he or she cannot determine the action or direction of the dream; it unfolds as though one were watching television or a movie. As in neural damage and drug anesthesia, the inputs essential for volition, and the feeling of freedom, have been removed. All of these observations suggest that volition (the experience of freedom) is, at least in part, derived from the inputs arising from intended and actual movements, and they suggest that, when sensory inputs from a part of the body are terminated, that part is excluded from the autocept.

CHAPTER SUMMARY

The increasing number and sensitivity of sensory systems that developed during evolution introduced an adaptation dilemma: more potentially pertinent information was available, but this very capacity introduced the problem of information overload. The evolutionary solution to this burgeoning complexity has been the emergence of simplification mechanisms that effectively reduces the intractable clutter without substantial reduction in response effectiveness.

As various inputs are repeated under different conditions and perspectives, such phenomena as percepts, motocepts, chronocepts, and autocepts emerge. At the same time, and congruent with the emergence of such phenomena, certain critical cues are being organized to provide the perception of spatiality, whether visual, auditory, or tactual.

Characteristics of the environment that are foundational to perceptual organization include edgeness, movement, and multiplicity. When these are buttressed by particular cues given in experience (in vision, such cues as parallax, retinal disparity, and others) the three-dimensional perceptual world populated with phenomena emerges. The perception of phenomena, rather than particulars, is achieved through fusion and the constancies, size, shape, and brightness. These fused representations include percepts, motocepts, chronocepts, and the autocept, all of which are vital simplification mechanisms.

The process whereby the perceptual world is organized and structured is remarkably plastic. Although there is a profound motivation for its retention (particularly in the autocept) the organization and structuring that results can be altered (at least for brief periods) if certain of the cues that have been unconsciously incorporated to provide such structuring are placed in opposition.

The autocept is a vital motivational system. It consists of the most consistent and recurrent inputs of an individual's life, finding its core in the first experiences of the embryo, expanding to include the vital current experiences of the neonate, and ultimately includes the principles and precepts of the individual's society.

The autocept makes a number of contributions to efficient adjustment, but its most important role is motivational. As such, it often enters actively into behavior as both the occasion for particular choices and the drive for their realization. Thus, volitional behavior, rather than being disregarded or banished, is viewed as being derived from each individual's particular autocept. Since each autocept is the lawful outcome of the individual's most important and persistent experiences, the behavior resulting from its influence is likewise lawful and is typically appropriate for that individual's welfare.

Volitional behavior (i.e., those acts that appear to the person as originating from the autocept) is not due to some hypothetical freewill agency. It arises from the sensory reverberations from visceral activities, ongoing emotional processes and inputs from actual and intended movements that serve as background for the variable flux of other input. The apperceptual monitoring of such enduring reverberations constitutes an essential source of each person's experience of having the ability for free choice. Support for this conclusion is most clearly derived from a consideration of dreams. In dreams (as discussed at length in chap. 8) the sensory reverberations of such body processes are excluded from awareness because of the raised sensory and motor thresholds that occur during sleep; in dreams the action typically unfolds as though we were watching a movie.

The related notion of intentionality is also explicable in this context. In the dream events just happen. The individual, though present, does not decide to do something and then do it. Rather, because the raised thresholds block out the inputs that during the waking state are monitored for selfness, things happen to the dreamer rather than being caused by the dreamer. Both volition and intentionality are given emphasis by the fusion process implicit in the emergence of the autocept. When a person tries to reconstruct why he or she made a certain decision the particulars that singly or taken together actually caused the action are often hidden, fused inextricably into the totality.

It is suggested that perceptual structuring provides a better-than-adequate portrayal of both the external environment and of the critical and recurrent aspects of experience. Thus each of the phenomenological unities discussed (i.e., percepts, autocepts, motocepts, and chronocepts) provides effective summarization and simplification without substantial loss of information. The three-dimensional perceptual context within which such phenomena reside excludes aspects of the physical environment that were of little or no import, while emphasizing others with greater adaptive significance.

6 Learning–Memory

As the title of this chapter suggests, learning and memory are inextricably bound together. Perhaps the intrinsic dilemma (see chap. 2) finds its clearest expression here: Without memory there could be no learning and vice versa. There is no easy solution to unraveling the constant overlap and interaction that exists. The simplest approach, and the one that I follow in this chapter, is to treat them separately; not because two separate processes are involved, but because researchers and theorists in our field have tended to separate them. The first part of this chapter deals with acquisition or learning (the topic of greatest interest to the behaviorists), while the last part deals with storage or memory (the side of the coin emphasized by cognitive psychologists). As we proceed we should keep in mind the fact that only one (learning-memory) process is involved. By viewing both acquisition and storage as a fundamental event in information processing, and by considering the evolutionary origin and present adaptive contribution of this event, a synthesis is achieved.

The selection pressures responsible for evolution have gradually resulted in organisms that are efficient information-processing systems. Learning–memory is viewed as the manner in which information about both internal cycles and environmental concurrences is catalogued and stored so that, if the same or similar situations are again encountered, more effective behavior can result.

LEARNING

The evolution of learning and its adaptive contribution were both directly dependent upon the existence of short-term concurrences in the external environment. With the advent of learning, organisms were no longer bound by inherited mechanisms, which allowed response to long-term regularities, but could now make adaptive responses to short-term concurrences that in many cases took place during a small portion of the individual's own life span. "If it happened once, it may happen again," and the great number and variety of such "if A, then B" concurrences in the environment suggests the importance of learning as an adaptive process and the importance of association as a basic principle of such learning. It also lets us surmise that the strength of an association may be a direct function of the number of such concurrences encountered by an organism.

The various ways in which appropriate response to repeated concurrences abets the survival of organisms is considered at length later in this chapter, but a simple example here may help the reader appreciate the functionality of such learning and to anticipate the manner in which associations derived from concurrences are involved in most of the learning experiments that psychologists have performed.

There is a large sycamore tree on my small farm, standing alone in the center of the pasture. Other than providing shade in the summer, it has little significance for the cattle that I raise. I placed a block of salt under this sycamore tree, the short-term concurrence becoming "if tree then salt." Within two days all of my cattle were taking advantage of this new-found regularity in their environment, with resulting improvements in their health and survivability.

Perceptual Structuring and Learning

In the last chapter, I suggested that perceptual structuring provides the stage upon which ordinary experience takes place. Since perceptual structuring appears early in an individual's life, the ordinary flux and flow of experience that takes place thereafter primarily involves phenomenological unities located within the confines of three-dimensional space. Although perceptual structuring occurs early, the plasticity of the process should be kept in mind. The visual field can be altered dramatically, at least for a short period (Goodson, 1981), if the circumstances are provided for such change (see chap. 5). Furthermore, the learning of skills (i.e., development of motocepts) may continue throughout the individual's entire life, and the autocept, though its nucleus may emerge early, probably continues to change until the individual reaches adulthood.

Perceptual structuring is intimately related to the kinds of ordinary learning that are considered in this chapter. Once the perception of phenomena within three-dimensional space begins to occur, such structuring is automatically imposed thereafter, and the memory encodes representing the individual's day-to-day experience will fall within the frame provided. Even though a city child may never have seen a cow, perceptual structuring will ensure that the first cow he or she sees will be perceived as a unitary phenomenon within three-dimensional space. As he or she encounters more and more cows a fused-memory residual, that is, the concept cow, will come into being. This fused composite (concept) develops not only because many similar cows are experienced, but also because the visual field has already been structured to occasion the perception of phenomena per se.

Levels of Approach

The topic of learning and memory can be examined on at least three levels. First, one might be concerned with the environmental conditions that brought it about and provided the basis for its evolution and for its present adaptive contribution. In general, this approach involves presenting various concurrences or regularities to the organism, and then determining the extent to which behavior is modified to adapt to them. All behaviorists from Meier to Skinner have followed this paradigm: Impose certain consistent operations on the organism and then determine the degree to which behavior is changed in some way.

The second approach centers on the physiological changes that are assumed to occur when learning takes place. When it can be demonstrated that a rat has learned a maze, a pigeon has learned a discrimination, or a human being has learned his telephone number, it is assumed that a change has occurred somewhere, presumably in the nervous system. The question of what happens in the nervous system has not been resolved, but theorists have been ingenious in their efforts to explain what might be taking place. Such hypothetical constructs as trace, neural bonds, synaptic knobs, and engrams have been used to suggest how memory encodes might be physiologically represented. Since the discovery of DNA and RNA and the determination of how genetic memory is represented, it has been hypothesized that learning may be recorded in extremely plastic molecules, either working alone or in conjunction with others. Aoki and Siekevitz (1988) presented data and reviewed findings suggesting that the phosphorylation of sites on a large protein (MAP2) may be involved in the process whereby experience brings about the changes in the nervous system that are demonstrated in learning and memory. Needless to say, the person who finally isolates the particular physiological mechanism that accounts for learning and memory will make a most notable contribution.

Finally, the third level for evaluating both learning and memory is the experiential, and it is here that I place my emphasis, although not to the exclusion of the other two levels. As indicated in chapter 4, sensation is assumed to be the evolutionary solution to the input problem in complex organisms. It provides the manner in which the changes taking place both without and within the organism are represented, and, after being ordered and deployed by the perceptual-structuring processes discussed in the last chapter, it provides the material that accounts for most learning. My emphasis upon experience as an appropriate basis for discussing both learning and memory does not diminish the importance of the other two levels. Data from the behavioral, the physiological, and the experiential level must all be systematically integrated if we are to achieve a total picture of the way organisms work.

History

The first efforts to study learning were centered in the experiential level. Locke thought that the mind was initially a tabula rasa, and that learning brought about the establishment of ideas and their association in the mind. Many philosophers (e.g., Berkeley, James Mill, and John Stuart Mill) and most of the early psychologists, (e.g., Wundt, Mueller, and Titchener) were associationists with a primary concern with how ideas got hooked together through either simultaneous or successive presentation.

The experiential level provided much more than speculation. It was the level involved in the most systematic and precise series of experiments that have ever been performed on learning and memory. I am, of course, alluding to the work of Ebbinghaus (1885), which resulted in basic techniques and vital data on the learning–memory process. There has been a resurgence of involvement with the experiential level, as implied by the term, observation learning pioneered by such psychologists as Bandura, Ross, and Ross (1961) and Hillix and Marx (1960). Although this resurgence is certainly timely (as discussed at length in a later section), it should be remembered that Ebbinghaus was the first researcher in observation learning and that Hull (1940) and Underwood (1949) followed him in this tradition.

The sudden advent and massive repercussions of behaviorism brought about a hiatus in the study of both learning and memory at the experiential level. Watson's (1913) blanket indictment of mind as legitimate subject matter and his emphatic rejection of introspection as an acceptable approach were primarily responsible for this shift in emphasis. The demands of objective behaviorism established the widespread use of animals in the search for general principles and laws of learning. In this context, we recall the cats of Thorndike, the dogs of Pavlov, and the rats and pigeons of Skinner. We also recall the great system builders of the 1940s and 1950s such as Guthrie, Tolman, and Hull, and finally, in the 1970s with cognitive psychology, the reemergence of concern with the experiential level.

The differences between early experiential psychologists and behavioral psychologists were not as great as might appear at first. Both groups were associationists, and both groups used many of the same principles (e.g., similarity, repetition, and contiguity) as the cement between the components being studied. These associationists, whether experiential or behavioral, were ridiculed by psychologists in the Gestalt, humanist, existential, and psychoanalytic branches of the field, who took a more dynamic and global view of the organism. They used such derogatory phrases as brick and mortar, empty box, and brass instrument to register their unhappiness with the particularism and elementarism that they saw in the associationist approach.

Although the mainstream in psychology was first represented by those concerned with mind and then shifted to those concerned with behavior, there was a third approach, the physiological, that was remarkably different from either mentalism or behaviorism and that provided data that was eagerly accepted into the mainstream without any question concerning its legitimacy.

Those with a physiological focus have been concerned with two interrelated problems: the localization of function, and the determination of process in the nervous system. Research on these issues has a short history, probably because a number of heuristic discoveries and equipment developments had to take place before it could be adequately pursued.

The first experimental localization of function—perhaps separation of function would be more accurate—came with the work of Bell and Magendie in 1840. They established different exit locations from the spinal cord for sensory and motor nerves. The work of Flourens in the 1850s, using extirpation of parts, localized certain general functions of the brain and also established the finding, which is still good doctrine, that the brain has both unified and particular functions. Then, in 1860, Fritsch and Hitzig, using electrical stimulation, discovered the motor center and the somesthetic areas of the brain. Shortly thereafter, Monk and Ferrier, also using electrical stimulation, discovered the visual centers and mapped the auditory and olfactory centers as well (see Boring, 1929/1950).

While these discoveries were taking place, research into the nature of the nerve impulse gathered momentum. Its speed was first discovered by Helmholtz, both by stimulating a frog's muscle and by measuring the difference in reaction time to hip and toe stimulation in human subjects. This was followed by the work of Bernstein and Ostwald, who finally established the membrane theory of nerve conduction in 1920, a theory that with but slight modification, remains good doctrine at the present time (see Boring, 1929/1950).

A search for the locus of learning began with the work of Lashley (1929). He first trained rats to run a maze, and then extirpated various parts of the rats' brains. After many replications and much frustration, he finally concluded that the silent areas of the brain work as a whole and that each part appears to have potentially the same function as any other part. Thus, Lashley's principles of mass action and equipotentiality were developed, which when combined are remarkably similar to Flourens' 1870s principle of action commune (Boring, 1950). After the work of Lashley, there was a long hiatus in physiological research as it might relate to learning.

The accumulated research from the past (Boring, 1929/1950) and that of modern times (Frith & Friston, 1997) suggest that both Flourens and Lashley wrote better than they knew. There are specific centers responsible for particular activities and functions but there are myriads of interconnections and redundancies. The seemingly intractible complications between action propre and action commune still highlight controversies that are taking place, and the silent areas still retain their mysteries. The human brain has 10 billion neurons, each connecting with 60,000 to 100,000 other neurons. When we consider that the number of possible synaptic connections exceeds the number of positively charged particles in the known universe (Pally, 1997), that all of this complication resides within the confines of the skull, and that all of the activity we are interested in ceases with death, the enormity of the problem is staggering.

Sometimes a casual observation can initiate a chain of events with ever-widening repercussions that leads to a major trend. In 1980, Roy and Sherrington surmised that activity in particular parts of the brain would cause increases in blood flow.

Years of research have confirmed that increases in activity in particular areas of the brain are correlated with increases in blood flow in that particular area, and have also indicated that the increase in blood flow is greater than that required for normal metabolism, that it is the result of specific neural activity (Frith & Friston, 1997). This robust finding is foundational to both MRI (magnetic resonance imaging) and PET (positron emission tomography) research.

MRI research rests on certain basic characteristics of blood; blood is mostly water and water consists of both hydrogen and oxygen atoms. If a particular frequence of the electromagnetic continuum (in the range for radio waves) is transmitted through the brain at the same time that the brain is being subjected to a strong magnetic field the hydrogen atoms will resonate, and they will resonate differentially depending upon the strength of magnetic field gradients. Such resonation can be detected in a receiver coil, and a detailed, layer by layer (i.e., three dimensional) image of the brain can be obtained, and if particular neural action is taking place that activity can be determined and recorded (Frith & Friston, 1997; Goodson & Goodson, 2001).

Research into brain locationalization and function using MRI has resulted in great progress during the past few years. A sampling suggests the variety and scope of such research: on willed action (Frith & Friston, 1997), face recognition (Gauthier & Logothetis, 2000) working memory (Paulesu, Frith, & Frackowaik, 1993), mental rotation (Richter, Georgopoulos, Ugurbil, & Kim, 1997) conscious awareness (Duzel et al., 1997), memory retrieval (Kapur et al., 1995), and encoding and retrieval (Nyberg, Cabeza, & Tulving, 1996). Many studies have been performed, and the pace quickens as research centers and universities acquire the necessary equipment.

Yet recently, demonstrating the self-corrective nature of scientific involvement, this enthusiasm has been tempered by a sobering reassessment. Bub (2000) indicated some of the methodological and analytical problems inherent in all such research. In his timely review of *Human Brain Function* edited by Frackowaik, Frith, Dolan, and Mazziotta (Eds.) (1997) he suggested that "cognitive scientists are remarkably divided on the immediate potential of functional imaging—as regards the modular organization of the cognitive system" (Bub, 2000, p. 467). He noted the extreme difficulty of controlling head movement, the diffuse nature of blood flow, the latency of blood flow response (six to eight seconds), the intrusion of extraneous brain activity (i.e., background neural noise), and the dynamic interaction of all cortical systems. Highlighting the difficulties is the seemingly intractable obstacle that in certain cases researchers are trying to measure neural processes that take place in milliseconds with blood flow shifts that are diffuse and that occur some seconds after the critical condition has been imposed. It is, as one of my colleagues observed, somewhat like trying to examine the intricate workings of a Swiss watch with a crowbar.

These difficulties are not insurmountable. As is often the case, the ingenuity of researchers rises to meet the difficulties encountered; many are obviated by careful instruction and head clamps, rigorous demands on instrument placement and operation, multiple measures of each event both within and between subjects, the averaging of data, and the application of appropriate statistical procedures.

In all such studies a subtractive procedure reminiscent of that introduced by Donders and utilized by early researchers on the timing of cognitive acts (Boring, 1929/1950) is utilized. An experimental condition (or multiple conditions) is imposed, measurements of the control image is taken and the values are subtracted from that of the experimental image. Yet resonance images are much more difficult to quantify than simple reaction times: Repeated measures, averaging, and the assumption of the normal distribution of data is required.

The second major method for measuring changes in blood flow in particular areas of the brain involves the use of radioactive tracers. A tracer is injected into a vein, becomes mixed in the blood, and for a short time positrons are emitted that can be detected and recorded. PET has two major disadvantages, possible danger to human subjects and the high cost of necessary equipment; but careful monitoring and a few well-funded research centers have performed many experiments, particularly with animals. Frith and Friston (1997) gave a detailed description of the PET procedure and summarized some of the research.

The great puzzle of how memories are represented in the brain, has made little progress since the theoretical consideration of Hebb (1949) many years ago. To him memories reside in cell assemblies that are gradually established as a function of practice. This notion may be correct, as far as it goes, relative to the establishment of habits. Yet much learning requires but a single observation: The encode appears to be established in a fraction of a second, little time for the development of synaptic knobs. We can only surmise that memories may be represented in remarkably plastic coding mechanisms within neurons or neuron matrices, possibly similar to the genetic encoding that accounts for maternal behavior in many species and for web spinning in spiders where complex designs and the acts to produce them are somehow represented in small genetic packages. Evolution is opportune and parsimonious and such a transition might be expected during evolution as many instinctual behaviors become replaced by learning.

As previously suggested, psychologists have been concerned with three different levels of data, the physiological, the behavioral, and the experiential. My focus is on the experiential level, not because it is necessarily more important but because it more clearly represents my background and interest. The other two levels are of equal importance and ultimately, as our disciplines mature, data and theory from all three levels will become unified into a comprehensive explanation of the way humans work as lawful derivatives of the evolutionary process.

As this synthesis is taking place it will help us to examine data and theory both at and across these levels. We know, for instance, from cognitive studies that memories exist in associational constellations, giving credence and support to Hebb's (1949) notion of neuron webs, from studies on observational learning pioneered by Bandura and Walter (1963) that encodes can be established in one exposure of brief duration, and from our examination of dreams and eidetic imagery that encodes can be markedly similar to sensory inputs. Such cross level and within level examination should abet the previously mentioned synthesis so that we can understand ourselves as dynamic adaptive systems.

Such evaluations may also provide direction and suggest limitations in our search. For instance, in our search for centers and processes in the brain we should keep in mind that both learning and apperceptual shift can occur in milliseconds and that somewhere, somehow, such facility and plasticity in process must exist in neural systems; a sobering conclusion, and one that profoundly questions whether a technique as diffuse and prone to methodological difficulties as the measurement of shifts in blood flow as in MRI and PET will be appropriate in the study of such fleeting processes.

Bub's (2000) analysis should be carefully read by all who are interested in the determination of location and function in the brain, but it is not a prescription for despair. His treatment of the promise and pitfalls of such research is an appropriate description of the learning by doing implicit in all research into complex systems, whether in astronomy, subatomic particles, or the genetic code, where progress is often the outcome of many coalescing factors: the conjoining of equipment development, seemingly disconnected observations and discoveries, shifting paradigms, the leavening of past discoveries, contentious claims, and sustaining commitment. Yet this, after all, is the way science works. Progress, particularly where issues are fundamental, is typically gradual, marked by small increments, occasionally by rapid advances. This gradual unfolding will no doubt prevail in our study of the human brain. Such enlightenment will come as new methods for observation, measurement, and analysis are brought to bear on this fascinating problem, the unraveling of that most complex of all unraveled knots. The human brain is awesome in its complexity and capacity, but it is also the lens now turned inward upon itself, the instrument of its own evaluation, which is reason enough for optimism.

FOUR KINDS OF LEARNING

An inventory of the many hundreds of experiments that have been performed on acquisition since the turn of the century indicates that four different kinds of learning appear to have been involved. Although my treatment is sequential, none of these kinds of learning occur alone. Indeed, in human beings all four types are represented.

Cue Deneutralization

As was noted in chapter 3, a paradoxical problem for adaptive behavior was introduced when organisms became capable of reacting to a cue rather than to the critical situation itself. When effective reaction to a cue takes place, the behavior occurs before the critical situation is encountered. Thus, in avoidance behavior no direct activation can be involved. When organisms became capable of responding to cues, and this is particularly the case where learning is concerned, a secondary activation system was a necessary prerequisite. In escape behavior, no such problem exists. The critical circumstance itself is directly imposed and can provide the impetus for the behavior in question.

Pavlov (1927) pioneered the investigation of cue deneutralization with some of the most precise and definitive research performed in psychology. After pairing the sound of a metronome with meat powder for a number of trials, Pavlov found that a dog would salivate to the metronome, even though no meat powder was presented. This was a straightforward demonstration of cue deneutralization; prior to conditioning, the metronome was neutral as far as salivation was concerned, but, after consistent pairing (a short-term environmental concurrence) of the metronome and meat powder, the association was established. Here we have a clear example of the learning of a functional association (i.e., a cue instigating a response that is important for the ingestion of food).

Many replications of this basic research using animals as diverse as rats, chimpanzees, and human beings, have been performed. Findings from such research lead to what may be given as a basic law of positive cue deneutralization:

> Any previously neutral cue that repeatedly occurs in conjunction with equilibrium-trending activity will acquire the capacity to occasion preparatory or approach behavior.

The avoidance side of cue deneutralization has also been actively researched. Some of this work was completed in Pavlov's laboratory (Bekhterev, 1907), but it was Miller and Dollard (1941) who first gave a convincing demonstration of both its development and function in behavior. After repeatedly pairing a stimulus with the onset of shock, they found that rats would tend to avoid that stimulus thereafter, and in certain cases, would even learn a completely new response in order to get away from it. Numerous other studies have been performed since such classical experiments, and, in general, they all support the same finding. These results may be stated as a general law of avoidance cue deneutralization:

> Any previously neutral cue that repeatedly occurs in conjunction with increasing imbalance will tend to produce avoidance behavior when it is encountered thereafter.

Both approach and avoidance cue deneutralization confer remarkable adaptive benefits: The organism need not come directly into contact with the critical situation, but can respond appropriately before it is encountered. The capacity for cue deneutralization evolved because it allowed organisms to respond adaptively to short-term environmental concurrences, and it finds its present functionality to the degree that it effectively reflects such concurrences in each individual's life.

In cue deneutralization, the associations that are established reflect short-term concurrences between some cue and either a state of increasing imbalance or a state of balance reestablishment. Two overlapping and complementary processes, preparation and activation, are involved. Thus, in Pavlov's experiments, the metronome came to elicit salivary responses preparatory to eating, and it also came to activate behaviors toward the food source. Likewise, in avoidance situations, the cue brings about such responses as the release of adrenalin into the blood stream, as well as increases in blood pressure and heart rate, all of which have preparatory functions, and it also brings about the introduction of the secondary activation input, fear, which initiates and sustains the avoidance behavior.

Thus, cue deneutralization involves the development of associations that are critical to the maintenance of optimal balance in the various physiological systems of the body. As a function of such learned associations, the adaptive facility of the organism is remarkably increased. It can respond to the cue, and thus either approach or avoid circumstances that are vital for survival. In all such examples of association establishment, the learning of a new stimulus is involved. In the next example of learned associations—which, for historical purposes, we call instrumental learning—a new response, instead of a new stimulus, is incorporated into the individual's repertoire.

Instrumental Learning

In retrospect two apparently different approaches may come to be seen as very similar, if not identical. This is true of the research and concepts of Thorndike (1898), who began his work with chicks, dogs, and cats in the puzzle box as early as 1896, and that of Skinner (1938) who began his work with pigeons and rats in versions of the Skinner box, in 1937.

Let us review some of the major features of Thorndike's approach as a way of introducing the nature and function of instrumental learning. Thorndike built a primitive box consisting of a square frame, approximately three and a half feet on a side, covered with wire mesh. Into this puzzle box he placed his subject, a hungry cat. On the outside of the box, near the door, he placed a saucer containing some fish. It was the cat's problem to learn that pulling a string (or activating some other triggering device) would release a catch that allowed the door to fall open, permitting it to emerge and eat some of the fish. After a considerable number of trials, the cat was able to learn the short-term concurrence between string pulling and accessibility of fish.

The work of Skinner and his associates and that of Thorndike are remarkably similar in both method and adaptive implications. The Skinner box is an enclosed chamber, approximately two feet on a side, containing a bar that, when pressed, activates a dispenser that releases a food pellet into a food tray. When a hungry rat is placed in the box, it immediately begins to indulge in what Thorndike would call trial and error behavior. Sooner or later the rat (as was also the case with Thorndike's cat and the string) accidentally hits the bar and releases a food pellet into the food tray. After a number of such short-term environmental concurrences, the association between the bar pressing and the pellet in the tray is established. In this example, also, the learned association is a clear reflection of a relatively short-term concurrence in the organism's environment, and, after learning has taken place, appropriate behavior relative to particular body imbalances is much more likely to take place.

This brief overview suggests that the traditional difference between classical and instrumental conditioning is largely a function of the emphasis of the researcher. Pavlov and his coworkers, and those who have followed in his tradition of studying cue deneutralization, emphasized the manner in which new stimuli develop the capacity to prepare and activate, while researchers such as Thorndike and Skinner have fixed their attention upon the establishment of new overt behaviors. A moment's thought would suggest that Pavlov's dogs exhibited many overt behaviors, such as head turning and lowering, which could well fit into the instrumental paradigm. Likewise, it appears obvious that Thorndike's cats and Skinner's rats also exhibited both preparation and activation, as involved in salivation and various visceral changes, which are clearly within a classical conditioning interpretation.

Regardless of the degree of overlap involved in these two approaches, the development of the associations is dependent upon the existence of short-term concurrences in the organism's environment, and both approaches demonstrate the vital contributions that such learned associations make in ongoing homeostatic behavior.

Insight Learning

Although the Gestalt psychologists have been primarily concerned with perceptual organization, they have also, from the very beginning of the Gestalt movement, tried to integrate learning into their point of view. Also from the beginning, they have been more inclusive in their approach than other researchers. With varying degrees of success, they have tried to integrate the three levels of psychological data into a comprehensive view. In Wertheimer's (1945) book on problem solving, he emphasized the experiential level in human beings. In Kohler's (1917) work, although the central concern was on the experience of the chimpanzee as the reorganization of the perceptual field took place, the precise description given emphasizes the importance of the behavioral level. Finally, with the development of the notion of a coercing brain field (see Kohler, 1920) and the residual trace, we see an effort to deal with the manner in which the nervous system participates in both learning and memory.

Let us review an example of Kohler's work with chimpanzees. As some readers will recall, Kohler was sent from Berlin in 1912 to take charge of the ape station at Tenerife, one of the Canary Islands off the east coast of Africa. He later was isolated there because of the outbreak of World War I, and thus had the opportunity to work with chimpanzees for four years without interruption. His general approach was to place a hungry chimpanzee in a six-foot-square cage with bars across the front. He then would place a food object, usually a banana, on the outside of the cage just beyond reach, and various tools, typically sticks of varying lengths, on the inside of the cage with the chimpanzee. Very precise observations were then made on the manner in which the chimpanzee solved the problem of using one or more tools to secure the banana.

Kohler's most dramatic example of problem solving was provided by a remarkable chimpanzee named Sultan. Sultan was placed in the cage with two sticks, neither of which would reach the banana on the outside of the cage. The following is a summary of Sultan's caretaker's observations. First, Sultan looked around the cage and apparently observed the sticks on the inside, and the banana on the other side of the bars. He then picked up one of the sticks and thrust it through the bars as far as he could, apparently trying to drag in the banana. He then reached for the other stick and tried to drag in the banana with it, but it was also too short. He kept trying different things, such as pushing one stick toward the banana with the other stick, but no solution was found during the entire hour of the first trial. Yet during the second trial, Sultan solved the problem. He telescoped the two sticks together, dragged in the banana, and ate it.

Kohler decided that learning is not a slow trial and error process, as Thorndike had postulated. Rather, it happens in a burst of insight when there is a sudden reorganization of the perceptual field into a meaningful pattern. Furthermore, according to Kohler, once such insight learning has taken place, it is difficult, if not impossible, to extinguish.

Does learning take place by slow increments, or does it take place in only one trial? The argument continued for a number of years, with thoughtful adherents on both sides of the issue. It was gradually, though some researchers feel not adequately resolved, with the obvious conclusion that Thorndike was using a relatively stupid animal in a situation where the various features of the problem were hidden, while Kohler was using an intelligent animal in a situation where the various features of the problem were easily discernible; thus, slow learning would be expected in the first case, while rapid learning would be predicted in the latter.

It now appears (see the following section) that Thorndike's cat was involved in new learning, while Kohler's chimpanzee was involved in the conceptual manipulation of learning that had taken place prior to the experiment. In both cue deneutralization and instrumental learning, we are dealing with the establishment of new and functional associations, while in problem solving, we are dealing with the manipulation of fused encodes (i.e., concepts) that have emerged from prior experience. However an association logic applies in problem solving more profoundly than in the other two types of learning. It is as if the individual brings an "if A then B" learning set into the problem situation. This associational frame of reference is another reflection of the short-term regularities that are common occurrences in the environment.

Observation Learning

Although the initial research on learning and memory (Ebbinghaus, 1885) was concerned with the variables that influence observation learning, this approach was strangely neglected during the period from about 1915 until about 1960, at which time Bandura et al. (1961, 1963) began to emphasize its importance. Why was observation learning ignored for so long? Perhaps because it is so universal that it was taken for granted. It is implicit whenever a professor gives a lecture and assumes that his students ingest at least a portion of it. It is also implicit every time a movie is presented to a class or, for that matter, any time two people have a conversation. When I asked one of my colleagues why he was lever training a rat in a Skinner box, he made the declaration that he was doing so in order to show his class that the most important principle of learning was reinforcement. When I pointed out that he was taking a much more important principle for granted in his demonstration, that the sheer observation of what he was doing would result in some ability to remember what he had done, he was astonished and subsequently changed his orientation towards the relative importance of the different kinds of learning.

A few simple examples should clarify the nature of observation learning. If a person hears a sequence of numbers such as 2-3-6-8 he will immediately be able to repeat them. Or if she observes a number of cards being presented—say, a king, three, ten, and four—she also will be likely to remember them also. If a person glances out of the window for just a moment and then closes his eyes, he will be able to recall a great deal of what he just saw. Indeed, as an individual indulges in daily activities, a memory replica of much of what is encountered is automatically established. Thus, an individual can reconstruct, with some precision, a great number of the experiences that occurred during any particular day. As previously mentioned (chap. 3), it is this capacity for the immediate establishment of an encode that allows the development of the internalized world, which is the foundation for effective foresight, planning, and thinking.

Observation Learning Update

Ebbinghaus (1885) would have had no trouble interpreting his findings in the light of cognitive processing. He was, after all, a member of the structuralist school, which viewed the study of the mind as exclusively appropriate for psychologists. But with the advent of behaviorism, the focus shifted, and, as far as the mainstream of the discipline was concerned, mentalistic interpretations became unacceptable. In spite of this, evidence for observation learning seeped in through crevices exposed when researchers were ostensibly involved in demonstrating other, presumably more objective, principles. Tolman (1948), while insisting that he was a legitimate behaviorist, had a theoretical position and a research orientation that was compatible with a cognitive interpretation, thus creating an accepting environment for its resurgence.

One study that Tolman published (Tolman & Honzik, 1930) should be remembered here. In this experiment, rats were allowed to wander through a multiple-unit maze without receiving reward in the goal box. As might be expected by someone holding an S-R reinforcement theory (Hull, 1943), little learning was demonstrated. Yet after 10 trials of free maze wandering, food was introduced in the goal box. The rats immediately performed as well as another group that had received food on every trial. Apparently the observations made by the rats as they wandered about were sufficient to provide a functional cognitive map of the maze. Yet the Tolman–Honzik experiment was not designed to demonstrate observation learning; rather, it was designed to demonstrate that latent learning (i.e., learning without obvious reinforcement that remained latent until incentives were introduced) was possible.

A more obvious example of observation learning that occurred while the experimenter was involved in demonstrating another principle comes from the early work of Tinklepaugh (1928). Monkeys and chimpanzees were allowed to observe the experimenter place a piece of banana under one of two identical cans. A screen was then placed between the subject and the cans and a less desirable food (lettuce, carrots, and orange slices) was substituted for the banana. When the subjects were allowed to respond they almost invariably went to the can under which they had observed the experimenter place the banana. This was a clear example of observation learning. Yet it was not designated as such. Rather, since the concept of expectancy was an important issue at the time (because of the Tolman vs. Hull contrast), the emphasis was placed on the fact that the subjects typically displayed emotional responses and failed to eat the less preferred food when it was found where they expected banana. Indeed by their jumping and shrieking the subjects seemed to be saying, "Where's the banana?"

In an interpretation of the same study by Restle (1975) the emphasis is, I believe, correctly placed on memory rather than short-term habit or body orientation. But the experiment's obvious implications for observation learning is once again overlooked.

Another study with animals where observation learning seemed to be involved, although not directly specified, was performed by Solomon and Turner (1962). These experiments paired a tone with shock to a dog's paw a great number of times. Yet no response to the shock could be made because the dog's leg was paralyzed with curare. In spite of this lack of response and thus of reinforcement during training, a strong association between tone and shock was established; that is, after the effects of the drug had worn off, the subject withdrew its paw each time the tone was sounded. Apparently observing the two contiguously presented stimuli (tone and shock) was all that was necessary for the learning to take place.

The imitation of behavior as a function of observation (as well as an example of animal discovery and cultural transmission) was reported by Kawai (1965). The Japanese macaques of Koshima Island learned to wash their food about 40 years ago by simply observing the behavior of others, beginning with the remarkable discovery of the functionality of such washing by a genius female macaque. The pioneer work of Kohler (1917) on the island of Tenerife during World War I, is a solid demonstration of the efficacy of observation learning. It was necessary for Kohler's chimpanzees to conceptually manipulate the perceptual aspects of each problem situation, but the observation of the field was an essential precursor to such manipulation.

More examples of observation learning could easily be given. Every reappraisal of the many experiments performed in learning, from the classical conditioning studies of Pavlov to the learning set studies of Harlow, suggests that observation learning had entered the picture, often in a fundamental way. Yet as ubiquitous as observation learning now appears in retrospect, its importance was not generally accepted until the research and practical applications of Bandura. Of the many studies performed by him and his colleagues, as well as the many inspired by his approach, one series is outstanding, both as examples of his work and of its implications. Bandura et al. (1961, 1963) allowed children to see movies of adults doing certain things, such as striking a large doll with a hammer. During the testing phase, when the children were exposed to an actual doll with an actual available hammer, a remarkable and disturbing number of them picked up the hammer and struck the doll with it. Such studies have been replicated under varying conditions with children of different ages and have created much concern in adults worried about the unsavory repercussions of allowing children to watch violence on television.

In some accounts of observation learning, the term imitation learning is used as if the two were essentially the same. This identification, I believe, is inappropriate and misleading. Imitation may be a fair indication that learning has taken place, but it should not be confused with the conditions or processes that were involved in the learning in the first place.

Most human learning, and probably much of that in other species, takes place when input is apperceived. That we overlooked this most obvious type of learning for so long is a testament to the myopia that can hinder even the most astute and conscientious observers when they become caught in a particular frame of reference with a fixed set of preconceptions. Although the term observation learning seems appropriate, it should be emphasized that the apperception of input (as explained later) is actually the operative process that results in the establishment of an encode.

In observation learning, such principles as contiguity and need reduction, which are sometimes stated as essential requirements for learning, are secondary or auxiliary in their contribution.

Apperception and Encoding

Contiguity has been the most widely advanced principle used to account for associations, from the English empiricists to the stimulus-response theorists of the behaviorists. Contiguity is certainly crucial in the establishment of associations between encodes, but it does not make an important contribution to encode establishment per se. The contiguity of two chalk marks on a blackboard may affect their relationship, but the variables that account for their being there are quite different. Their presence reflects the nature of the chalk, the pressure with which it was applied, and the color and texture of the board. Similarly, the contiguity of two inputs may be crucial in establishing an association between them, but the process that occasions their encoding is quite different. I do not mean to deemphasize the importance of associations, reflecting as they do the "if A then B" concurrences in the environment that make such learning critical to survival. Yet before such associations can develop, the A and B must be observed (i.e., apperceived) so that encoding can take place.

Need reduction (the many variants of which range from general satisfaction to more specific drive stimulus reduction) appears to be involved in observation learning in an even more indirect way. Where is need reduction involved in the encoding of a number of playing cards? It may play a role in the establishment of encodes, but only insofar as the termination of the need (or the pleasure input that is perhaps correlated with it) gives emphasis to the encoding situation.

Although need reduction and contiguity, as well as a number of other principles to be discussed later, may play peripheral or secondary roles in observation learning, one fundamental principle appears to be both necessary and sufficient for such learning to take place—apperception. Apperception, as briefly described in chapter 3 and as explained at length in chapter 7, refers to the momentary focalization of either memory encodes or sensory inputs.

Most human learning results from the apperception of inputs that are introduced when observations are made. The principle of observation learning is given in the following statement:

> Any input (sensory or memory) that is apperceived will result in the establishment or strengthening of an encode.

If apperception constitutes the necessary and sufficient condition for observation learning to take place, one might wonder whether the two processes are distinguishable. There is much overlap, but there are also important differences. In the next chapter, we discuss the variables that determine which input (sensation or memory) will be apperceived. These variables, insofar as apperception and encoding are inextricably bound together, will affect observation learning. In this chapter, we discuss observation learning, while holding the process of apperception constant. In other words, if apperception is full open, what are the other variables that might enter the picture to influence the establishment or the strengthening of an encode? This leads us to a second principle of encode establishment:

> If all other factors are held constant, the more often a given input is apperceived, the more durable the resulting encode will be.

Although repetition has generally fallen into disrepute as a principle of learning, this low status seems completely unjustified. Certainly the repetition of a stimulus or an activity provides no assurance that learning will result; the mere presentation of a circumstance may or may not occasion apperception. Yet if apperception is taking place, repetition of input will strengthen the encode. Hull (1943) was essentially correct in stating that strength of habit is a function of the number of reinforced trials, but he was, I feel, correct for the wrong reason. It is not the drive-stimulus reduction that strengthens the encode, but the high probability of apperception that the balance reestablishment circumstance provides.

Much data from overlearning studies supports the view that repetition is an important factor in observation learning (Ebbinghaus, 1885; Krueger, 1929; Underwood, 1949). If two groups of subjects learn the same list of nonsense syllables until they reach the same facility, overlearning by either group will establish the encoded series even more firmly. If practice makes perfect, extra practice will ensure that perfection lasts longer.

Three other variables, in addition to repetition, can influence the encoding process in observation learning: discreteness, intensity, and duration (Hull, 1943; Underwood, 1949). Once again we confront the overlap that exists between apperception and encoding.

When I discuss the variables that influence observation learning, I assume, as previously mentioned, that apperception is taking place, and that it is being held constant. For example, if we wanted to determine whether or not a bright, clearly drawn triangle would produce a better image on film than a dim, fuzzily drawn triangle, it would be necessary to hold the shutter setting, the size of the lens opening, and the position of the camera constant. One general statement will suffice to cover the variables of intensity, discreteness, and duration:

> If all other factors are held constant, apperceived input that is more intense, discrete, or enduring, will produce a more lasting encode.

A number of experiments support the implications of this statement. Hull's (1943) research on stimulus-intensity dynamism, as well as studies by Underwood (1949) and others, indicated that the three variables mentioned are important in the establishment of encodes. There is a problem in most of these studies, however. Are the positive results due to the apperceptual mechanism or to the encoding mechanism? That is, was apperception being held constant when the stimulus material was being presented?

The conditions that affect apperception and encoding appear to be so intertwined that decades of precise experimentation will probably be required to assess how much a particular variable contributes to each process. Take duration, for example. An input that remains longer in apperception than another should occasion a more durable encode. Yet since apperception is an intermittent, periodic process, the more durable encode may result from the occurrence of a greater number of apperceptions during a particular time interval.

Concept Formation

In chapter 5, it was suggested that, as a function of perceptual structuring, the various items in the individual's environment come to be represented as phenomena (i.e., percepts) located in three-dimensional space. When such percepts are encountered again and again, their common features become fused into summary encodes or concepts. Thus, from apperceiving many different houses, the summarizing concept, house emerges. The same thing happens with any phenomenon that is repeatedly experienced. After apperceiving many different apples of varying shapes and sizes, a memory encode representing all such similar experiences emerges. Yet apples are not only seen; they are also smelled, felt, and tasted, and the inputs from these various modalities also become fused into the emerging concept. Thereafter, whether one feels, sees, tastes, or smells an apple, the encode representing all such input may immediately be recalled. The emergence of concepts, that is, this intrasensory and intersensory summarization, occurs quite automatically. (See Saltz, 1971, for an excellent review of research and theory on concept formation.)

In the early stages of concept formation, these summarizing encodes are almost exact duplicates of phenomena. Yet as learning proceeds, and as more and more examples of a particular phenomenon become included, the pictorial character of the concept tends to dissipate. After the learning of language (McNeill, 1979), even greater dissipation takes place, so that in thinking the individual may use the encode of the written or spoken word rather than the encode of its physical referent.

Thus, concepts become less directly representative as the individual grows older, and for some people, large segments of experience may become fused into blobs that have little, if any, direct similarity to the separate phenomena that have become fused into the concept. The capacity for abstract thought is, at least to some degree, a function of the extent that inclusion has taken place. Individuals capable of such internal locomotion may use very little pictorial representation and to some degree may have lost the capacity for it.

It is not accidental that eidetic imagery (i.e., pictorial memory) is greatest in children and in individuals who have not been educated. Further, we may surmise that the summarizing encodes of animals remain relatively undiluted by symbolic representations and by the more refined fusion mechanisms postulated for human beings.

Although there is a tendency for precise pictorial imagery to dissipate as the individual grows older, it is not necessarily a progressive phenomenon, nor is it equally generalized throughout all aspects of an individual's memory. Yet the more such inclusion has taken place, the less the individual will be able to recall similar new experiences separately. When we meet another person, we have much more difficulty in remembering the specifics of his or her features than in recalling the wart on his or her nose. All human beings have features that are essentially similar, and aspects of the new person that are common to the recurrent input from countless meetings with other people tend to become fused into the generalized representation. The wart, on the other hand, does not fit. It stands out (pun intended) and can be recalled with remarkable clarity.

This somewhat whimsical example suggests that a fused encode (i.e., concept) may provide a basis not only for generalization but also for discrimination. The fused encode is a composite summary of a series of similar percepts. As such, it provides a basis for generalization when a similar percept is imposed again. It also represents a norm from which even slight deviations may stand out, thus providing the basis for discrimination.

As is the case in the emergence of percepts, there is a loss of specific information in the summarizing process that results in concepts. However, this loss of detail is offset by the advantages conferred by simplification. The complexity of encoding is reduced, the facility of recall is augmented, and the rapidity of the thought process is increased.

Associations Among Encodes

Encodes are established primarily via the repeated apperception of a given input. The association of inputs also requires apperception, but another factor is heavily involved: the extent of prior associations between such input and previously encoded material. Each encode that is established becomes a component of a much larger associational matrix.

Ebbinghaus (1885), the great innovator in the study of learning and memory, was keenly aware that the study of encoding appears to bring with it two interrelated problems, (a) that all new learning must take place within a matrix within which varying degrees of association have already taken place, and (b) that many inputs are similar to some that have already been encoded. Memory is comprised of an intricate web of associations, and very few, if any, inputs are totally unfamiliar.

How, then, can one study the establishment of novel associations? We can do so, Ebbinghaus believed, by using items (inputs) that are totally unfamiliar. To this end he invented nonsense syllables such as BOT, DUZ, and ZOG and the memory drum, the first teaching machine, and proceeded to perform, with himself as the sole subject, many definitive experiments.

One, of course, may question whether a completely novel nonsense syllable can ever be devised, but the list that Ebbinghaus used seems remarkably free of prior associations, and his work, along with that of others who have used them, provides some of the most precise data on the establishment of associations.

What conditions are responsible for the development of associations? As suggested in a previous section, apperception is always a requirement. Although many other conditions have been emphasized by various writers, contiguity appears to be the one that is most universally appealing, and it is also the one that I find most appropriate, as expressed in the following statement:

> The strength of association between two encodes will be a direct function of the number of contiguous presentations of two apperceived inputs.

According to this statement, contiguity is required for associations to be established, and repetition is necessary for associations to be strengthened. It should be emphasized that there must be a time lapse between the two components; otherwise, fusion rather than association will take place. And, as we see later, the optimal interval is approximately 0.5 seconds. If inputs are widely separated, other components may intervene and interfere. Contiguity is thus important primarily because it tends to lessen the probability that extraneous encodes will disrupt the association.

The previous discussion is directly pertinent to the relationships established in serial learning. Any number of inputs that are sequentially presented will tend to become associated to the extent that the series is repeated and that extraneous inputs do not intervene. The same conditions that affect the learning of a maze by a rat also account for the encoding of a list of nonsense syllables or the refinement of a motor skill requiring a series of sequential acts. In all such examples, a series of inputs (whether specific bits or fused phenomena) becomes associated through contiguity and repeated presentation.

Many variables can, of course, affect the development of such encode chains, though apperception is always a requirement. These variables include the number of components in the series, the intensity at which they are presented, their similarity to prior encodes, and the similarity among the various components in the series. The effects of most of these variables on serial learning have been thoroughly examined by many investigators, including Ebbinghaus (1885), Hull (1943), and Underwood (1949).

Biological Constraints on Learning

From the previous discussion it might be inferred that learning is a ubiquitous process that, regardless of species, provides the basis for appropriate behavior if "if A then B" concurrences are encountered in the environment. This inference, I believe, is essentially correct. Yet a question immediately arises. Since the evolutionary history of each species is at least to some degree unique, might there not be aspects of learning that reflect the particular behavior and process potential of each species?

As a matter of fact, both the assumption of ubiquitous principles and the possibility of unique intrusions have long been of concern to psychologists, although it must be admitted that the unique intrusions, at least until quite recently, have often been viewed as bothersome contaminants. The search for a few general principles that would subsume the data from all learning studies, regardless of species, has been the commitment of most theorists and researchers. To this end they have fashioned many types of apparatus: the puzzle box of Thorndike, the maze of Small, the discrimination box of Yerkes, the delayed reaction apparatus of Hunter, the obstruction box of Moss, and the Skinner box of Skinner (See Munn, 1950, for detailed descriptions and drawings of such equipment). All were designed in the hope of getting at generalized learning principles.

Most of these researchers were well aware that each animal brings instinctive, innate, propensities (both behavioral and morphological) into the experimental situation. But they were viewed as idiosyncratic annoyances that had to be discounted or ruled out if the pure principles of universal learning were to be exposed. Thorndike, Watson, and Pavlov all shared in the commitment to generalized learning principles. They believed (see Timberlake, 1983) that such principles could best be derived by isolating the subject from the natural conditions of its existence and by ruling out contaminating inherited proclivities.

Find an apparatus that is equally appropriate for all species, and universal learning principles in all their purity will be exposed, seemed to be the operational commitment.

The Skinner box (or one or more of its modifications) seemed to be the universal apparatus. The Skinnerian approach seemed to be the solution to both method and terminology; a method that could be universally applied and terms that were universally applicable. Skinner's general research view of organisms as empty boxes also has a cast of universality. Genetically based idiosyncracies in various species immediately become reduced to zero. Behavior simply occurs and can be appropriately or malappropriately shaped as a function of reinforcement. "The total act ... is constructed by a continual process of differential reinforcement from undifferentiated behavior, just as a sculptor shapes his figure from a lump of clay" (Skinner, 1953, pp. 92–93).

The first experimentally derived news that the box was not empty came from two different sources, (a) research on aversions acquired through the pairing of ingested substances with nausea, and (b) research demonstrating species-specific (presumably inherited) modes of behavior that interfere with the demonstration of general learning principles.

As examples of the first, Garcia and Koelling (1966) demonstrated that a number of animals, particularly rats, can easily learn the relationship between the taste of a particular food and a subsequent illness, although taste and illness may be separated by considerable time intervals. This Garcia effect seems, at least to some investigators, to obviate general learning principles in that the relationship can be established so easily and the time interval between cue and outcome can be so considerable. It also appears that species-specific factors may either facilitate or constrain the learning, that is, taste aversion can be easily established with nausea but not with shock. A buzzer will not become an aversive stimulus when paired with nausea (Seligman 1970). The Garcia effect does not manifest itself in an all-or-none fashion; certain flavors seem more readily associated than others, the more novel a flavor the more easily it can be associated, the shorter the time interval between ingestion and illness the stronger the association, and the greater the illness the stronger the resulting aversion. (See Bolles, 1975, for an extended treatment of the variables that influence the Garcia effect.)

Something reminiscent of a reverse Garcia effect has also been reported. Hogan (1973) found that three-day-old chicks had no initial preference for small seeds over grains of sand. He then exposed one group of chicks to a pile of sand and another to a pile of seeds. After an hour, when both groups were placed back in the choice situation, the sand chicks still showed no preference, but the seed chicks had developed a decided preference for the seeds. Apparently, at least partial digestion had to occur before the positive effects of the seed could be demonstrated, and digestion of food takes time.

Do such findings indicate that idiosyncratic factors may be sufficiently represented in certain species to override or nullify general learning principles? Perhaps. Learning capacities may be more highly refined relative to certain concurrences than to others. There have been long-term selection pressures in rats regarding poisons. Not only have humans been trying to eliminate rats for hundreds of years, but also the nature of their existence as small omnivorous scavengers has imposed further hazards from poisonous food items. Thus, we might expect the evolutionary emergence of highly refined poison discrimination mechanisms. Since there is typically some delay between eating and toxic effects, there should be very little surprise that experiments confirmed that aversive behavior can be demonstrated across considerable intervals between food ingestion and the occurrence of nausea. Eat a little, wait a while, and if it makes you sick, do not eat anymore is the strategy that rats have developed. Nausea has become a functional discrimination mechanism. Since there is a considerable time lapse between the ingestion of noxious food and the onset of nausea, Garcia and his colleagues discovered what evolutionary theorists should have predicted all along.

Does the long delay demonstrated between the conditioned stimulus of taste and the unconditioned stimulus of nausea require principles of learning different from those usually offered by learning theorists? I do not think so. The tastes involved are typically unique and the occurrence of nausea usually quite rare. Thus an interference explanation, as offered by Revusky (1971), provided the simplest and most traditional explanation of what is happening. When a unique taste is followed by the rare nausea input, very little input of sufficient similarity typically intervenes to interfere with the crucial association. It is as if the animal were put to sleep during the interval between taste presentation and nausea occurrence.

Of course, it may well be that natural selection has resulted in highly sensitive association mechanisms where taste and nausea are involved. After all, the short-term (approximately one half second) interval that many studies indicate produces the optimal association between most stimuli and most responses did not happen by accident, but by the adaptive functionality conferred.

Further, it should be noted that investigators are not surprised when sense modalities that have had special or unusual roles in the evolution of a species are either highly developed or correspondingly retarded. Bats have exceptional hearing and poor sight; both are reflections of a long-term evolutionary habitat. Hounds have highly sensitive olfactory systems, and hawks have exceptional eyes. Obviously such animals will be able to make finer discriminations in those areas where their modalities have greater acuity and poorer ones where their modalities have been blunted. Yet this does not mean that different principles of vision, audition, olfaction, etc., are involved.

A second series of investigations that both piqued and puzzled learning theorists was published by the Brelands (1961), a husband and wife team who trained various animals for advertisers. In one study they tried to train pigs to place a wooden disc in a piggy bank, a cute and appropriate metaphor. The pigs could easily learn to place discs in the slot in the bank, but as training continued curious interfering behaviors began to occur. The pigs began to toss and root the discs, and the bank behavior dissipated. The Brelands also tried raccoons and ran into a similar problem. The raccoons could perform on a continuous reinforcement schedule but when placed on a 2:1 ratio (i.e., two discs for one food reward) their behavior became erratic; they began rubbing the discs together and dipping them, and generally behaving as if the discs were food.

In both examples, the animals' idiosyncratic behaviors were apparently interfering with learning how to use discs in exchange for food. A number of investigators (Bolles, 1975; Breland & Breland, 1961; Shettleworth, 1983; & Timberlake, 1983) suggest that biological constraints derived from a species' own unique history may be at work in such cases. The Brelands (1961) stated, "Moreover, it can easily be seen that these particular behaviors to which the animals drift are clearcut examples of instinctive behaviors having to do with the natural food-getting behavior of the particular species" (p. 683).

Species-specific factors may be at work in the cited examples, but it is also possible, as Moore (1973) suggested, that past learning is providing at least a part of the interference. Prior to being tested most animals have had extended experiences in finding and preparing food typical of their environmental niche in ways that are appropriate to their morphological structures; pigs root, chickens scratch and pursue, raccoons wash, etc. The Japanese macaques on Koshima Island have learned to wash their food before they eat it, something they never did until recently (Kawai, 1965). It is very likely such washing behavior would now intrude into any token reward study that might be undertaken with them. Certainly to conclude as the Brelands did that animal "misbehaviors represent a clear and utter failure of conditioning theory" (1961, p. 683) seems premature, if not unjustified. All the cases they cite could be the result of past learning that interfered with present acquisition or could simply be each species' special mode of demonstrating general principles. Perhaps we should reread the Brelands' first report (Breland & Breland, 1951) in which they argued convincingly, on the basis of training hundreds of animals, that general principles derived in the laboratory could be applied without considerable alteration to a large number of different species.

It seems obvious that the precision and facility of learning may vary, depending upon the evolutionary history of an organism. As a case in point, rats can learn 16-unit alley mazes faster than humans can learn a comparable 16-unit finger maze. This does not mean that rats have better learning capabilities than humans; it does suggest however that the evolutionary history of these two species has been different, that the facility of learning varies, depending upon the task. Some of our earliest rat psychologists were aware of the importance of evolutionary habitat as a factor in designing appropriate apparatus. In his use of a maze for studying learning in the rat, Small (1901) pointed out the similarity between the maze and the rat's "propensity for winding passages" (p. 208) and suggested that, "experiments must conform to the psycho-biological character of an animal if sane results are to be obtained" (p. 106).

Thus, we should never lose sight of the fact that biological factors (constraints, boundaries, limits, and intrusions) may profoundly interact with each learning situation. Every organism inherits a capacity for learning, and this capacity is certainly biologically based. Every organism inherits mechanisms of discrimination, and these modalities (such as vision, hearing, taste, and smell) are either dull or sensitive, depending upon biologically based mechanisms. There are even innate intrusions in every consumatory process. Human beings, for instance, do not have to learn to swallow food or to suck a nipple. Even the chewing of food, though learning may refine the process, is couched in biological (both morphological and behavioral) structures. The most obvious of all activities, walking or running, involves biologically based feedback loops from balance and kinesthetic receptors, as well as reciprocal innervation among various muscle groups. In many animals, entire muscular neurological patterns are prewired in at birth, and, though perhaps refined by learning, remain remarkably stable through life.

Very recently there has been an upsurge in the questioning of, if not rejection of, general learning principles. According to the view implicit in the work of many evolutionary psychologists, learning is not a blanket category; it is shot through with particulars: Biological (evolutionarily derived algorithms, modules) constraints and directives that enter into the when, the why, and the how of learning, whenever it takes place. It may vary not only with the species being considered but also with the maturation level, gender, or hormonal condition of the individual. For example, a number of investigators hold the view that social learning in humans is hedged in by such directives, that selection pressures have insured that certain particulars of social exchange will be influenced by biological predispositions. Each individual must indulge in behavior that benefits others, often at personal cost; the mother nurses the infant, the hunter shares the meat, and the warrior protects the group. Social living has thus resulted in a biologically based tendency for reciprocal altruism, a module of give and take, of cost and benefit.

In an effort to clarify this imperative, writers occasionally use explanatory metaphors. Much has been made of the prisoner's dilemma, a situation that exists when two individuals who have been involved in the same crime are questioned in separate rooms by interrogators who promise that if each will rat on the other, he or she will be given a lesser sentence. The dilemma for each is: "Should I rat on my friend and get a lesser punishment or shall I stonewall with the possibility that my accomplice will also stonewall and that we may both get away with it?" I mention the prisoner's dilemma because it carries such a heavy explanatory burden (R. Axelrod, 1984; Boyd, 1998; Cosmides & Tooby, 1992) and to emphasize that it may obscure rather than clarify. It is true that when any individual gives, he or she is not sure that other individuals will reciprocate. It is also true that altruism could not have evolved except through the survival implications of give and take, but the metaphor is so burdensome and its applicability so dubious that it detracts rather than helps. A complicated rationale is used to suggest the rather obvious point that any genetic combination that codes for an inheritable characteristic (behavior, propensity, predisposition, etc.) that increases the likelihood of its own survival will tend to be incorporated into the species.

Regardless of how one may view the appropriateness of the prisoner's dilemma as an explanatory metaphor, the give and take requirements of group living not only resulted in many biologically based expressions of altruistic behavior, but also, according to some researchers (Cosmides & Tooby, 1992), occasioned the corollary development of cheater detector modules, that is, special evolutionarily derived techniques designed to ferret out those individuals who tend to take more than they receive.

During the past few years the search for predispositions (modules, algorithms, etc.) has centered on many aspects of social interaction in human society. Here of course we encounter again (see prologue) the ubiquitous problem of cultural overlay, that is, the extent to which a given attribute is a product of learning in a social context rather than derived from selection pressures encountered during evolution.

To some degree researchers have managed to control for this ubiquitous complication and to determine that much learning is coerced, directed and, in some cases, restricted by inherited predispositions—zones of sensitivity or emphasis that lie within the more inclusive process. Thus, social interaction is influenced by cheater detection mechanisms (Cosmides & Tooby, 1992), foodsharing allocation by equity modules (deWall, 1989), mate selection by preference algorithms (Buss, 1987), what is good and bad by an intrinsic morality (Alexander, 1987), language acquisition by modular structures (Pinker, 1994, 1989; Pinker & Bloom, 1992), the beautiful and the functional by an intrinsic aesthetics (Kaplan, 1987), cognitive facility by genetic predispositions (Plomin & DeFries, 1998), what is funny by humor dispositions (Weisfeld, 1993), human language interpretations by gossip algorithms (Barkow, 1992), and the list goes on.

In the summary of their chapter on cognitive adaptations for social exchange, Leda Cosmides and Tooby offered a statement demonstrating both the remarkable extension of the computer metaphor (see prologue) and the contemporary segmentation of mental processes. "On this view, the human mind would more closely resemble an intricate network of functionally dedicated computers than a single general purpose computer" (Barkow et al., 1992, p. 221).

The search for biologically based predispositions within the context of more inclusive learning, whether individual or social, is thus a primary concern of many contemporary evolutionary psychologists. In this arena Garcia and Koelling (1966) and the Brelands (1961) were truly pioneers. They found that learning was not a general, global process but was rife with particulars reflecting unique species-specific encounters. Lorenz (1970) was also an innovator in this regard; imprinting reflects a maturational sensitivity of the very young that is reminiscent of the facility with which human toddlers learn language (Chomsky, 1986; Corballis, 1992; Hurford, 1991).

The questioning of the domain-general commitment is a positive happening. Blanket concepts generally obscure, but there are many universal conditions that all organisms share. Where selection pressures have overlapped common adaptive solutions should be expected. Thus we should search for particulars that reflect selection pressures peculiar to a given species, but parsimony, which is the watchword of science, demands that we emphasize general principles. Studies on classical conditioning are central to our understanding of cue utilization in all species. Instrumental conditioning studies, particularly those emphasizing schedules of reinforcement, describe the seemingly erratic behaviors of all organisms in their search for sustenance: sometimes there is A then not B! Those involved in observation learning (Bandura, 1961, 1963; Ebbinghaus, 1885; Hillix & Marx, 1960) provide information on the development of the internal environmental replica that is central to foresight and planning in humans and, at least to some degree, in many animals.

All of these approaches to learning emphasize conditions that have molded all of us and, as such, domain-general principles would be the predictable outcome.

Summary and Discussion

The universal survival imperative for all organisms is maintain or regain balance or perish, and all activity, as suggested by the postulate of process, is geared to the service of this imperative. In every species there are inherited behaviors, varying in specificity and complexity from the coughing reflex of the human infant to the migratory behavior of the Pacific salmon. And in most species there are learned behaviors. These are invariably cradled in a context of inherited mechanisms. All coordinated movement involves innate feedback loops; for example, eating and drinking involve innate swallowing and peristaltic mechanisms. All stimulation, to be effective, activates sensory mechanisms that impose innate and qualitatively unique sensory inputs. When learning takes place it invariably occurs within the context of process systems that are characteristic of each species, and will be influenced (either restricted or facilitated) by the relative acuity of the equipment involved.

The obvious fact that every species may have different sensory, integration, and response capabilities does not force us to conclude that a generalized evaluation of learning is inappropriate. Species do have different evolutionary histories, but they also have overlapping, long term, evolutionary commonalities, including those relating to the evolutionary emergence and function of learning. Short term (if A then B) concurrences, pertinent to homeostacing activity, have existed in the environments of all species, regardless of their seeming uniqueness. Individuals of each species must respond in terms of the equipment they have available, and such equipment can constrain, channel, facilitate, etc., learning in seemingly idiosyncratic ways. However, this does not mean that there is a different type of learning for each species. When the special biological overlays are removed or discounted there will be found, I believe, a generalized process of learning reflecting universal evolutionary encounters with the selection pressures imposed by responding effectively to short-term concurrences, as they have been present during the evolutionary development of all species.

Since 1971 comparative psychology, my initial area of commitment, has undergone a remarkable, and to me disquieting, contraction. It has apparently become a victim of the cognitive resurgence, away from behaviorism; it is very difficult to get at the cognitive operations of animals; they do not speak and, as far as we know, they do not introspect. However, the contemporary involvement in evolutionary thinking does not diminish the importance of behavior or reduce the importance of animal studies in our efforts to understand ourselves. Behavior is the output component in every biological information-processing system. The viewpoint of Watson, that the only thing we can know about another organism is its behavior still has the ring of truth. Behavior is the source of inferences about the cognitive operations of others, whether humans or animals. It constitutes the manner in which the balances essential for life are maintained and regained, and it constitutes a critical source of psychological data, both historically and at the present time.

A timely defense of comparative psychology and its role in the emerging evolutionary synthesis was given by Wasserman (1997). Not only are the data from animal studies justifiable for their own sake, but may profitably be culled for insight into the nature of human processes and their evolution. The researches in paleontology, particularly those on hominids, are eagerly examined for indications of evolutionary development in humans. The bones of all animals reflect such universals as gravity and the requirements of locomotion. The skulls of hominids reflect the increasing cranial capacities that are presumed to culminate in human beings. Everywhere we look evolution expresses the presence of universals in the environment: eyes, lightwave reflectivity; ears, vibrations emanating from objects; noses, the presence of molecules in the environment. We do not marvel, but take for granted, that bones, eyes, and ears not only reflect universals but also are honed to reflect each species' particular environmental niche, so we should not be surprised that learning, and indeed all the attributes involved in information processing, should demonstrate species variation. There are environmental universals, and these are represented in characteristics common to all species; there are environmental niche particulars, and these are represented in the unique attributes of each species.

MEMORY

An encode (or memory) might be defined as a residual resulting from the apperception of a supraliminal stimulus. Such a definition would be too broad, however, because there are two types of residual, particularly in vision, that should not be classified as memory. Both are derived from peripheral receptor effects. The most fleeting is the positive afterimage, which may last for about 500 milliseconds. It results from a continuation of the experience that was introduced. For instance, if a bright light is flashed into the eye and then turned off, the experience of brightness will continue for a brief interval. Immediately after the dissipation of this positive afterimage, the negative afterimage occurs.

This experience is apparently a function of the recruitment of the receptors and is opposite to the positive afterimage. Thus, if imposed stimuli resulted in the experience of white, red, or blue, the resulting afterimage would be black, green, or yellow, respectively, and vice versa. These images fade slowly and, if the viewing situation is appropriate, may be observed for a full minute after the initial stimulation is removed.

Do such afterimages qualify as memories? Only in a very technical sense. Under ordinary circumstances negative afterimages are not noticed by the observer and, on those rare occasions when they are, they probably interfere with the central encoding (memory) processes that will be described later. Positive afterimages, however, may be functional in the perception of movement as well as in the fusion thresholds for both vision and audition (see chap. 4).

Immediate Awareness and Its Functions

There is now general agreement that there are three different levels (if not distinct types) of memory. The most basic and most fleeting type apparently lasts for about 750 milliseconds (Blumenthal, 1977). Although the term memory (as implied by such terms as immediate, iconic, echoic, sensory, etc.) is typically used in contemporary literature to represent this brief type of representation, the identical process was discussed in the earlier literature under the name immediate awareness, and this latter term now seems more appropriate for reasons that I give later.

Both of the psychologists who started our first laboratories, Wundt and James, accepted the notion of a brief experiential representation. This was Wundt's *blickfeld*, the field of awareness, which included representations from all sense modalities that, taken together, comprised the sensory side of the ongoing process of consciousness. To Wundt (see Boring, 1929/1950), this blickfeld was preapperceptual, that is, it was the field relative to which the focalizations (*blickpunkt*) of apperception occurred.

The viewpoint of James (1890), at least in its general outlines, was very similar. James was concerned about the problem of the present, suggesting that it could not be infinitely brief, with each segment separated from the others. Rather, in order for the various psychological processes to be integrated, a time span or stream of consciousness had to exist.

The positions of both Wundt and James are, I believe, essentially correct. There does appear to be a field of awareness that progresses as a unified and continuing stream. The reader may check this out for himself or herself. The field of awareness is filled with inputs from the sense modalities, and from the various activation and affective systems of the body, and this influx is occurring simultaneously. Thus, sounds, smells, sights, pressures, fears, and pains, etc., are all represented.

How wide is this span of our immediate awareness? Much research, some of which will be summarized later, suggests that it is approximately 750 milliseconds. This is the width of our experienced now or what James called the specious present. It is apparently a direct representation of what researchers have designated by such names as sensory memory, buffer delays, echoic and iconic memory, stimulus trace, etc.

What kind of process accounts for our experienced now? Perhaps it is due to a wave of overlap between the lingering central reverberations of past receptor activation and new ones that are continuing to occur. As long as stimulation is above threshold, new sensations will be introduced, but the sensory effects from just-previous stimulation will also still be present. Sensations arising from each new stimulation combine, I believe, with those arising from dying reverberations to constitute immediate awareness (the specious present, the experienced now, etc.). Since stimulation (except under laboratory conditions) rarely shifts abruptly, sharp cleavages or disruptions are the exception. As we interact with our environment, sensations shift gradually, melding slowly into one another, providing the stream of consciousness that James talked about.

Thus, during any particular waking moment sensory inputs are typically being continuously introduced, and as such, they constitute the moving belt of our reality. Even when abrupt changes occur, as when a horn blows or when we hear a scream in the night, such inputs simply add a proportional increment to the general background of the continuing stream.

This stream of consciousness (immediate awareness) is preapperceptual. It constitutes the matrix within which the perpetual flux of apperception takes place. When apperception occurs, that is, when any one of the multitude of inputs within immediate awareness becomes focalized, it becomes represented in the next level of retention, traditionally called short-term memory.

In the late 1950s and early 1960s, Sperling (1960) performed a landmark series of experiments that not only helped define the nature and duration of immediate awareness but also marked the resurgence of interest in cognitive processing. Sperling's research was a refinement and extension of experiments initially performed by Wundt and his students in the 1880s and 1890s, emphasizing the 60-year hiatus cognitive psychology endured during the tenacious ascendancy of behaviorism.

In a series of experiments, Wundt (1899) exposed subjects to a large number of simultaneously presented stimuli for very short intervals. He then found that the subjects could report only a small portion of the total array. However if signaled to report a particular part of the array, the subjects could report that segment almost perfectly. Sperling's experiments are essentially replications and refinements of Wundt's early studies. In Sperling's (1960) simplest and perhaps most convincing study, subjects were exposed to three rows of letters, four letters to a row, for 50 milliseconds. After this brief exposure he found that the subjects could report less than 50% of the total array. Yet if the exposure was followed by an instruction tone, either high, medium, or low, representing the corresponding row, the subjects could report the letters in the designated row with nearly 100 percent accuracy. However, when the instruction tone came as long as one second after the exposure, partial reports were no more accurate than total reports.

These results indicate that immediate awareness can hold large numbers of items, but that it dies out even as reporting is taking place. Sperling's studies (a number of different variables such as size, color, upper vs. lower case, and general location were examined) all confirmed that immediate awareness lasts only about 750 milliseconds. Directly after exposure, the entire array of sensory input is briefly available, but it immediately undergoes decay, to be completely lost if it is not enhanced by apperceptual focalization.

Since Sperling's remarkable studies, replications of his design examining immediate awareness in other modalities have been reported (e.g., Bliss et al., 1966 on tactual input; and Norman, 1969 and Treisman, 1964 on echoic input). In all cases there is remarkable agreement; sensations tend to linger for about 750 milliseconds, and during this interval any part of a larger matrix may be focalized in apperception.

What is the adaptive function of immediate awareness? In any complex organism, such as the human being, energy changes are continually taking place both on the outside and the inside of the body, and many of these changes activate sensory systems of the body. Thus, at any moment hundreds of information bits are represented in sensations and retained momentarily in immediate awareness. However no system can respond effectively to a number of information components simultaneously. Directional and functional activity depends upon a single channeled system; and apperception, as defined in the present theory (see chap. 7) is the single channeled focalizer that makes appropriate adaptive behavior possible. In short, immediate awareness constitutes a buffer delay system that holds potentially vital information in abeyance until the single channel of apperception can be cleared. Indeed, apperception flits among the hundreds of components momentarily available in immediate awareness, depending upon their adaptive priority (see chap. 7).

In spite of the fluctuations of apperception within the mass that it constitutes, the field of immediate awareness remains continuous. Thus, the material from audition, vision, feeling, touch and pressure, as well as transient and persistent hungers, etc., continues to flow within and to constitute the field. All of this input, as Wundt insisted, is preapperceptual; it is automatically and nonvoluntarily introduced.

From an adaptive point of view, the automatic and persistent imposition of the sundry inputs from all the receptor and need systems of the body is imperative. Such input represents the external world, with all its potential hazards and benefits, and it constitutes a continuing read out of the condition of our internal process systems as well. Thus, our avenues of knowing must remain open to any and all information that might affect our continuing existence. They do. At this moment I am being bombarded with the sound of music from a colleague's office. It is extremely irritating and disruptive. Yet the only way I can keep it from intruding is to ask him to turn it off or to put something in my ears. The same is true for all the inputs from the various sensory and need systems of the body. They are being automatically and continually introduced. Our sensory systems are perpetually vigilant. They let in all the available information uncensored, leaving the determination of appropriateness to other process systems (primarily apperception, as we shall see).

One of the most difficult problems for homeostatic theory is encountered at this point. If all activity is homeostatic, as stated in the postulate of process (see chap. 2), how can we explain the introduction of such inputs as hunger, anxiety, or pain? That is, since pain is so disequilibrating, why experience it at all? The answer is simple. We have simply been shaped by the selection pressures of evolution so that such input is automatically introduced. Responding to the critical circumstances that pain represents was not only the occasion for its evolution but for its immediate introduction. The windows of our sensory systems are open and we cannot fail to experience what comes through them. To do so would be to override the way we are structured. It is true, as discussed at length in the next chapter, that we may not apperceive such disequilibrating input when other powerful inputs are competing for focalization.

An important function of immediate awareness is to provide continuity and stability. Our experienced world is not given in disconnected slices, but as a smooth flowing stream without abrupt transitions, a stream that is continually variable as new situations are encountered, but one where the variable aspects meld into one another. This melding provides the continuity that makes the ceaseless interplay between experience and behavior both appropriate and possible.

Why is each temporal segment of immediate awareness so brief? Its fleetingness may be a byproduct of the limits of physiological reverberations, but it is also obvious that the sheer amount of material momentarily included would argue for its rapid dissipation. If the totality of information available at each moment were continuously accumulated, the processing systems of the individual would soon be inundated, and the capacity for adaptive behavior reduced to a minimum. Brevity in immediate awareness serves its function admirably; the interval (750 milliseconds) is long enough for inputs of adaptive importance to be apperceived, to influence behavior, and to be represented in the next stage of retention, short-term memory, while inputs without significance can disappear and take their clutter with them.

Is immediate awareness (sensory memory, echoic memory, iconic memory, etc.) the same thing as Wundt's blickfeld and James' stream of consciousness? I believe so, but certain issues need to be clarified. Perhaps stream is an unfortunate and misleading metaphor. It implies an undifferentiated flowing process confined within easily discernible boundaries. However the stream of our immediate reality does not consist of an undifferentiated mixture of sensory noise. Rather, it consists of unitary phenomena within three-dimensional space, and this order and structure remains remarkably stable through time. If an individual closes her or his eyes and then opens them for very brief periods, the visual field that is introduced during such fleeting glimpses is not just a two-dimensional mosaic; it consists of phenomenological unities located within three-dimensional space. Whether such organization is inborn or learned (an issue discussed at length in chap. 5) is incidental to this point. The stream of our visual reality consists not only of sensory inputs but also of perceptual organization. The same seems to be true of echoic awareness. Many experiments in our own laboratory with subjects in a circular sound cage indicate that the echoic representation comes with spatiality automatically imposed. Blindfolded subjects subjected to 100-millisecond clicks on a horizontal plane can judge direction with 95% accuracy, and they make such judgments immediately after hearing the clicks. This discussion suggests that the term sensory memory, which is in common usage (Blumenthal, 1977), is too restrictive. Certainly sensation is involved but perceptual structuring is also automatically imposed. The world is not restructured each time we awaken, with the various items of our experience set in their appropriate places. Perceptual structuring, although it may be modified by placing cues in opposition (see chap. 5), lasts a lifetime.

Is apperception a necessary precursor to the order and structure of immediate awareness? Probably not. The cues that are operative in the development of perceptual organization are usually not even noticed; indeed, many of them cannot be apperceived even when we try to isolate them. Thus, the inputs arising from convergence and accommodation are so subtle that they cannot be focalized, and even more obvious ones such as movement parallax, overlap, and retinal disparity are typically so tenuous that we are seldom aware of them. Von Helmholtz (1924) noted that perceptual organization is given as an unconscious inference from specific cues. By this, I believe, he meant that perceptual organization is not dependent upon apperception for its ongoing manifestation, nor is it apparently dependent upon apperception as it develops.

Immediate awareness thus has many functions. It constitutes the span of the continuous belt of our reality, it provides the field within which apperception takes place, and its rapid dissipation helps alleviate the confounding and confusing clutter of incidental sensory noise. Although it is not a stream, as this term implies an undifferentiated flow, it is continuous, at least during the waking state. As such, it has no gaps and rarely any sudden or abrupt transitions; rather, it is a smooth progression of the phenomenological unities of our experience.

Immediate awareness is not simply a representation of raw sensory input per se. It includes the organization and structure (see chap. 5) that the perceptual processes impose on all of our experience. It is a brief representation that lasts approximately 750 milliseconds unless it is elevated into the next stage of buffer storage by the process of apperception. It is preapperceptual, it is automatically and nonvoluntarily imposed, it constitutes our immediate reality, and it provides the material for the next stage of retention (short-term memory).

Short-term Memory and Its Functions

Immediate awareness, or the field of awareness, lasts only about 750 milliseconds and then fades to become lost unless some component within this field is apperceived. As soon as apperception takes place, short-term memory is activated. Short-term memory is thus post apperceptual and is a direct outcome of the apperceptual process. However, short-term memories also fade and in turn become lost unless renewed by further apperceptions, which tie them into a long-term memory code or structure.

Many experiments, beginning with Daniels (1895), demonstrated that short-term memory lasts approximately 15 seconds. These studies all presented a series of items to a subject and then immediately involved that subject in an interpolated activity that prevented apperceptual renewal. Remarkably consistent results were obtained: Pepper and Herman (1970) with kinesthesis, Phillips and Baddeley (1971) with tones, Wickelgren and Norman (1966) with touch. Indeed the loss curves are almost perfectly overlapping, dropping rapidly at the outset with all recall absent at approximately 20 seconds. The rapid loss that takes place in short-term memory is common in the experience of everyone, most notably in the use of telephone numbers. Look up a number, use it immediately or it will be gone, necessitating another troublesome search through the phone book.

Does interference or decay, or possibly both, account for this rapid loss? Certain investigators (Conrad & Hull, 1966; Norman, 1966) believe that interference is primarily responsible. They found a relationship between the number of interpolated items presented following the presentation of lists of items and the loss of such material. Also, emotional input can apparently disrupt the material in short-term memory. When I meet people, invariably an emotional situation for me, I can never remember their names.

In spite of these examples, it seems doubtful to me that interference accounts for much dissipation of short-term memory. The apperceptual process is so rapid, as high as 20 blips per second (as we discuss at length in the following chapter) that a retention of all resulting memories until interference took place would seem both unwieldy and adaptively inappropriate.

Long-term Memory

Immediate awareness is preapperceptual and, as such, provides the field within which apperception takes place. When a component within this field is apperceived it is immediately transferred to short-term memory, where it in turn fades, if it is not reinforced with new apperceptions and fitted into enduring associational structures. Thus the material in long-term memory consists of a cumulative record of experiences that were of sufficient pertinence to occasion the indelible "fixing" implicit in "if A then B associations" and in the development of each individual's internalized world.

Long-term Memory Update

Research on long-term memory was formally initiated by Ebbinghaus (1885), who, as we have seen, invented both the nonsense syllable and the memory drum (the first teaching machine), and who used himself as subject. Ebbinghaus wanted to study memory as a pure phenomenon, uncontaminated by prior learning, and the use of nonsense syllables minimized the effects of prior associations. In so doing he effectively deemphasized what many contemporary researchers find most fundamental in memory: The fact that memories do not take place piecemeal in a vacuum but within a context of associational structures, some functionally isolated but many overlapping in varying degrees of complexity and inclusiveness. The many experiments by Ebbinghaus gave us the basic, and still applicable, data on the laws of primacy and recency, the relative influence of distributed versus massed practice, and curves of both acquisition and forgetting. Because of the precision and thoroughness of the research, Ebbinghaus's data provides a dependable foundation for work on memory today.

The fact that memories are organized into particular categories and strategies, rather than existing as isolated tidbits, was first demonstrated by Watt (1904). He presented subjects with two numbers, such as a 5 with a 2 beneath it. The subject would respond with either a 7, 3, or 10. However if, prior to the presentation, the subject was told that the problems would be in addition he or she would accurately respond with 7, and the reaction time would be much faster. It appeared that the instructions to add made an addition strategy immediately available.

All long-term memories appear to be sorted into appropriate categories, whether we call them strategies, structures, or lattice works of association. Thus if we tell a subject that we are going to ask her to give us the capital of a certain country, she will be able to respond much more rapidly if we also give her the continent in which the country is located. Such narrowing down the scan to an appropriate category increases both the speed of recall, and the speed and accuracy of response. In hunting and gathering societies where the appropriateness and abundance of food items vary as a function of time of year, time of day, and presence of other plants or animals, it would be functional for the members of a foraging group to have a searching strategy that would be appropriate (Cashdan, 1989). Contemporary children, and adults for that matter, have similar strategies when something is lost. They invariably ask how big, what color, what did it look like, etc., in order to narrow the range of their search.

A number of thinkers have postulated processes, in some cases innate and in others learned, that limit the scan to items appropriate for a given hunting strategy or background need. Lorenz (1970), for instance, talked about a searching image that certain animals, particularly birds, have in their pursuit of prey; and both Freud (1935) and Hull (1943) suggested that individuals may be born with a predisposition to seek out items that are appropriate to a particular biological imbalance. Thus in Freud's object of the primary process the hungry infant seeks for an objective duplicate of an internal image of milk, and in Hull's first postulate the hungry individual has an innate tendency to search for and to consume food. Such searching images also seem implicit in the expectancies of Tinklepaugh (1928) and Tolman (1948).

Information concerning the manner in which human beings work as information-processing systems can sometimes be gained from observing their seemingly trivial activities. Take the game Botticelli for instance. This version of twenty questions begins with the person who is "it" stating that he has something or someone in mind that begins with a particular letter, say T. We note that the statement of the letter already limits the scan. The other players must earn the right to further limit the scan by stating that they have a T that, say, was a tribe native to India. If "it" cannot think of Todas the other player has earned the right to ask a further limiting question that typically is: Is it animal, vegetable, or mineral? Thus as the game proceeds the players gradually limit the scan, both for themselves and for "it" until they either exhaust their repertoire or limit the scan to something or someone, which might turn out to be anything from tigers to Teotihuacan. In the game of Botticelli, the notion that memories are contained in categories is taken for granted. It makes us wonder why it has taken so many years for psychologists to come to the same conclusion.

Perhaps the first formal research designed to intentionally break with the isolated tidbit approach was that of Bartlett (1932), who was concerned with how information is organized in memory. In a typical experiment, an isolated subject was told a story. This subject then repeated the story to another subject, and so on until the story had been told and repeated a number of times. Invariably Bartlett found that memory is a fluid, active process, during which certain events drop out and other events are added. According to Bartlett, there is a tendency to fit memories into meaningful and familiar constellations.

In the thirties and more extensively in the forties, Tolman (1948) emphasized what he termed cognitive maps as a way of conceptualizing memory in both rats and people. Cognitive maps were not static replicas of the organism's environment, but were seen by Tolman as dynamic representations that not only mirrored the environment but also were influenced by the intentions and purposes of the individual as demands (i.e., needs) shifted in predominance. Cognitive maps, however, were primarily visual, a reflection, no doubt of Tolman's many experiments on place learning in rats. More recently, Bransford (1979), Minsky (1975), and Newell and Simon (1972) include such concepts as problem-solving strategies, memory schemas and memory frames of reference to designate memory categories. Also, there has been an effort to broaden the functional implications of Tolman's cognitive maps to include strategies for problem solving, as demonstrated by Harlow's (1949) learning set experiments, Bundy's (1975) analysis of the manner in which individuals solve algebra problems, and Greeno's (1978) and Larkin's (1979) analysis of steps taken in solving problems in geometry and physics.

From an evolutionary viewpoint, long-term memory is remarkably functional. When vital "if A then B" concurrences are again encountered, the associations in long-term memory ensure that rapid and appropriate behavior will take place, whether such behavior represents the preparation and activation demonstrated in classical conditioning or the toward or away from behavior portrayed in instrumental conditioning.

The internalized world, which represents an accumulated replica of the individual's transactions with his or her environment, is even more functional. As our ancestors went out on the various hunting and gathering forays that were essential for their existence, the progressive encoding of salient features of the environment allowed the individual to know where he or she was at each moment, and provided the wherewithal for a return to the home camp after the day's activity was over. Obviously, during our evolutionary history individuals who were more proficient in world internalization would tend to survive, while those less efficient would be selected out and take their duller genes with them. So after millions of years of selecting in and out, the capacity for world internalization with all its significance for subtle nuances of behavior achieved its high degree of functionality. It remains perhaps even more functional today. Without the accumulated replicas of their ongoing interactions with the tangles and complexities of modern highways, modern cities, and sprawling suburbia, people would never reach their destinations or find their ways home again.

Long-term memory, then, does not consist of a hodgepodge of discrete elements haphazardly represented. It consists of constellations of encodes, many of them summary concepts, highly organized and related to provide rapid and appropriate readout of information pertinent to each survival situation.

CHAPTER SUMMARY

The organization that is imposed during perceptual structuring (see chap. 5) provides the functional stage upon which the flux and flow of ordinary experience take place. Four different kinds of learning provide the basis for encoding important features of such transitory experience.

Observation learning, the type that accounts for most human learning (and probably the learning of most mammals) is dependent purely upon the process of apperception. If the apperception of a particular input takes place, a replica of that input, that is, an encode, will automatically be established and strengthened by repetition.

The overlapping interdependency between apperception and encoding presents difficult problems of classification, but the functional intimacy between the two processes would be predicted from an evolutionary orientation. The apperceptual process follows the path of greatest equilibrium, that is, the input most immediately pertinent to the welfare of the organism is the one that is apperceived. Since pertinency has already been established in terms of that which is apperceived, it seems not only functional, but parsimonious, that encoding would be a direct outcome of the apperceptual process.

In observation learning (i.e., encode establishment) we find the highest refinement of the learning process. This type of learning, which I term world internalization, provides a cumulative functional replica of the organism's experience that, as we see in the next chapter, serves as the basis for planning, foresight, and creative thinking.

Cue deneutralization and instrumental learning (as represented by the work of Pavlov and Skinner) demonstrate how associations come to represent important short-term concurrences in the organism's environment. Insight, as demonstrated by Kohler, involves the use of previously established encodes and their associations for the rapid solution of problems.

At least where mammals are concerned, observational learning, as dependent upon apperception, is foundational to the other three types of learning. Through observational learning the world is internalized, and the short-term concurrences contained within this world become represented in associations. Such associations are implicit in cue deneutralization, instrumental behavior, and the potential for problem solving that can sometimes take place when organisms confront situations similar to those encountered in the past.

In brief, although many secondary variables are always involved, the establishment of encodes is dependent upon apperception, the development of associations upon contiguity as well as apperception, and the emergence of percepts and perceptual structuring upon the simultaneous occurrence of inputs within the field of awareness—each temporal segment that lasts approximately 750 milliseconds.

These observations are applicable to human learning. In other animals, apperception may play a less critical role. As we move down the philogenetic scale, although some kind of focalization is always required for effective and directional behavior, less refined modes such as eye fixation and positioning, head directioning, or postural orientation may play more important roles. The capacity for apperception, even as other attributes, evolved by degrees. In humans and probably other primates, although the influence was no doubt reciprocal, the capacity to focalize sensory input may have been refined as a process bonus by the rapid and facile shunting requirements implicit in the encode manipulations necessary for effective planning and foresight and for coping with the complexities of social interchange.

The resurgence of interest in information processing, inspired by the computer revolution, has resurrected (from its 60-year eclipse due to behaviorism) cognitive psychology with its enduring concern with memory. When tied in with evolutionary theory, the steps in cognitive processing can be understood as techniques that have been selected into organisms for coping with the ever-shifting cycles within their bodies and the ever-changing energies in their environment. Thus, immediate awareness (lasting approximately 750 milliseconds) constitutes a continuing readout of both internal and external reality; short-term memory (lasting approximately 15 seconds) represents those features of this reality that are of sufficient pertinence to occasion apperception; and long-term memory (enduring for prolonged intervals) consists of a lasting storehouse of those features with sufficient portent to have occasioned multiple apperceptions. In this fashion information is filtered for adaptive pertinence through the buffer stages of both immediate awareness and short-term memory before it is finally encoded in long-term storage. There it remains as functional latticeworks of "if A then B" associations and an accumulated replica (internalized world) of the individual's past experience. Such encoded information can be utilized for coping with present problems and for predicting happenings in the future, as implied by such terms as foresight and planning.

A commitment to inherited predispositions, as represented by the modules and algorithms of contemporary evolutionary psychologists, gathers momentum. According to this view, the tabula rasa of Locke is not blank, but marked by zones of receptivity and exclusion. The empty box of the behaviorists is not empty, but filled with special capacities and proclivities derived from each species' evolutionary history. The optimistic environmentalism espoused by Watson, that any human infant has the potential to become "butcher, baker, or candlestick maker," may seem hollow now. A remarkable shift in only 50 years.

However such shifts are healthy; renewed commitment and honest contention add energy to our discipline. Nativism, as represented by evolutionarily derived predispositions, may appear to be in the ascendancy, but environmentalism, though perhaps diluted and attenuated, is not dead. Of all the outcomes of the long travail of our evolutionary history, the capacity to learn is perhaps our most adaptive attribute. The fact that all humans have a common evolutionary history and to some degree share evolutionary histories with all other living things, suggests that in certain areas domain-general principles will be characteristic of all of us. Species vary in their capacity for vision, hearing, and odor detection. Why should not we expect variations in the capacity to learn? Searching out and clarifying variations both within and between species relative to more inclusive categories has long been the commitment of comparative psychologists and those who study individual differences. The new emphasis helps us determine where to look for special abilities, and when they are found, provides a theoretical framework which makes them meaningful.

7

Apperception

In this chapter we confront, in detail, the most important step in information processing: apperception. It has already been encountered in the overview (chap. 3) and in the treatment of both short-term and long-term memory. It provides a readout of vital events in both our internal processes and our external environment. It is responsible for most of the encoding that takes place. It is the process whereby those memories that are appropriate to each survival circumstance are retrieved. It is the ultimate intermediating mechanism: Between the input (memories–sensations) and the output (behavior) comes apperception. In short, it signifies the pertinence and appropriateness of each available bit of information so that efficient adaptive behavior can take place. The ongoing course of apperception is completely lawful: It is a focalizer, not a selector or planner.

In chapter 4, the view was defended that various transmitting mechanisms (whether specialized receptors, free nerve endings, or fluids) are being activated during every moment of an individual's life, and that many of these mechanisms introduce functionally pertinent sensory inputs into the field of awareness. Because reacting to all these inputs would involve unresolvable complexity, we observed in chapter 5 how such multiplicity is simplified by the constancy and fusion mechanisms that contribute to the development of phenomenological wholes and their orderly arrangement within three-dimensional perceptual space. With the advent of learning (see chap. 6), organisms were confronted with added problems of complication in the form of vast numbers of memory encodes. Not only are fused sensory components (i.e., percepts) being continually introduced, but also thousands of encodes are perpetually available.

Simultaneous reaction to all such inputs (whether memory or sensory) would be impossible and completely nonadaptive. A situation of information overload would exist and directional behavior could not occur. The evolutionary solution to this intractable multiplicity was the emergence of a mechanism that would allow focalization upon only one component at a time—focalization, as implied by the postulate of inference, on that component that is most significant for adaptive response at each particular moment.

Apperception is the term that seems most appropriate for this focalization mechanization. The term attention has long been used by psychologists to represent the selection process that appears to be taking place. I prefer the term apperception because attention has typically been restricted to focalization upon sensory input alone. In my view, this is only half the process picture. Sensory components are involved, but memory encodes are also available. Indeed, the mental life of the individual consists of a continuing interplay between sensory and encode focalization. During any given moment there may be a rapid shift between these two kinds of information input. A sound may be focalized, for instance, and then a series of related encodes may follow in apperception. A scream in the night may conjure up all kinds of unusual possibilities, with each encode that is focalized perhaps activating an even more bizarre and disquieting sequence.

Apperception is the most important information-processing operation that takes place in complex organisms, particularly in human beings. Without it, learning could not occur, and thinking would be impossible. Furthermore, it provides the immediate occasion for most, although not all, overt behavior. It is the ultimate simplification mechanism, without which an organism's behavior would be reduced to chaotic and nonfunctional oscillation.

Apperception, as I define the term, is the process whereby inputs from sensation, memory, or both are brought into focus during any given moment. It is assumed to be both completely automatic and completely lawful, and is without doubt the most highly refined homeostatic mechanism that has emerged during evolution. The following statement summarizes the previous discussion and relates apperception to the postulate of process given in chapter 2:

> That which is apperceived at any given moment constitutes the most homeostatic resolution of the interaction among inputs from both sensation and memory.

The tendency to view apperception as a selector, as if it were a little mobile creature with a volition of its own, is sometimes present in the writing of psychologists, particularly cognitive psychologists. Most often this tendency is not admitted or possibly even recognized. However occasionally it is spelled out with remarkable candor.

> It would seem that a higher agency of mind, call it the executive agency, has available to it the proximal input, which it can scan, and it then behaves in a manner much like a thinking organism in selecting this or that aspect of a stimulus as representing the outer object or event in the world. In short, cognitive theory at its essence incorporates a homunculus concept. (Rock, 1983, p. 39)

Whether apperception is itself a causal agent or is the lawful outcome of coalescing variables has been an issue of much and prolonged contention, as reviewed by Johnston and Dark (1986). They pointed out that viewing apperception (attention) as a causal agent invariably involves a homunculus and an infinite regress.

If we plead a homunculus as an explanatory concept, where does it end? We have explained nothing, but have merely pushed the problem back a step; once started, further steps can be taken, back and back into an infinite regress of little men within little men.

Fortunately, we need not assume such a self-defeating and pessimistic view of cognitive activity. Apperception is a lawful process. It is a focalizer, not a selector, and the path of its lawful operation can be determined inferentially and in certain cases experimentally. Even James (1890), who accepted freewill on pragmatic grounds in his philosophy, was a determinist in his treatment of apperception (attention).

THE FIELD OF AWARENESS AND APPERCEPTION

The field of awareness is the moving belt of our experienced reality. The inputs from all sensory systems of the body are simultaneously and irresistibly represented; it is thus a multi-parallel system. As long as a sensory system is being stimulated above threshold, the sensory resultant is displayed within the field of awareness. It is within this field that the flux of apperception takes place.

All of this can be readily appreciated by considering a simple example. At this moment, the field of awareness of the reader consists of the overlapping and simultaneously presented input from the various sensory systems of his or her body. Inputs from audition, vision, olfaction, gustation, pressure receptors, and perhaps pain receptors coalesce and become an aspect of the experienced now. If a person is hungry, thirsty, anxious, or nauseated, these inputs become simultaneously added. Apperception is in continuing flux within this matrix:

Inputs from any sensory system or particular components within a given system may be focalized.

From an adaptive frame of reference this is entirely plausible. The field of awareness consists of events in the external and the internal environment, transformed (see chaps. 4 & 5). It is the functional representation and organization of those energy changes, both internal and external, that have been of enduring importance in survival and evolution.

Certain sensory systems are more sensitive than others, depending on the evolutionary history of the species, as is olfaction in the hound, audition in the bat, and vision in the vulture. Certain segments of a dimension may be more sensitive than others, as expressed in the Weber-Fechner function. However once the threshold of a sensory system is reached, the resulting input is displayed within the field of awareness. All inputs have equal, parallel, and simultaneous representation.

However what about such variables as intensity, pervasiveness, size, and movement, etc.? These variables are important, as are others to be discussed, but because of their influence on apperception, not from any priority for inclusion in the field of awareness.

There has been much discussion concerning the relative status of the field and the point of awareness. Wundt (1886) insisted, and I agree with his position, that the field of awareness or blickfeld is preapperceptual. It is the context in which apperception (or blickpunkt) takes place. Neisser (1967) suggested that there is a preattentive segregation of the field of awareness according to certain Gestalt principles such as figure ground. I agree that the segregation is preattentive (preapperceptual) but feel, following Helmholtz, that subtle cues given in experience, as it takes place within the field of awareness, account for perceptual organization, and that once such organizing and structuring has taken place it is automatically imposed. Thus, the visual aspect of the field of awareness is presented as phenomena displayed within a three-dimensional context.

ENCODES AND THE FIELD OF AWARENESS

Encodes are a crucial aspect of mentation. They are instantly available for focalization, and rapidly and easily become a part of every sensing-thinking sequence. Yet these memories are seldom (except in dreams and in eidetic imagery) represented in the field of awareness with any degree of clarity. Indeed, they become apparent therein as fleeting and tenuous images when we daydream or when we try to summon up some past experience. Sensory input is simultaneously and continuously displayed in awareness, but memories just seem to leap into consciousness from some subliminal realm where they are perpetually and appropriately available, but where they are hidden until they leap.

Boring (1929/1950) reviewed the early efforts to clarify the elusive images involved in thought. Kulpe, of the Wurzburg school, perhaps represents the most noteworthy victim of the ephemeral nature of encodes. He and his students searched for the conscious contents of thought for many years, but failed to find them. Watt devised a technique called fractionation with which he tried to zero in on the ephemeral process. He instructed subjects to respond verbally to a stimulus word as rapidly as they could. He found that he could divide the mental operation involved into four periods: (a) the preparatory period, (b) the presentation of the stimulus word, (c) the scanning of memory for the reaction word, and (d) the utterance of the reaction word. He thought he would find the conscious content of thought in the third period, but nothing was found. The reaction word simply leaped out without any discernable conscious precursor. It was, as Boring stated, as if "... one does one's thinking before one knows what he is to think about; that is to say, with the proper preparation the thought runs off automatically, when released, with very little content" (1929/1950, p. 404). All this led Watt to emphasize set (or *aufqabe*) as the critical factor in thought.

In other research, Watt discovered that memory operations apparently reside in separate clusters. For instance, if subjects are presented with two numbers, say a 6 and a 4, and asked to respond as rapidly with a solution as they can, they will respond either with a 2, a 10, or a 24, depending on whether they subtract, add, or multiply. However if given the set for any one of these operations, say to add, they will respond more rapidly and with fewer errors. Watt's findings not only gave support to the view that the process of recall has little representation in awareness, but also indicated that operations can be encoded separately, to be rapidly demonstrated when required.

However human beings can recall particular memories that appear in the field of awareness as faint replicas, as Hume pointed out, of sensory input. Such replicas may be narrow and restricted or broad and inclusive; thus a person may have a particular image of the sausage they ate for breakfast or a scene of their neighborhood. In both cases the image is typically vague and amorphous.

Thus, memories may enter consciousness and as such become a part, however fleeting and indefinite, of the field of awareness. Yet the process of recall and the process involved in moving from one memory image to the next (as implicit in thinking) have no representation in consciousness and enter into awareness only as tendencies that somehow automatically happen.

The amorphous character of memory images and the even more nebulous process of transition from one image to the next are typical of mentation in all of us, but are not universal. In eidetic individuals the process of retrospection can proceed with remarkable pictorial vividness, and dreams often achieve the verisimilitude of an actual sensory experience, with like richness and clarity. It is also likely that the hallucinations of mental illness often achieve the clarity of immediately given sensory experience.

Yet these are the exception, and even when they occur the individual can typically, except in psychosis, distinguish between the world as remembered and the world as directly experienced. This discrimination is critical to survival. If memories and sensations were indistinguishable in the field of awareness, a nonfunctional and debilitating increase in complexity would occur. The selection pressures of evolution have worked efficiently to keep the world of memory and that of sensation separate, and for obvious reasons. Sensation represents the condition of the organism and the organism's environment in the present moment. Hazard or benefit exists only in the present, not in the past.

For adaptive behavior to occur, it is essential that all aspects of sensation be represented in the field of awareness, so that priority can be given (as apperception fluctuates) to those aspects that are most vital at each moment. It is also essential that the backlog of the individual's experience with this or like vital situation be immediately available. Memories reside in a preconscious reservoir (outside the field of awareness) unobtrusively potential, but readily available when the circumstances in the immediately given sensory world make them adaptively appropriate. This facility is a testament to the evolutionary honing that has gradually produced the vital interchange between the problems of the present and the accumulated wisdom from the past.

This discussion gives some indication of the complex set of processes that are implicit in even the simplest reaction situation. It underscores the amorphous character of the memory image and the impalpable process whereby potential encodes suddenly and automatically become focalized. The memory image is an indistinct replica, often barely discernable in consciousness, and the process that propels it into consciousness has little if any representation in awareness at all. Small wonder that the Wurzburg school was called the school of imageless thought, and the chapter on thought was missing in Kulpe's much anticipated book (see Boring, 1929/1950).

Narrowing the Scan

The field of awareness, including as it does the inputs from all sense modalities and activation systems of the body, is remarkably complex.

When we add to this complexity the potential multitude of memories, the sheer number of items that are available for focalization becomes enormous. Fortunately for efficient and rapid processing, such information bits, whether sensory or memory, are not arranged randomly and haphazardly. They reside in composites, operations, and associational constellations (see chap. 6). If one of these subsuming factors is isolated prior to focalization, the speed and facility of information processing will be increased.

The studies by Watt, just mentioned, emphasized that set is a critical factor in the speed of retrieval. In other research, Posner and Cohen (1984) found that the speed of response was facilitated when subjects were precued to the area of the visual field within which a target would appear. Egly and Homa (1984), in a similar study, demonstrated that subjects could be trained to expect targets in three different areas of the visual field (consisting of three concentric rings located at 1, 2, and 3 degrees from the point of eye fixation). Valid precueing as to the ring in which the target would appear facilitated accuracy of response, while invalid precueing inhibited accuracy.

Johnston and Dark (1986) reviewed many studies on priming effects, where the presentation of one stimulus prior to the presentation of another influenced the processing of the later stimulus. When the set is given that the target will appear in a particular modality (called modality priming), response to targets presented to that modality will be facilitated, while those presented to other modalities will be inhibited. If a given word is read prior to the presentation of the same word during a threshold test (called identity priming), the recognition of that word will be facilitated. If a word such as butter is read prior to the threshold presentation of a related word such as bread (called semantic priming), the response will be facilitated. If a subject is primed with the word mineral (called schematic priming) responses with items within this category will be facilitated. Indeed, as Johnston and Dark stated in their first empirical generalization, "All levels of stimulus analysis can be primed for particular stimuli" (1986, p. 47). In a later section of this chapter, we see that all of the variables that influence apperception may be viewed as ways in which the component being focalized establishes the set (as occasioned by either inherited or learned associations) for the one that follows it.

THE PHYSICAL MECHANISM

Although specific inferences about the physical process responsible for apperception are undoubtedly premature, certain general hypotheses, which the reader may consider as questions about its operation, can be stated. It appears that the process is cyclical, somewhat analogous to the sweep of a radar scope, but much more rapid. Accordingly, there may be a relationship between the upper limits of the speed of apperceptual fluctuation and the frequency of the hypothetical cycles. Yet what kind of cycle? Perhaps a reverberating neural circuit similar to Hebb's (1949) notion. If it could be isolated from the other potentials emanating from the brain, some evidence for its existence might be gathered from brain wave recordings. Research is moving in this direction.

Many studies have been involved with measuring those electrical potentials from the scalp that are correlated with various information-processing operations, such as encoding, apperceptual shift, discrimination, and choice. The delicate problems of measurement and the ephemeral nature of these potentials (or ERPs, as they are called) is emphasized by the fact that they are never clearly indicated in any single brain wave recording. The ERP is typically so small and variable, and the background noise so considerable, that reliable signal–noise ratios can be obtained only with computer averaging over many stimulus presentations. Also, studies using magnetic resonance imaging (MRI) and positron emission tomoography (PET) are making progress in isolating the physiological events that are correlated with the different stages in cognitive processing. These studies parallel the choice reaction time experiments initiated by Donders during the last century, and they mark at least a beginning of the effort to isolate the physiological events that are correlated with cognitive activity and perhaps eventually to the resolution of the elusive mind-body problem discussed in chapter 4 (see Bub, 2000; Harter & Aine, 1984; Hillyard & Kutas, 1983).

If the physical mechanism accounting for apperception is cyclic, might it be related to the processes accounting for the lower absolute thresholds for pitch and flicker fusion? This does not seem likely. The lower thresholds for pitch and flicker fusion are both responsible for the emergence of continuity in sensory input, whereas apperception involves the focalization upon inputs that are already present in the field of awareness. The difference is between how input is introduced in the first place and which variables determine whether or not it will then be brought into focus.

We might also leap to another equally incorrect conclusion, that apperceptual fluctuation is tied to eye movements. There is typically a relationship between the position of the eyes and the component in the visual field that is being apperceived, but these two factors can vary independently. Although the eyes may be directed toward an object straight ahead, some component in the periphery of the visual field can be centered in apperception. Indeed, an input in the periphery may remain in focus while the eyes are moving through a relatively complete arc. That apperception and eye movements may vary independently has been demonstrated in a number of studies, beginning with Helmholtz (as reported by James, 1890) and more recently by Posner and Cohen (1984), Remington and Pierce (1984), Sperling and Reeves (1980), and Tsal (1983).

The prior analysis of apperception (chap. 3) as a vital simplification mechanism suggested that it may be single-channeled so that if any particular information bit is being focalized, all others will be excluded. Reaction to two or more information components simultaneously would obviate the possibility of effective directional behavior; thus the selection pressures accounting for the emergence of apperception as an adaptive mechanism would have worked against this eventuality. Yet, sometimes it seems that apperception may vary from the very specific to the quite general, or even be split, as seemingly demonstrated in the previously mentioned study by Egly and Homa. At times only one item appears to be in focus, while at other times a blob composed of many different items appears to be involved.

Is apperception a single-channeled system, but so rapid in its fluctuation that it seems to include a variety of components simultaneously? As any one component is brought into focus at a given moment, are all others blocked out? Although the evidence is still coming in, it appears that both of these questions may be answered in the affirmative, as we see in the following section.

DURATION OF THE BLIP

As long as man has been thinking about his own nature, the notion of a selector mechanism as an aspect of mental activity has probably been commonly held. Aristotle considered it at length and concluded that attention was a single-channeled system. Leibnitz (see Boring, 1929/1950) first used the term apperception to emphasize varying degrees of consciousness, and Herbart (1882), in an analysis similar to my own, put forward the view that the process of focalization involves a dynamic resolution of a multitude of different components competing for dominance.

How many items can apperception include? Several researchers, beginning with Jevons (1871), investigated the range of attention without preconceptions about an absolute refractory period. Jevons threw a number of beans into a tray and found that he could take in as many as eight in a single glance. Using an early version of a tachistoscope, Cattell (1886) concluded that the human being can attend to six, and possibly eight, dots at a time. Fernberger (1921) believed that there is an intrinsic tendency for subjects to group dots as the number grows larger and the time limit grows shorter.

Behaviorists like Hunter and Sigler (1940) tried to study the effects of other variables on attention span and concluded that training, knowledge of the number of items, size of the field, and viewing distance are all related to the number of dots that can be recalled after a single exposure. The process of subitizing was first named by Kaufman, Lord, Reese, and Volkman (1949). They claimed that a subject can recognize six or fewer items immediately. When more than six are presented, the subject will become less accurate and tend to estimate the number.

Mowbray (1953) took the view that, as the period of presentation becomes shorter and shorter, attentional fluctuation between inputs becomes progressively more difficult so that, at least when complex cognitive phenomena are involved, focalization is unitary. However Mowbray (1954) suggested that when very simple inputs (e.g., letters, numbers, and three-letter words) are involved, attention may possibly be multi-channeled. Gardner and Long (1962) made a list of the variables that they supposed might affect attention span. These variables include duration of presentation time, the amount of information that is permanently stored, level of attentiveness, spatial grouping, intensity of items, size of visual field, viewing distance, character of input, absence of recent inputs, sense modalities used, and number of sense modalities involved. Adams (1962) introduced the variable of set, concluding that as expectancy increases, the period of fluctuation decreases to the point at which no more than one item can be brought into focus at one time.

How rapidly can apperception fluctuate? Is apperception a single or a multi-channeled system? These overlapping questions are fundamental. If it can fluctuate very rapidly, a number of different items might be subsumed, giving the appearance of a multi-channeled system, while only a single channel is actually involved. Research on the speed of apperceptual shifting probably began with the work of Bessel (1823), inspired by miscalculations at the Greenwich Observatory. Although there were various notions to explain the personal equation, apperception doubtless accounts for at least a part of it. In 1885, von Tchisch performed the first psychological experiments specifically designed to investigate the speed of apperceptual shifting. He discovered that a person cannot tell at which point on a clock face a moving pointer is located when a click is sounded. The sound apparently jams out the visual input so that the person tends to report the number indicated by the pointer just before the click, rather than during it.

In a study performed in our own laboratory, Brussel (1968) was able to replicate von Tchisch's finding. Brussel's experiment was much like von Tchisch's, except that the face of his clock was divided into colored sections and responses were timed in such a way that backward displacement between auditory and visual stimuli could be determined to the nearest millisecond. In addition, the speed of the pointer on Brussel's clock could be varied. He found that as the speed of the pointer increased, the amount of backward displacement increased proportionately. The total mean delay between the auditory and visual stimuli was 57.49 ms—or approximately 1/29 of a second. These data suggest that attention (apperception) can shift about 20 times per second. However, there is one fundamental flaw in both this study and the one just mentioned by von Tchisch. Since the subject was following a pointer with his eyes, the reported backward displacement could be due to visual lag.

In another study, Davis (1957) obtained a similar result although he pursued the problem from a different angle. Davis presented subjects with a visual stimulus (light) and an auditory stimulus (click) that had variable intervals of time between them. The interval varied randomly from 50 ms to 500 ms. The point at which subjects were most markedly unable to respond to both the visual and auditory stimuli was at 50 ms—approximately the same as the speed of attentional shifting found by Brussel.

These results suggest that apperception is a single-channeled system; when any particular component (whether sensory input or memory encode) is brought into focus, all others are blocked out. There is some physiological evidence that supports this view. In 1956, Hernandez-Peon, Scherrer, and Jouvet implanted an electrode in the cochlear nucleus of a cat and found that a noise produced near the cat's cage produced a discernable spike in recordings from the electrode. They then presented the cat with inputs from other modalities. In one instance the cat was allowed to observe a glass beaker containing live mice, and in another the odor of fish was blown into the cage. Under these conditions, the cat immediately became alert, and, at the same time, the brain wave spikes from the continuing sound decreased in magnitude. Apparently, input from the visual or olfactory modality was jamming out the auditory components. Although Hernandez-Peon's interpretation of his results have been disputed (see Horn, 1960; and Hugelin, Dumont, & Paillas, 1960), his research constitutes a pioneering effort to clarify the as yet unknown neural mechanisms underlying apperception.

Broadbent is probably the leading exponent of the single-channeled view. He presented a mechanical model for human attention that has had a profound impact on both research and theory since 1961. His model consists of a Y-shaped funnel standing vertically with the converging branches separated by a hanging flap. A number of small balls represent bits of sensory input. If a ball (sensory input) coming down one branch strikes the flap, it will move over and momentarily close the other branch before it swings back to the vertical. Thus although more than one ball (information bits) may descend the branches, only one can enter the lower stem of the funnel at any given moment. This single-channeled model of attention (apperception), albeit simple, makes sense out of a variety of experimental findings, as cited by Broadbent (1957) in his delightfully written account. These include studies that support his predictions on the influence of such variables as intensity, earliness in time, absence of recent inputs on the channel, position of input on the hierarchy of channels, and the type of information in permanent store. In Broadbent's model and his inferences from this model, we see, as recounted in detail later, a clear coalescence of the thinking of cognitive psychologists and inferences from evolutionary theory.

We have known for many years that set is an important factor influencing apperception. Referring to the many arguments about the personal equation that took place during the latter part of the last century and the early part of this one, Boring stated, "attentive predisposition favors earlier clear perception. If you are expecting the bell, listening for it, then the sound comes into consciousness more quickly than does *a simultaneously presented visual stimulus*" (Boring, 1929/1950, p. 146, italics added). Goldstein and Fink (1981) found that subjects who were told to attend to one of two overlapping but different colored line drawings revealed very little processing of the irrelevant drawings. Apparently the set established by the instructions effectively kept the relevant stimuli in focus for the entire 0.75 seconds of immediate awareness. However, overlapping objects can be selectively processed if the set (instructions) encompasses both. Becklen and Cervone (1983) instructed subjects to respond when a specific event occurred in a relevant game (a hand game) that was simultaneously presented on a screen with an irrelevant game (a ball game). There was no performance decrement, and subjects had no awareness of happenings in the irrelevant game. Apparently the set effectively blocked the apperception of the irrelevant game. Yet when subjects were instructed to attend to both games, although a severe decrement in performance (reflecting the time taken in split scanning) took place, subjects were able to process information from both.

A remarkable study by Posner, Boies, Eichelman, and Taylor (1969) gave, in my view, the most sensitive (since it was so simple) measure of the effects of complication on the speed of apperceptual flux. Subjects were presented with letters in rapid succession and asked whether the second letter was the same as the first. The experimenters found that the response "same" was 80 to 100 ms faster when the letters were visually identical as in A and A, than when they were semantically identical as in A and a. This study also provides a very sensitive measure of the duration of the field of awareness. The superiority of the visually identical over the semantically identical letters ceased to exist when an interval of 1.5 seconds intervened between the presentations. By then the first image apparently fades so that immediate image-matching cannot occur.

Do these and similar studies provide concrete evidence concerning the speed of apperceptual flux? They may indicate the range within which such activity occurs, but several operations are invariably involved. Say we expose a subject to a matrix of 15 items for 50 ms and find that he or she can report 5 items during the 750 ms that the matrix holds in the field of awareness, thus apperceiving at least one item every 50 ms. Is not this a fairly accurate measure of the speed of apperceptual flux? No, because other operations, such as discriminating items yet to be reported, and the report itself, also take up time.

It is probable that new equipment and more subtle experimental designs may help us more nearly approximate the speed of apperceptual flux, but that we can ever achieve a completely accurate statement seems doubtful. The processes involved are so ephemeral, fleeting, and in many cases overlapping and interdependent, that a literal determination of the minimum period of apperceptual focus suffers from an intractable complication reminiscent of the Heisenberg indeterminacy principle in physics. Yet, we can say that the minimum period is exceptionally brief and that no available substantive evidence indicates that more than one input component can be apperceived during this minimum period.

This discussion of the minimal period of apperceptual focalization emphasizes a complication that has without doubt confounded many of the experiments on information processing. Apperception can apparently fluctuate up to 20 times per second, while the field of awareness lasts approximately 0.75 seconds. Thus components in the field of awareness can be apperceived numerous times before the representation fades.

As a case in point, consider an experiment by Allport, Antonis, and Reynolds (1972), which purportedly disproved the single-channel hypothesis. They found that "people can attend to and repeat back continuous speech while at the same time taking in complex unrelated visual scenes, or even while sight reading piano music" (Allport, et al., 1972, p. 225). However it turns out that the overlay material was presented for 1.7 seconds, giving ample time for considerable apperceptual shifting between the shadowed (the one the subject was focalizing and responding to) and the overlay task.

When we consider the duration of the blip, we are invariably talking about sensory input. Yet what about memory inputs? They constitute a large proportion, perhaps as much as 80%, of the material that comes into apperceptual focus. Why are there no studies on the duration of the encode blip? The answer probably relates to difficult, if not insurmountable, problems of experimental design and control. We can measure the reaction time of recall without difficulty, as say, when we present a stimulus such as night and measure the time interval until the subject says day, but the measurement of the amount of time that the encode day remains focalized, and the minimal interval of such focalization, is quite another matter. We simply have no way to control the duration that any encode remains focalized or any way to get at the duration of an encode blip. The duration of the sensory blip, on the other hand, can be controlled by manipulating the duration of the stimulus. The only way we can even begin to exert some control over the encode blip is through the subtractive procedure (see chap. 3), which invariably introduces intractable problems of presentation and measurement. Until appropriate research is forthcoming we may assume, and this will probably not be far wrong, that the encode blip is approximately one twentieth of a second, as is the case with the sensory blip.

INCLUSIVENESS OF THE BLOB

The question of how rapidly apperception can fluctuate is also related to the very difficult problem of ascertaining the size of an input. This problem, in turn, is tied to the extent and number of items that have been fused into the particular phenomenon in question. Although the percept *chair* may be a composite of many discrete sensory elements, it has assumed unity as far as apperception is concerned. The same is true for all percepts and, indeed, for all concepts. In both cases, various numbers of individual components have become fused into unity so that more economical and rapid processing can take place.

The issue of the emergence of percepts is still controversial. The Gestalt psychologists insisted that nativistic organizing principles account for our perception of phenomena (percepts). They also insisted that wholes are perceived prior to their parts (Kohler, 1930/1971; Wertheimer, 1925/1950). Empiricists agree that there is an innate tendency for phenomena (percepts) to emerge in perception, but they maintain that cues derived from experience are necessary for such perceptual wholes to develop. Once the individual components have become fused into the totality, they also hold that global percepts have perceptual priority over the parts that make them up. Thus we perceive such phenomena (percepts) as cars, tigers, and chairs and can separate out the subsumed components only with difficulty (see chap. 5).

The fusion implicit in concept formation is only one example of the condensation that occurs in memory. Long sequences of events may likewise be fused and simplified. There is a well-worn joke that demonstrates this process. A group of individuals was isolated in a dungeon. Time passed. They entertained themselves by telling jokes—the same jokes over and over again. To save energy they finally gave numbers to the jokes. One day a new guy was thrown into the dungeon. He sat there with his back against the wall in the semidarkness until he was startled to hear one of the other prisoners shout out the number 36. To his astonishment he heard everybody burst out laughing. Bewildered, the new prisoner asked a person sitting next to him what was going on. The situation was explained to him and after pondering for a bit he asked if he could shout out a number. On being assured that he could, he shouted out the number 37. However no one laughed: not a sound. Again bewildered, he asked his companion what was wrong. The answer, of course, was, "Well, some people just can't tell a joke." My point in telling this joke, that we shall call number 37, is that a long series of inputs had become fused into an associational unity that had been tagged so that common understanding and reaction was made possible.

Such fusion, condensation, and tagging is typical of the encoding process and is foundational to the use of language, the cognitive mapping that occurs when people interact with the environment, and the planning and foresight essential for adaptive behavior—conditions that provided the selection pressures for the evolution and function of this very capacity. Appropriately tagged encode constellations, with their embedded associations and affective overtones are fundamental in the internal locomotion (thinking), which has long been and yet remains the basis for human adaptation and survival.

This analysis simply hints at the complexity of the remarkable web of association and affective overtones that are involved in human mentation. Whatever their source or extent of inclusion, sensory, intersensory origins, or burden of affect, such constellations may become fused into an intricate blob—a blob that is typically tagged: a tune, a gesture, a face, a taste, a symbol such as number 37, or $E = MC^2$.

In certain individuals (children, the uneducated or the mentally handicapped), the encodes involved in the thought process may be literal and representational, while in others, more capable of abstract thought, mentation may involve amorphous blobs of varying degrees of inclusiveness. To the extent that the thought process consists of such blobs, the speed of mentation can be remarkably increased, but the possibility of error is correspondingly elevated and failure of communication more likely. This can be readily appreciated when we consider the inclusive vagueness of such terms as paradigm, theory, democracy, or Christianity. Thus, inclusive blobs may simplify and abet the thought process by providing the basis for conceptual integrations, or they may serve as the focus for the encapsulation implicit in logic-tight compartments and prejudice.

Such associational fusion is typical of all human mentation: the logos and slogans of corporations; the epithets of group insult; the flags and symbols of nationalistic identification; the symbols, practices, and precepts of religion; and, as we see in the final chapter, the symbols and structure of language. In all such cases the affective loadings may vary from neutrality, through bland subtlety to sizzling prepotence, depending upon the circumstances of their development and the number of people involved. Regardless of their associational inclusiveness, emotional overtones, and action-instigating power, all such mentations are reflections of the selection pressures that provided the basis for their emergence and their present use.

As might be expected, such inclusive categories develop as the child grows older, no doubt as a function of both maturation and learning. Piaget (1929), who pioneered the understanding of the emergence and use of concepts, believed that all children go through four developmental stages: (a) reflexive reaction, (b) object recognition, (c) encode function transfer (e.g., as in using a toy wagon for an airplane), and (d) the ability to manipulate concrete encodes (i.e., to actually think). All these stages are precursors to finally achieving the capacity for abstract thought, that is, the manipulation of inclusive concepts. Although he and his followers have designed many ingenious experiments to demonstrate such development, there is now general agreement that these stages are not as separate and distinct as Piaget supposed (Kuhn, 1992; Spelke, 1991). Yet there is agreement that all children gradually move from the specific and concrete to a capacity for the manipulation of more inclusive, less directly pictorial encodes as more and more events and operations are funneled into inclusive concepts, and that the highest levels of mentation, as evidenced by educated and gifted individuals, may achieve greater speed and efficiency while utilizing such representations.

TOTAL RESPONSE TO PARTIAL CUES

In many cases, the input from a sense modality represents only a small portion of the potential information about an encountered circumstance. For instance, an individual may encounter only a part of a human face or a part of a motor car. When such partial presentation occurs, there will be a tendency to apperceive the fused encode (i.e., concept) of such partial cues immediately thereafter. This tendency to perceive a total residual following a partial exposure depends upon the extent of fusion that has taken place.

The adaptive function of total response to partial cues is considerable. Both predator and prey rarely present a total configuration. More likely, only certain parts of the head, eyes, or body are exposed. Every item always has a number of sides, may be viewed from many different perspectives, and is always changing in size, brightness, and shape, depending upon distance and angle of presentation. The apperception of the fused encode allows efficient response to take place, a response that is appropriate to the total survival significance of the item. Total response to partial cues appears to be a universal characteristic of all reaction, and there should be little surprise that the shaping processes of evolution have developed mechanisms for its refined expression.

This forest versus the trees issue, as it has been called, has a long history, beginning with the form-qualities of Ehrenfels (see Boring, 1929/1950) and extending into more recent research and theory (Broadbent, 1977; McClean, 1979; Miller, 1979; Navon, 1977, 1981). Most of the foregoing investigators report the precedence of the global over the particular, giving evidence for the priority of the percept and providing empirical support for the argument (given in chap. 5) that such functional simplification (for both apperceiving and encoding) facilitates information processing.

FUSION AND THE SPEED OF APPERCEPTION

Once a percept or a concept has developed, a remarkable amount of functional simplification is provided. Thus, as mentioned in chapter 5, when we encounter a tiger we do not apperceive the multiplicity of inputs (colors, textures, shades, shapes, etc.) that signal its presence, but rather the unified phenomenon. The concept tiger is a functional summary of a multiplicity of inputs plus a modal representation of all of our experiences with anything to do with tigers. After such fusion has taken place, either the spoken or written word tiger or its picture or a partial presentation of an actual tiger, may occasion the apperception of the fused encode with all attached survival implications. Thus to respond appropriately the various inputs that represent the tiger need not be separately focalized, only the unified phenomenon. This remarkable simplification speeds up the information processing necessary for effective behavior to take place.

Once fusion has occurred there is an enduring tendency for it to remain intact; thus, meanings and attributes tend to accrue to certain familiar items, and it becomes difficult to separate them from such items. For example, it would be tedious to think of black milk, white coal, or green blood. This observation has been supported by a series of experiments dealing with the Stroop effect. Stroop (1935) found that it takes longer to respond correctly to colored words when the color does not match the word, as for instance when the word green is written in red. Apparently the word green and the color green have become fused into a composite encode; when the word green is written in red, an apperceptual complication arises that takes time to resolve, resulting in increased reaction time.

Variations of the Stroop design might be used to study the degree of fusion, association, or both that has taken place between components of many different kinds. Thus, it would be predicted that the simultaneous presentation of the spoken word pear with the visual presentation of the picture of an apple (or the written word apple) would result in longer reaction times than when the simultaneously presented components are congruent with past learning. Indeed, it would be expected that two simultaneously presented components, whatever their nature, should result in longer reaction times to the degree that there is association or fusion incongruity.

In a seemingly contradictory study, however, Lewis (1970) found that reaction times were longer when simultaneously presented words were similar than when they were not. Lewis used a dichotic listening design to present different words simultaneously to the two ears. The subjects were to carefully attend to (i.e., shadow) the words presented to one ear and to immediately repeat that word aloud as soon as possible. It was found that reaction times were longer when the unattended words were semantically related to the attended (shadowed) words. For instance, when the simultaneously presented words were synonyms, the reaction times were significantly longer than when the words were unrelated.

Do these findings run contrary to those found in the Stroop studies, where greater incongruity resulted in longer reaction times? Do these results also indicate that there may be some processing through an unattended channel? I do not think so. It turns out that the attended words were always given to the same ear. Thus the greater the semantic similarity between the two words, the more difficult it became for the subject to discriminate that he was responding to the components presented to the attended ear. That is, the flux of apperception (up to 20 blips per second) simply took more time as the subject tried to resolve this increasingly difficult discrimination problem. The increased reaction time does not indicate, as Lewis concluded, that the unattended message was being processed at a deeper (semantic) level. Almost incidentally, it would seem, Lewis reported results that are among the most convincing evidence I have found that attention (apperception) is a single-channeled system. "No subject was able to report any word from the unattended message while maintaining errorless shadowing" (Lewis, 1970, p. 227).

At present, we can only try to answer the inclusive question in a general and seemingly circular fashion. The inclusiveness of an input (whether percept or concept) is indicated by what is experienced as unity in apperception. Regardless of the circularity implicit in this statement, it offers some possibility of empirical test. If a human subject is briefly exposed to a room filled with items and is later asked what was seen, he or she will be able to recall such phenomena as chairs, tables, and pictures, suggesting that he or she is recalling totalities, rather than individual particulars. Yet the staggering complexity of the problem is revealed even in this simple example; not only does the extent of fusion determine what the person sees, the context is also of critical importance, and the intrusion of language complicates both encoding and recall.

Questions concerning the nature of the blob have had a long history. Mill held a view called mental mechanics that assumed that all the particulars contained within the blob could be extricated and identified, as such, with analytic introspection. J. S. Mill, his son, called his version mental chemistry, and assumed that the individual components are lost in the melding process; the blob is an emergent that is different from and more than the constituent parts. Wundt accepted J. S. Mill's view under the term creative synthesis, as did Titchener, and made it central to structural psychology (see Boring, 1929/1950, for a detailed treatment of this issue).

Deciding which point of view concerning the formation and nature of the blob is correct seems a moot issue at the present time. Certainly a blob can involve the fusion of many components in intricate combinations and juxtapositions, melded together to emphasize a unitary functionality that can be understood in terms of its adaptive contribution to information processing.

The Conditions of Fusion

Is apperception a necessary precursor to fusion? I have posited in a number of places that apperception is a single-channeled system. I have also stated that fusion is dependent upon simultaneous, as well as repeated presentation of inputs. However fusion by its very nature involves the melding of a number of inputs into unity, implying either that apperception is not required or that strict simultaneity of focalization is not necessary.

This suggests that simultaneous representation in the field of awareness rather than simultaneity in apperception is the requisite condition for fusion. This hypothesis is supported by the fact that the cues involved in perceptual organization, particularly in visual depth, are represented in apperception, if at all, as barely perceptible intrusions. What, for instance, are the experiences derived from convergence and accommodation? Further, the experiences representing such cues as retinal disparity, movement parallax, linear perspective, and overlap can be isolated, but only with difficulty. Typically they are not even noticed. Once perceptual organization has taken place, all such cues apparently manifest themselves automatically, irresistibly, and unconsciously as Helmholtz insisted. Thus it may be that neither apperception nor simultaneity is required for fusion and the other processes involved in perceptual organization. These issues may seem esoteric but they are important if we are to disentangle the overlapping processes involved in information processing. Although apparatus of extreme sensitivity and experimental designs of great subtlety will be required before the role that each process plays can be teased apart from the others, I am optimistic that it will some day be accomplished.

VARIABLES INFLUENCING APPERCEPTION

It has been previously hypothesized that apperception is a lawful process that works on homeostatic negative-feedback principles. Yet many variables enter the picture to determine which of the thousands of potential inputs (both sensory and memory) will be brought momentarily into focus. Since (as discussed at length in the following chapter) apperception is a critical mediating mechanism between input and output, the development of a predictive system about behavior must depend upon an understanding of the manner in which variables coalesce to determine the path of focalization.

Inherited Predispositions

Since the apperceptual mechanism is the outcome of selection pressures imposed during evolution, a consideration of important survival variables would be profitable for inferring which factors influence focalization. For instance, movement (see chap. 5), whether of prey or predator, has had enduring survival significance during evolution; therefore, we might predict that anything shifting in the visual field would have high priority for focalization.

Likewise, since there has been an enduring relationship between the proximity of an object (a factor critical to survival), and the amount of energy emanating from it (whether light wave reflections, pressure fronts, or molecular density), it would be predicted that intensity of stimulation (see chap. 2) would be an important variable influencing focalization.

Since the approach of the predator or the nearness of prey typically produces a number of similar sounds, odors, or light wave reflections, repetition would also be predicted to have an important influence. However repetition as a variable has limitations; repetitive stimulation may have short-term survival significance, but over an extended period it may mask other vital information. Fortunately, sensory adaptation (see chap. 8) helps free the apperceptual mechanism from the dominance of prolonged repeated stimulation.

Finally, discreteness or contrast may be an important factor. Other organisms, whether predator or prey, have always been critical to survival, and, as their edges set them off from the background, the variable of discreteness should influence which components in the field of awareness will be more readily focalized (see chap. 5).

We may summarize this discussion with the following general statement:

> Those characteristics of sensory input that have provided the most vital information during the evolution of a species will have the greatest relative potential for apperceptual focalization.

In line with this statement, we may infer that certain categories of sensory input will be more readily focalized than others. Pain should be more readily apperceived than thirst, and thirst more readily than hunger. It is even probable that certain specific inputs within a given modality may have a greater potential for focalization than others. Thus the color of blood is probably more readily brought into focus than any other color.

The discussion on sensory pervasiveness in chapter 4, and that on movement cessation gambits and breaking up the edges gambits considered in chapter 5, might well be recalled at this point. Certain inputs have been of such crucial import during the evolution of species that mechanisms of both concealment and detection have been reciprocally sharpened in response to the selection pressures involved. The greater tendency for certain inputs to be apperceived is simply an expression of the sharpening of a vital detection mechanism; that is, the greater the survival importance of a bit of information, the more effectively and rapidly it should be processed.

Learned Predispositions

In our discussion of observation learning (chap. 6), an intimate interplay between such learning and the process of apperception was suggested: Apperception is both necessary and sufficient for encodes to be established. The interdependence of apperception and learning demonstrates a remarkable parsimony. Since apperceptual focalization is most responsive to those components that have had enduring survival significance, it is both appropriate and predictable that encoding will result when components are focalized. In this manner, apperception works as a filtering mechanism for learning. Though hundreds of components may be potentially available in the field of awareness, only those that are most crucial are apperceived, and, in turn, only those that are most crucial are encoded.

Learning depends upon apperception; and that which is learned affects the course of apperception thereafter. In the adult human being, we find a complex network of associations that have been established by the contiguous occurrence of repeated sensory inputs. Once a particular if A then B association has been established, the apperception of one component will tend to bring about the apperception of the other immediately thereafter:

> The probability that an input (sensory or memory) will be focalized is a direct function of the number of past apperceived contiguous occurrences between that input and the one immediately preceding it in apperception.

This statement covers a wide range of seemingly different input circumstances. If a stimulus such as night is focalized, the memory input day will tend to follow in apperception. If a series of items such as the alphabet has been learned, the apperception of each item will trigger the next item in the sequence. If the activation input hunger is focalized, encodes representing familiar food items will tend to follow in apperception. If the secondary activation input fear is focalized, memories of past experiences having to do with reacting to or coping with fear will tend to follow in apperception.

Indeed, sequential focalizations are characteristic of all aspects of mental life. Apperceptual fluctuation is a continuous process. Associated encodes follow one another, often to be disrupted by sensory inputs that may initiate another encode chain. At times there are rapid alterations between encodes and sensory inputs, and, in many cases, sensory components may follow one another in rapid succession.

Apperceptual fluctuation follows the most equilibratory path among the myriad components from memory and sensation that are dynamically interacting during every waking moment of an individual's life. Associations between inputs (sensations and encodes) may be viewed as equilibratory paths, with each path determining the course of apperception until some new input disrupts the chain to initiate another.

Apperceptual flux typically follows the same sequential order that prevailed when the input was imposed, reflecting the directionality that exists in the environment. This directional isomorphism between environment and process has fundamental adaptive implications. The environment consists of relatively permanent concurrences between cues and critical circumstances. Thus, even as "if A, then B" is characteristic of the environment, it is also characteristic of both the associations which develop in learning, and the lawful sequentiality that is characteristic of apperceptual fluctuation. It is much easier to say the alphabet forward than backward or to repeat the days of the week or the months of the year in the same order in which they were initially presented. Similarly, familiar numbers such as one's phone number flow easily when the preestablished directional sequence is followed, but this becomes a tedious task when the process is reversed. With the learning of language, the directionality implicit in the way words are hooked together ensures that the order in which encodes follow one another will become even more fixed.

As the fusion of various inputs into percepts takes place, and as such composites become represented by concepts, facility for abstract thinking develops. Yet the path of apperceptual flux, however abstract, is profoundly affected by the directionality that is present in the environment.

Novelty

A familiar situation is, almost by definition, less hazardous than one that is unfamiliar. The sheer fact that the organism has lived sufficiently long in a situation for it to become familiar provides logic for this inference. Hazard lies in the unfamiliar, and when some novel input intrudes into the field of awareness, that input will tend to be focalized. Thus, the following statement:

> Holding other factors constant, the greater the novelty of a given input, the greater will be the tendency for that input to be apperceived.

Activation Input

Rare is the moment when an individual is not experiencing some nuance of activation input. Hunger, anxiety, fatigue, subtle moods, and pervasive fears provide but a part of the background for the succession of inputs from the ears, eyes, and other modalities. Depending upon its persistence, intensity, and pervasiveness, such background input will influence the direction and speed of apperceptual flux. This would be predicted from an evolutionary frame of reference: activation input (i.e., background) indicates that certain process systems of the body are moving toward hazardous imbalance. Since the immediate welfare of the organism is threatened specifically in the area that the input represents, it is essential that this input affect the course of apperception and, thus, behavior. For example, when an organism is deprived of water, fluid proportions within the body move away from balance, and the activation input thirst is introduced. Since thirst is an outcome of growing imbalance within a vital homeostatic system, the organism is structured so that this background input will become the predominant factor influencing apperception and determining action.

What establishes the relationship between various activation inputs and particular sensory and memory components? Although innate factors may dispose apperception to move in certain prescribed channels, learning undoubtedly plays the predominant role, at least in human beings. The input representing a given state of imbalance is not very different—except in terms of quality and persistence—from that derived from more obvious sense modalities. Therefore, the same principles of learning previously discussed (chap. 6) can also explain the associations that develop between activation inputs and inputs derived from other sources. After such associations have been established, the apperception of an activation input will occasion the tendency to focalize related encodes and pertinent sensory components.

All biological information-processing systems, and this is certainly true of cognition, are geared to the service of maintaining and regaining vital balances within the body. The entire course of evolution and the shaping implicit in natural selection have determined that this homeostatic priority should exist. The various ways in which background inputs coerce the direction and speed of apperceptual flux is but another manifestation of the remarkable manner in which even the most highly refined example of information processing is functionally subservient to the survival requirements of balance reestablishment.

It should be reemphasized that the individual responds to the activation input rather than to the circumstance that produced the imbalance. Thus, the spur that results in action—whether overt or covert—is neither the number of hours of deprivation, as the behaviorists maintain, nor some underlying physiological state, as assumed by physiological psychologists. Deprivation may produce the imbalance and such imbalance may be represented as a physiological state underlying experience; but the effective drive, spur, or activator (call it what we will) of behavior is the apperceived experiential input. An increase in food deprivation may be correlated with an increase in general activity, and such activity may be potentially related to underlying physiological variables, but the actual determinant of action is the experiential input hunger.

This discussion leads to the following hypothesis:

> The more intense the activation input, the greater will be the tendency for previously associated inputs (encodes, sensations, or both) to follow in apperception.

For example, as hunger increases in intensity, not only will it tend to be focalized more readily, but also memory encodes and sensory inputs of food items will have a greater tendency to follow in apperception.

Conversely, since apperception is a singular process, the hungrier an organism is, the less will be its tendency to apperceive extraneous inputs of any type. Thus, the following hypothesis:

> The more intense an activation input, the greater will be the tendency for that input to jam out materials that are unrelated to the critical imbalance state.

The need domination of apperception implied in the statements just given leads to a third and related hypothesis:

> The more ambiguous the sensory input, the greater will be the likelihood that such components will be apperceived as related to the existing state of imbalance.

Accordingly, a hungry organism will tend to apperceive food items in an ambiguous circumstance, whereas an organism that is sexually deprived will tend to apperceive sexual items in the same situation. This tendency to apperceive even seemingly extraneous inputs as related to critical imbalances may reach remarkable extremes. A person desperately needing to urinate may see every rock or bush along the road as a potential situation for relief, the desperately hungry person may see food items in the ambiguous context of an inkblot, and a person desperately in love may construe neutral statements as supporting his or her involvement.

Although activation input will typically trigger the focalization of associated material, if such input becomes extremely intense, it may completely flood apperception and jam out relevant encodes, sensory components, or both. The continuing presence of a hot iron on some part of the body, the sensory repercussions of orgasm, or the demobilizing anxiety that sometimes occurs in neurosis are but a few examples in which activation input may be sufficiently intense to exclude other materials from apperception. Under such circumstances, little rational behavior can occur, simply because the encodes essential for it cannot be apperceived.

What is the evidence that background input influences the course of apperception? Here again a fundamental difference in terminology comes to the fore. Most researchers and theorists on this problem talk about the influence of central determinants (to use the term coined by Bruner & Goodman, 1947) on perception. To me the use of the term perception in this context is inappropriate. Motives, needs, moods, sets, etc., do not influence perception. Perceptual fusion and structuring are fixed and remain relatively stable regardless of transitory central states. The influence is on apperception, not perception.

Let us consider some of these studies. In general, they all follow the same format: The experimenter establishes the subjects' operant level of apperception under normal conditions and then determines how much some need, set, or mood alters this operant level. In one of the first studies of this type, Siipola (1935) demonstrated that subjects who were given the set to see words having to do with animals or travel tended to see both when ambiguous words were flashed for one-tenth of a second on a screen. In another series of early experiments, Sanford (1936) found that hungry subjects would make food-relevant words out of word stems (for instance ME would be seen as meal or meat) more often than nonhungry individuals. Leuba and Lucas (1945) found that different moods (whether anxious, happy, or critical) established in hypnotized subjects significantly influenced their responses to quasi-ambiguous pictures.

There have been a number of studies dealing with the influence of personality variables on apperception. These typically involve some verbal test, such as the Allport-Vernon used by Postman, Bruner, and McGinnies (1948); a space orientation test, as developed by Witkin et al. (1954); or some projective device, such as the Rorschach or the Thematic Apperception Test. In general, all of these studies support the view that background input—whether need, value, emotion, or any number of ephemeral personality variables—may influence the course of apperception. The influence of such central determinants continues to be investigated, and numerous studies have been published on the topic (see Bartley, 1958). I would like to reemphasize that the term perception is a misnomer in this context. The only central determinant that can influence perception is prior learning or inherited predispositions (chap. 5). Background input influences the course of apperception but has little if any effect upon either fusion or structuring.

Influence of the Autocept

One of the most important motivational systems, at least in human beings, is the autocept. The reader will recall from our previous discussion (chap. 5) that the autocept is a phenomenological unity that has a profound tendency to maintain its organization and structure.

The input representing the autocept, and that which arises when the integrity of this system is threatened, always remains in the background within the field of awareness. The inputs from other need systems fluctuate as a function of relative deprivation or surfeit, but inputs representing the autocept are perpetually available. As an individual interacts with his physical and social environment, many different circumstances may impose varying degrees of stress upon the integrity of the autocept. Whenever this happens, volitional activity may be the immediate result.

Since the autocept constitutes a functional summarization of the experiences that have been most immediate and recurrent, such volitional behavior is optimally designed to support the individual's welfare. It is true that a person may, on occasion, behave in a manner that seems contrary to his or her own best interest, but when this takes place, it is likely an expression of other activation systems that have momentarily achieved priority.

Much of the behavior that an individual would construe as volitional is probably instigated by systems other than the autocept. The inputs characterizing the autocept are always in the background, so when the source of the motivation is obscure or amorphous, the individual may simply state, "I did it because I wanted to," mistakenly ascribing the autocept as the source of the motivation.

Although the activity arising from the autocept is termed volitional, the notion of freewill is not applicable. The resulting behavior is an outcome of the structure of the autocept, but this structure represents the fusion of the most critical and recurring experiences in an individual's life. Thus an individual can behave the way he or she wants, but what is wanted is invariably an outcome of the kinds of experiences that have been encountered and fused into the phenomenological unity of the autocept. The autocept enters into behavior as does any other activation system, and the behavior resulting from it is no less lawful. In chapter 8, I give detailed consideration of the manner in which the autocept influences behavior, whether ordinary or bizarre.

The Light-Shift Model

A relatively simple model may help clarify the manner in which background input affects the ongoing process of apperception. Let us picture an ordinary room filled with many kinds of objects, each with particular reflection characteristics. Let us also assume that we can flood this room with light—either ambient white light from a constant source or light of any color that can be beamed from a projector. If the room is flooded only with white light (analogous to input from the autocept), items in the room will stand out to the extent that they have been of importance in the fusion and structuring of the autocept. Yet if we flood the room with red light as well (analogous to an input such as thirst), items having reflection characteristics peculiar to this particular wavelength will stand out while others will fade into the background. Each time the color of the light is changed (i.e., each time activation input is shifted), different classes of objects (specifically, those that have been associated with the particular activating input in the past) will become dominant, and others will be deemphasized.

This model may also help clarify the relative influence of the autocept and other need systems. That is, the more intense the white light, the less will be the influence of any given color that may be introduced. Thus, the apperception and consequent behavior of an individual with a highly integrated autocept should be less dominated by input from one of the other need systems.

MENTATION

Mentation refers to the sequential apperception of encodes, sensations, or both. It may involve goal-centered, specifically directed intentions, the interplay between sensations and encodes that occurs in casual thinking, or the free wandering sequencing common to daydreaming and night dreams. In all cases the sensory inputs and memory encodes may vary from simple pictorial distinctness to amorphous blobs representing associative webs of complex meanings and affective overtones.

The shadings and nuances of mentation have been dissected and classified in many treatments and subsumed by many terms, both in common usage and in psychological evaluations. I consider a few of the most common categories, which, though overlapping and fraught with ambiguity, may provide some indication of the subtlety of the information-processing machinery involved and the adaptive contribution conferred.

Thinking

In chapter 3, internal locomotion (i.e., thinking) emerged from our discussion as the highest and most functional rung in the evolutionary ladder of increasing adaptive facility. By locomoting within the harmless surrogate world of encodes, problems vital to survival can potentially be resolved without exposing the thinker to the hazards of external reality. During such locomotion all of the individuals' past experience with the same or similar problems is rapidly scanned so that appropriate behavior can take place. Projections into the future can also be made; these are readouts of the past that add efficiency and refinement to foresight and planning.

It was previously noted (chap. 4) that some kind of similarity must exist between the initial sensory input and the resulting encode; otherwise, the internalized symbolic world could not be functionally representational. This similarity allows for accurate and precise interplay between memory and sensation, yet it must be sufficiently different to keep these two categories of experience from being confused.

Although initial input may consist of discrete components, such sensory bits (to the extent that they are recurrently and synchronously presented) become fused into phenomenological wholes. As more and more of these percepts are imposed, a fused encode or concept emerges. When language becomes available, such fused entities (i.e., percepts and concepts) become associated with particular symbols (as discussed in chap. 8), both written and spoken. It is relative to all such components (percepts, concepts, and symbol encodes) that internal locomotion, or thinking, takes place. Thus, a person may think of a particular cow, the fused encode constituting the concept cow or the word cow.

However the expansion and inclusion does not stop here. Every concept also carries with it an affective burden: multiple nuances of emotional significance that accrue during the give and take of the individual's daily life. Thus each encode (concept) is couched in a complex of associations and counter associations. The emotional loadings and associational interrelationships may vary from the neutrality of an incidental and isolated encounter to a residual replete with affective significance within an intricate associational web. For me, the recall of the face of the person who checked out my groceries at the store yesterday is essentially neutral. The encode simply arises in the context of other thoughts. Yet the encode of my mother's face is fraught with emotional import and the accumulated burden of a multitude of encounters under hundreds of different circumstances. All of these coalesce into a vast internet of associations, an intricate blob haloed with affective overtones.

Many such encode constellations or blobs are unique composites, particular to the experience of each individual. Others are common to many individuals and are fundamental in communication; they include shared meanings and common emotional reactions. The encode of my mother's face is of particular significance only to me and my siblings, while the Stars and Stripes instigates a response common to most Americans. So it is with many encode symbols such as the Star of David, the Cross of Christ, the Rising Sun of Japan, and many rituals, practices, and precepts. Such collective representations are essential to the welding of many individuals into group thought and action.

Thinking may thus be defined as the fluctuation of apperception relative to available encodes and, as such, it is a completely automatic and lawful process. Thought tends to follow equilibrium paths representing past associations, but unless there is an enduring background input or some continuing problem, any given associational sequence rarely lasts for very long. New sensory input typically intrudes to disrupt the ongoing sequence and to initiate another encode chain. In certain cases, the focalization of particular encodes may introduce sufficient activation input, such as fear or guilt, to disrupt the series.

This brief account only hints at the complexity and facility of the thought process. It is perpetually shunting back and forth between sensory inputs and encode chains. It may be influenced by a transitory displeasure or coerced by a pervasive background input. It may be fleetingly affected by momentary threat to the integrity of the autocept, or it may be dominated by the high levels of anxiety and guilt that are derived from profound autocept disruption. As a homeostatic negative-feedback process, thinking is continually in the service of maintaining the welfare of the individual. Considering the perpetual cycling of the many critical balance systems of the body and the problems imposed by an ever-changing environment, this is no small service.

Imagining

It is remarkable how often language accurately reflects the nature and function of a process. This is particularly true of the type of internal locomotion that we term imagining or daydreaming. When we are imagining, we are literally indulging in a flight of images, representing various and sundry kinds of experiences that have been encoded in the past. Such imaging can be unusually flexible, original, and creative. Encodes of varying clarity and inclusiveness may become juxtaposed, separated, or interrelated in novel and unusual ways—ways that may be very different from the manner or order of encode establishment. I hope the reader will be tolerant as I try to produce an example of such free-wandering image construction:

> He was leaning back, legs propped up, relaxed in his overstuffed chair, staring at the fireplace when he first noticed it. It seemed to be a part of the moving flames, but it was somehow cold. As it grew, he felt that it was becoming a part of him. He shifted position, took a sip of coffee and looked deep into the flames. Now it was definitely inside him, a growing iciness, as if from some arctic place a draft had opened up, cold and

> growing colder. He looked about the room, drew momentary warmth from the familiarity of books in the bookcase and pictures on the wall and the greenish rug that, even as he gazed, seemed suffused by whitening frost. Then, as a flower might bloom rapidly into sight, the cold within him bloomed until the last thing that he could see above the burgeoning ice was the red flicker of the fire.
>
> Slowly, he became aware that he was somewhere on an endless plain where gleamed in utmost brightness myriad brittle shapes of many different sizes, each shining with an inner light. In the distance, a long low mesa, cut with craggy canyons, crowned with translucent topaz, defined the horizon. He was lost there on that endless plain, caught in the clutch of shapes that, even as he watched, began to move according to some implicit rhythm. Then, appearing over the edge of the topaz rim, a glimmer of red grew into a crimson tide, flowing over the cap rock, moving as blood might move, congealing slowly, oozing around the brittle shapes that by now had achieved a rhythmic frenzy. A sound came then. Beginning as a high shriek, it moved down the scale until it whimpered away into guttural bubbling. He knew that he had made this sound, and he also knew that it was his final claim to being.

Hopefully, this example is bizarre and original. It was an effort to see how much novelty in imagery might be achieved during a free-wandering example of image production. Actually, the term free-wandering is not applicable. In spite of my efforts to depict a novel circumstance, powerful constraints from past learning and past modes of expression contaminate the description. However some of the images are combined in ways that are completely different from any experience that I have had. This novelty potential is without a doubt possible in all imaginings, creative or otherwise. The total reservoir of encodes is apparently available for such exercises, and these may be combined in unusual or bizarre patterns. Yet there is no such thing as a random combination. Rather, the images that are used are directly representational of experiences that the individual has had. These images may be broken into bits and pieces and combined in novel ways, or the sequence may be scrambled or reversed. However even when this happens, thought is still lawful.

Whatever their motivation, the sequence and content of image constructions follow lawful equilibratory paths. An effort can be made, of course, to totally scramble images and words, and I have tried to do this on a number of occasions. Yet such an exercise is carried out only with great effort, and even then constraints from past modes of cognitive activity inevitably enter in. One must recognize the constraint in order to counteract it, and this very recognition becomes hard evidence for the constraint. Of course the rules of language implicit in this writing place further constraints on such expression.

Dreaming

Efforts to understand the meaning of dreams have had a long history: as spirits that travel forth on their own while the person sleeps, as a basis for prognostication for the ancient Greeks, and, with Freud, as an avenue into the workings of the unconscious mind and as a key to understanding our central selves. I also view dreams as a Rosetta Stone, not for the unconscious mind but for the conscious. In my view, dreams are natural expressions of mentation that occur when certain critical contributors to ordinary experience are removed. By examining dreams we can infer the function of such components during the waking state. Indeed, dreams provide a reversing lens through which some of the most fundamental aspects of human nature become exposed and clarified.

I consider dreams at length and in some detail, not only because they are universal in human experience and important to understand as such, but also because they shed light on the nature and function of ordinary waking mentation and, as we will see, upon the perennial issues of volition, freedom, and the self.

The following treatment of dreams has four sources: my own reflections on the topic, Freud's pioneer treatment, research data and dreams provided by my colleagues and students, and the writings and data from many others, both professional and laymen. In my treatment, I often make reference to Freud's theory pointing out areas of commonality, as well as disagreements. I also provide data on many dreams obtained from 150 of my students in introductory psychology classes, and will consider particular dreams, of my own and others, where they may help clarify or demonstrate my own position.

At the outset I believe it will be helpful to emphasize a major divergence of my own treatment from that of Freud. Where he perceived the often puzzling, chaotic, and bizarre aspects of dreams as due to the filtering action of censorship, I view the same characteristics as a predictable and natural outcome of certain fundamental conditions that are present in the waking state but either removed or reduced during sleep.

There are a number of such conditions, but certainly the one that is most obvious and theoretically important has to do with the observation that during sleep there is a raising of both the sensory and the motor thresholds. Since one of the adaptive functions of sleep is to provide a time for rest and restoration (as well as a state of minimum energy loss during darkness when activity would be both more hazardous and less profitable) such sensory isolation and motor quiescence seem predictable. Yet the implications of such isolation are often not appreciated. The raised sensory threshold either blocks or reduces a large number and variety of inputs, not only those from the obvious sense modalities, such as vision, audition, and olfaction, but also those derived from biological imbalances such as hunger and thirst, those representing emotional states, such as fear and anger, and most importantly, those that typically accompany movement and are central to the nature and function of the autocept.

In a prior discussion of the autocept (chap. 5), I suggested that it is fundamental in our experience of freedom and volition, and that the actual experience of volition and freedom is derived from the sensory feedback that arises from movement and intended movement. It was also suggested that the autocept is a critical motivational system that makes a vital adaptive contribution; that is, since it is a fused composite of those inputs that are most enduring, recurrent and immediate, resulting behavior is crucial to the welfare of the individual. When any circumstance threatens the integrity of the autocept, anger, fear, or jealousy, etc. (depending on the situation) is immediately introduced.

The question immediately becomes: How can we ascertain the role that the autocept actually plays? In my view an analysis of dreams provides this information. During dreams, because of the raised sensory and motor thresholds, inputs central to the autocept are essentially eliminated, and the role the autocept plays during the waking state is reduced to a minimum.

If we wish to determine the function of any aspect of a system, remove it and see how the operation of the system changes. The raising of the sensory and motor thresholds does exactly this. In dreams both volition and the experience of freedom are missing. The dream unfolds automatically as though one were watching a movie or television. This characteristic of dreams provides evidence that the experiences of freedom and volition are derived from the sensory feedback that is available during the waking state; that is, the input from movement and intended movement is blocked out because of the raised thresholds which occur during sleep (see chap. 5 for an extended discussion of this point).

It should be emphasized that a central aspect of the autocept still remains in the dream, namely the memory constellations that provide the individual's identity. The enduring presence of these encode constellations insures that the individual is still present in the dream. Only the volitional aspect and the experience of freedom are cut out. The dreamer knows who he or she is but cannot do anything about it. Dreamers are observers of what is happening, but they cannot make anything happen.

During sleep the individual is essentially isolated from inputs both from sensory systems and motor activities. Because of the raised sensory threshold, a more intense stimulus than ordinary is required for the apperception of such input to occur, and because of the raised motor threshold, a more intense than ordinary apperceived input, whether sensory or memory, is required for overt behavior to be initiated. The individual is essentially encapsulated within a context of sensory isolation and motor paralysis.

Mentation (i.e., dreams) during sleep is thus largely (but certainly not entirely) freed from the coercing influences present in the waking state so that encodes can follow one another or coalesce in bizarre or seemingly insignificant ways. This accounts for, I believe, the common finding that dream images often appear as composites in which seemingly incongruous items are overlapped or fused to provide a surprising and seemingly meaningless result. Once, when I was first writing this, I had the following dream: I was in a large bed snuggled up to President Reagan's back, who was somehow not President Reagan but my father. I looked down toward the foot of the bed as if to confirm the identity of the person and instead of seeing the legs and torso of a human being, saw freshly butchered sides of beef. Interpretation: Just before going to sleep I had seen President Reagan on television. The snuggling episode was a direct duplicate of an actual situation in which I, as a child, snuggled against my father's back when he was chilling with fever just before he died. The sides of beef? My father was a butcher and one of my clearest memories of him had to do with the sides of beef that would result when he sawed down the backbone. In this dream, experiences from the distant and immediate past are combined or juxtaposed. The principle of association still prevails, but the associations are not constrained by the coercion of the sensory input ordinarily present during the waking state, that is, the traffic that typically fills our day. Dreams are associational, but with the sensory and motor constraints removed the principles of similarity, contrast, and even opposites tend to predominate.

However dreams are not totally disordered as is seemingly the case in the neologisms of psychosis where both images and language may become scrambled and discordant. Dreams retain the logic implicit in those activities that have become automatic and unconscious as a function of repetition and the fusion, constancy, and structuring processes responsible for perceptual organization. The more habitual and automatic mental activity has become during the waking state, the more it will assume some semblance of order in the dream. Thus, in dreams of walking or talking those aspects that are automatic in the waking state are carried over into the dream: We may walk in strange places where things are aligned or juxtaposed in unusual ways but the process of walking itself proceeds normally. Or in dreams we may talk about unusual topics in bizarre ways but the rules of syntax and the order and sequence of words proceed much as they do in the walking state: Where encodes have become associated with repeated usage corresponding dream images tend to follow one another in coherent fashion.

Thus, there is an underlying logic in dreams derived from habitual patterns of thought and expression, and a basic organization derived from the principles of perception discussed in chapter 5. By and large, however, because of the raised sensory and motor thresholds, dream content and progression are remarkably varied and haphazard.

Freud ascribed the unusual nature of dream content and progression to:

> the free transference of intensities, and in the service of condensation intermediary ideas are formed.... The ideas which transfer their intensities to one another are very loosely connected, and are joined together by such forms of association as are disdained by our serious thinking.... In particular, assonances and punning associations are treated as equal in value to any other association.... Contradictory thoughts do not try to eliminate one another but continue side by side and often combine to form condensation products.... (Freud, 1900/1980, p. 450)

Freud saw condensation as resulting from the work of the censor upon a repressed wish dating from the infantile period of the individual's life, while I view it as a natural characteristic of mentation after the associational constraints, which are present during the waking state, have been removed.

In passing, it should be emphasized that, although Freud is a master at using association as his primary method of dream interpretation, he disregards it as a principle in dream formation. Whether this is an oversight or simply considered too obvious for him to mention is unclear. Yet the principles (i.e., of association) that I consider of paramount importance in dream formation as well as interpretation are of secondary importance to Freud. According to Freud, dream formation (i.e., the translation of the latent into the manifest content) involves three processes, condensation, displacement, and secondary elaboration, but these processes are outcomes of the filtering action of the censor. The censor is given responsibility for the fact that the associations in dreams are superficial, extralogical, and bizarre. "The current explanation for the predominancy of the superficial associations is the pressure of the censorship, and not the suppression of the directing ideas. Whenever the censorship renders the normal connective paths impassable, the superficial associations will replace the deeper ones in the representation" (Freud, 1900/1980, p. 387).

We come now to another critical difference between waking and sleeping mentation: The fact that dream images are universally reported as much more distinct and vivid than the images involved in thinking or remembering. Indeed this distinctness and vividness is so pronounced that many dreams have the clarity of situations experienced while awake. It is as if—and Freud defended this position—the part of the brain responsible for direct perception is reactivated by the encode. This reactivation, which Freud called regression (or retrogression), involves a curious reversal of process. During the waking state, our perceptual world is activated by the various receptor systems, but during sleep the activation (if Freud is correct) comes from processes already in the brain; the encode can activate the perception, rather than the perception being activated by the sensory system.

Why are dream images so vivid? If a dream sequence is nothing more than the sequential apperception of encodes, why are the images of the dream so much clearer and more distinct than the encodes that are typical in ordinary thought? Freud's retrogression explanation may seem fanciful, but there is some evidence, indirect though it may be, which gives support to his notion. In experiments in which human subjects have been deprived of stimulation for an hour or more, a dramatic increase in the pictorial clarity of encodes is often reported. Since sleep imposes a state of stimulus deprivation (because of the raised sensory threshold), the same kind of pictorial clarity might be expected. Yet why is the image clarity greater during stimulus deprivation? This question is very difficult to answer, but it may relate to lack of contrast. Since sensory input is missing, the apperceptual flux between such input and the fainter encodes can no longer occur, and the encode thus may appear clearer and more pictorial than it would under ordinary comparison circumstances.

Another possibility is that dreams often involve encodes established in childhood that are embellished and structured with repeated usage. During childhood, eidetic imagery with remarkable pictorial clarity is prevalent. Such images may become tied to activation inputs such as anxiety, guilt, fear, or hunger. Then if one of these inputs intrudes across the raised sensory threshold, the associated, highly pictorial images would tend to appear in the child's dream. Each time this happened, such memory constellations would become more sharply delineated and more closely associated with the activation input in question.

The following description of a recurring dream by a 64-year-old college professor represents such a constellation and also provides some insight into its origin.

> My first experience with the river occurred when I was six or seven years old. The water was only inches from flooding over the top. Standing near the edge of the bank it seemed that one was almost in the water itself. To me the whole scene was depressing, dreadful, menacing, especially under the dark overcast sky. In occasional dreams years later I am standing near or approaching the river with great fear and apprehension; nothing

> seems to be happening except the onrush of the grey-green water as it gets closer and closer to me. I always wake up to escape the river.

Another factor, which increasingly occurs as an individual grows older, probably contributes, though in a negative fashion, to the vividness of dreams. The pictorial clarity prevalent in childhood gradually fades when the individual reaches adulthood, as many experiences are funneled into abstract encodes in which specifics are subsumed by global concepts. Very likely, such concepts rarely appear in dreams. Indeed, since they are typically so abstract, there is little sufficient distinctness to be represented in a dream. Even if they did appear, there is little sufficient distinctness to occasion their recall when the person is awake. What is there to be remembered about a blob? Thus dreams may consist of concrete, pictorial imagery because the encodes persisting from childhood are of this nature and because much adult experience is subsumed within global concepts where adequate specificity for dream representation or for later recall is typically not present.

It should be noted that dreams are primarily visual, something that upon examination is unremarkable; in the waking state the visual, at least in human beings, also predominates. Inputs from other modalities and from the activation and affective systems of the body may, and often do, intrude, but the visual constitutes the major stream of waking sensory input. Only when we close our eyes or experience total darkness is the visual subordinated. Yet we cannot close our eyes in dreams. The dream, as it occurs, is all there is. The priority of the visual is probably the case in most mammals, depending, of course, on the evolutionary history of each species. If bats can dream, they probably dream in clicks.

The visual predominance of dreams may impose a curious kind of representation where concepts are involved. A kind of modal or additive fusion of the many items subsumed by a concept appears to be involved and may account for the high percentage of dreamers who experience unfamiliar scenes. Eighty-one percent of my student research group had dreamed about unfamiliar people, 63% about unfamiliar buildings, 41% about unfamiliar cities, and 52% about unfamiliar roads.

This does not mean that actual scenes, particularly if they are vivid and unique, cannot appear in the dreams of adults. They often do. However they enter as momentary additions to the action of the dream, sometimes in a logical way, but often incongruently. For instance, I saw a review of a Disney film, *Eo*, in which one of the protagonists was presented as a most striking and unique female. That night her face, or at least a reasonable facsimile of it, appeared in a dream that I often have, in which I have lost something and cannot find it. The face appeared for a moment, and then it was gone, being, as far as I could later infer, completely unrelated to other aspects of the dream.

According to my view, the pictorial nature of dreams, whatever its source, is emphasized when abstract concepts, ideas, problems, etc., are translated into dream content. In my student research group, there was a high incidence of dreams (reported by 74%) in which complex problems were represented by efforts to open locks, find one's way through a strange building, or magical equations. Some years ago I had a dream that I was trying to solve the ultimate problem, the solution to everything. I kept trying, to no avail, until the solution suddenly came to me. It was so satisfying and all encompassing that I, somehow realizing that I was dreaming, resolved to remember it so that I could apply it during my waking life. The solution to everything? $X = Y^4$. Absolute nonsense.

It is possible that unfamiliar items, that is, people, places, buildings, roads, caves, etc. may involve innate representations, as implied by Jung's archetypes and Freud's objects of the primary process. More likely, and certainly more parsimoniously, the unfamiliar image in the dream may be the visual expression of an abstract concept. It is also possible that an unfamiliar image may be caused by an overlapping of normally sequential associations, where items, which might be familiar and unremarkable if presented sequentially, achieve novelty and strangeness when simultaneously represented. Recently I had a dream about being bitten by a strange looking pig, and I felt pain in my hand. The pig had a sharp, pointed snout, white coarse hair, and short legs. When I woke up I pondered this strange pig dream. Slowly it came to me. The previous evening a half-grown opossum had appeared on our deck, and at that time I thought that it looked like a pig, and a rat. The bite and the pain? I had been bitten by a rat while running an experiment some 20 years ago, and the real rat bite and the dream pig bite hurt in the same way. Incidentally, people rarely dream a pain. Although 35% of the research group had dreamed of being shot or stabbed, 75% of these reported that they felt no pain. This is probably because they had never actually been shot or stabbed, or perhaps because pains have very little, other than locus and intensity, to distinguish them. The pain experienced by the 25% who reported it was probably displaced (Freud's term) from some other situation (as in my strange pig dream) or were actually feeling a pain of sufficient intensity to intrude across the raised sensory threshold, which indeed may have activated the dream of being shot or stabbed in the first place.

Yet another difference between mentation in the waking state and mentation during sleep has to do with present, past, and future. There is no past or future in a dream. The dream is encapsulated in the now; there is no planning for or anticipation of what's to come, no remembrance of things past.

Dreams follow associational paths, but since the sensory content that drives mentation during the waking state is blocked by the raised thresholds, dream progression becomes liberated, allowing seemingly trivial associations to predominate. At this point I should reemphasize major differences between my own and Freud's position. He viewed the seemingly haphazard progression of dreams as due to the actions of the censor, which I construe to be an explanatory homunculus, an explanation that goes nowhere. Freud occasionally subsumed the unusual mix that often occurs in dream images under the term condensation. A superficial examination of his theory would suggest that condensation happens automatically because of the imperious demands of the censor, as the latent content is transformed into the manifest content. Yet a more careful reading indicates that a number of coalescing associations may sometimes be responsible. Take his interpretation of his dream on the botanical monograph. He suggested that dream images result from the coalescence of a number of seemingly oblique associations. "Whatever dream I may subject to such a dissection, I always find the same fundamental principle confirmed—that the dream-elements have been formed out of the whole mass of the dream thoughts, and that every one of them appears in relation to the dream thoughts, to have a multiple determination" (Freud, 1980, p. 202). Yet Freud only talked this way occasionally and then only when interpretation (not dream formation) was involved.

Freud almost always explained the puzzling nature of dreams as an outcome of the action of the censor. Indeed it is the censor that accounts for the divergence between the latent and the manifest contents of dream. In my view, Freud did his dream theory a disservice with the censor concept. It is simply his name for an assumed process, a name that he then used as a tool of explanation without analyzing the nature of the process. This is perhaps Freud's greatest theoretical fault, the naming of a hypothetical process and then using that name in the service of explanation.

There is little doubt that both dream content and progress are heavily influenced by past learning, even though many associational constraints are removed during sleep. Dream images and situations may seem juxtaposed and bizarre, and dream progress weirdly fluctuating, but the images and situations are never completely novel. They always involve a mix of past experiences. People rarely dream of items with a very unusual color, such as green dogs or purple cats, simply because they have seldom experienced them. People blind or deaf from birth do not have visual or auditory images, respectively, in their dreams. Likewise people rarely have dreams in which the rules of progression derived from our daily encounters are broken. In dreams people rarely walk backwards or fall upwards. Cars run on roads and airplanes fly, and when they move it is typically in a forward direction.

Images may be fused, scrambled, and juxtaposed, they may involve incongruous combinations but always in spite of their strangeness or novelty, dream images typically reflect past experiences.

The influence of past learning is particularly obvious where (as previously mentioned) perceptual organization and structuring are involved. In dreams we almost never (1.5% of the research group reported two-dimensional dreams) interact within a two dimensional mosaic; rather, we inhabit a full-blown perceptual world filled with phenomena at varying distances from one another.

In summary, although mentation during sleep is remarkably liberated from the constraints of the waking state, a considerable amount of ordering and structuring remains. Such ordering and structuring, though based on innate capacities, reflect the influence of past learning, and as such account for the fact that some dreams, particularly if recalled immediately upon wakening, appear coherent and logical. Freud recognized that many dreams have an implicit order and structure that keeps them from being incoherent and illogical, and he coined the concept secondary elaboration to represent such organization. Freud, of course, believed that the jumbled and incoherent content of many dreams was related to the action of the censor, which worked upon the latent content to produce the bizarre effects. While I explain such bizarre contents as the predictable outcome of the flow of associations resulting when the constraints, described above, are removed. I should mention that Freud gave no explanation of the source of secondary elaboration, except to hint that innate processes are involved.

According to the associational theory of dreams, which I envision, a dream may be initiated and sustained by an activation input that is sufficiently intense to intrude across the raised sensory threshold. When this happens memory constellations that have been associated with this activation input in the past will tend to appear in the dream. I, for instance, often dream of tornadoes and, on at least three occasions, have dreamed of losing one of my children in a fire. Such dreams are common. Seventy-one percent of my research sample have dreamed of the death of a loved one.

Why do such terrifying sequences appear during sleep? Such dreams are not wish fulfillments as Freud would have us think. In my view, this occurs because the dreamer has gone to sleep while experiencing a high level of anxiety or fear, possibly from specific events, or perhaps from some combination of potential problems. If such anxiety or fear is sufficiently intense it may cross the raised sensory threshold. When this happens the dreamer will tend to focalize encodes that have been associated with anxiety or fear in his or her past. I came from an area of Oklahoma where tornadoes are frequent and where I was instilled with a high level of anxiety about them. Thus, when I am extremely anxious I tend to dream about tornadoes because tornadoes have been associated with anxiety in my past. If one goes to sleep with a level of fear high enough to cross the raised sensory threshold, one will dream about fearful situations, either from the past or, perhaps, from conjecturing about possible hazards for one's family or friends. One of the most frequent fear dreams reported by my students is of falling, 81%. An association theory of dreams would certainly predict this. Being hurt in a fall or being cautioned about the dangers of heights is common in the experience of everyone. Other common fear dreams in descending order of frequency are car wrecks, 40%; fires, 26%; tornadoes, 15%; and swirling water, 13%.

Emotion can enter the dream from two often overlapping sources. If an individual goes to sleep with a high level of fear, guilt, or anger, etc., such input may cross the elevated sensory threshold, and then triggers associated encode constellations. When these encodes appear in the dream they may arouse further emotion that becomes added to the emotional content already present. If this happens a positive feedback spiral may be established with each increment in emotion instigating even more prepotent encode constellations. Such reciprocal escalation may continue until terror is experienced in the dream. Positive feedback spirals, I believe, are typically present in nightmares and sleep-disrupting panic, from which the individual awakens to a less threatening world. Of course, panic dreams by their very nature are intense and dramatic—conditions that would tend to fix them indelibly in memory and give them priority for recall.

Another condition may be responsible for the emotional content of certain dreams. One of the more common features of dreams is the coalescing, fusion, and overlapping of images, typically in the visual mode. When each coalescing image triggers a minimal (i.e., subliminal) emotional input, it may combine with others to achieve sufficient intensity to cross the raised sensory threshold. Thus, an image that combines snakes, high places, rats, spiders, and failure, none of which would trigger an intense emotion by itself, appear together, the cumulative affect on the visceral changes accounting for the experience of emotion could be considerable.

However many, if not most, dreams are curiously devoid of affect; they proceed with little emotional content, essentially neutral. An individual, in remembering a dream, may be puzzled by the incongruity between the events that happened and his or her reaction to them. Situations that during the waking state would arouse dramatic emotional repercussions are strangely bland, as though one can have an experience without feeling it. In some dreams, particularly those involving social interactions, there seems to be the conditions appropriate for emotional reaction but nothing happens, a feeling that emotional reactions should be there, but are not, as may happen in dreams of embarrassment, achievement, or humiliation. On a number of occasions I have had the dream that I have gone to a class room to give a lecture and no one showed up. I simply waited there hopefully staring at the door. I am humiliated and concerned in these dreams but upon considering them in retrospect there is always the feeling that the emotional reaction was derived from an inference that I should have been humiliated rather than that I actually was.

This discussion suggests that there may be two kinds of involvement in all emotional reactions: Those that are primarily cortical and those that include the inputs from visceral repercussions. In dreams, because of the raised sensory threshold that occurs during sleep, the inputs from visceral upheaval are essentially excluded. Such exclusion may leave dream events emotionally neutral. The emotional reaction that does occur may be derived from the inference that we should be having emotional reactions rather than that we actually are; which suggests that they may be primarily cortical, particularly where social situations are involved.

However this may be, the events in many dreams transpire with little emotional content, a curious disparity between what happens and one's reaction to it. This may be due to the raised sensory threshold, the purely cognitive nature of much dream content or some combination of the two. It may also be influenced by the difficulty in recalling emotional reactions that are usually less discrete and pictorial than ordinary dream content. Indeed, emotion experiences, per se, are hard to remember even when they occur during the waking state. Yet we can remember that they did happen and that we did feel them, and we can also remember that they did not happen in our dreams.

I do not want to imply that emotions do not occur in dreams. They do. Sometimes intensely. In nightmares, horrible situations may arise with such intensity that the visceral repercussions will flood across the raised sensory threshold so that the individual may waken and lie there taking inventory of his or her heavy breathing, sweating body, and violent heartbeat, grateful that it was just a dream.

This discussion may have implications for one of the classic controversies arising from our efforts to understand emotion—the James-Lange versus Cannon-Bard theories of emotion. It suggests that certain emotions, those that instigate overt action such as fear and anger, may depend heavily upon inputs from visceral changes as stated in the James-Lange theory while others, those derived from social situations, such as humiliation, ego enhancement, etc. may be primarily cortical, as implied by the Cannon-Bard theory.

However if many dreams are devoid of affective content, as previously suggested, the question becomes, what drives them as they unravel in their multifarious ways? There is, I believe, an intrinsic perseverative tendency in mentation, which, once underway, keeps it going. Thus a dream may proceed, not because affect is continually intruding to energize it, but because associative paths of least resistance provide the basis for such mentation. During the waking state there is a tendency in everyone to achieve closure in mentation, whether in language or thought. Why should not the same tendency be present in dreams?

An associational logic also holds for dreams that involve apparently repugnant activity, such as having sexual relations with a family member (reported by 21% of the research group) or perhaps having a homosexual experience. If high levels of guilt intrude across the raised sensory threshold, encodes that had been associated with guilt in the past will tend to be focalized. Thus, when a man dreams about having sexual intercourse with his mother, it may not be that this is something he actually wishes (as Freud would maintain) but that mother incest had been associated with guilt in the person's past, either through learning or as an inherited predisposition.

Yet we can never assume that a dream sequence, whatever its theme, is always tied to an obvious set of associations: A dream of eating is not necessarily the outcome of an intrusion of hunger input across the raised sensory threshold. Consider the following dream (I present it without alteration as it was written) by a young woman who had lost 40 pounds during three months of dieting prior to the dream:

> Sunday night, September 21, 1986. I went to bed feeling guilty because I ate constantly all day long.
>
> Dreamed I was going through a buffet line. Everyone in front of me was getting a salad. I got mad because when it was my turn, they ran out of salad. So I got creamed corn and mixed ketchup and mustard in it—it turned blue, but I thought it tasted good. The person sitting with me had a Swiss Steak smothered in tomato sauce (my dream was in color). I only ate some of the corn. I then went and yelled at the manager for not having enough salad.

In my view,[1] a hunger-guilt[2] combination is the activating input that crosses the raised sensory threshold to instigate and sustain the dream. The young lady in question has had an eating problem for a number of years, and overindulgence has habitually been accompa-

[1]Dream interpretation is notoriously fraught with the potential for error. I offer mine, more to demonstrate a point of view than as a definitive statement.

[2]Such fusion would be predicted according to the principles given in chapter 5, that inputs that occur simultaneously become fused into unity. Such fusion of activation inputs may have important implications for understanding various kinds of aberrant behavior, as when sex-aggression, sex-guilt, or both become conjoined.

nied by guilt. Thus guilt and eating have become firmly associated, so we would predict that she will dream about eating when guilty. One guilt-expiating behavior strongly imbedded from her past is eating salad. Yet in her dream they ran out of salad. Only then did she take the creamed corn. To reduce her guilt, she put both ketchup and mustard in it. There is apparently less guilt derived from eating something unappetizing. Then, "it turned blue," a further unappetizing overlay. Then comes guilt reduction through invidious comparison. It was not so bad that she was eating creamed corn because the person next to her was doing something far worse, eating "Swiss Steak smothered in tomato sauce." In a further guilt-expiating maneuver, she "only ate some of the corn." Yet she did eat it, and this act was accompanied by anger, which was directed at the manager. "I then went and yelled at the manager for not having enough salad." Thus the manager became both a focal point of her displaced aggression and a further source of guilt expiation, since it was his fault they ran out of salad.

The main theme of the dream was instigated by the presence of hunger-guilt when the subject went to sleep, but various aspects of the dream produced even more guilt, which further influenced the progress of the dream. Here is a clear example of the positive feedback spiraling previously mentioned, in which the emotional repercussions rising from within the dream can influence the symbols and sequence of a dream. The young woman reported that she had no feeling of volitional control over the action or content of the dream. The dream unraveled automatically, with her being present in it but having no control over it.

When I came in the next morning I found the following dream from the same young woman shoved under my door:

> I dreamed about small dark chocolate doughnuts and large grape jelly doughnuts with tons of white icing on them. I remember I ate one chocolate one and really wanted a jelly one but didn't take it.

Although this dream is brief, it demonstrates guilt reduction through the indulgence in the lesser evil and autocept enhancement from victory over the greater temptation. It also clearly demonstrates the way in which adjustment gambits can automatically influence the content and progress of a dream.

Another happening reported by my students relates to certain curious influences on motor activity, either actual or as experienced in the dream. For instance, 70% of the research group had had a dream of trying to run away from something frightening only to find that flight behavior was curiously hindered; if they moved, they did so in strange slow motion sequences (53%) or could not move at all (48%).

A dream of motor incapacity, in my view, is a reflection of the raised motor threshold. This interpretation is supported by the common observation that, although a dreamer may actually try to move certain parts of his body or make a sound, it is very hard to do so. I have a recurrent dream that intruders have broken into the house and are coming toward the bedroom. On such occasions I somehow comprehend that it may be a dream, so I do my best to get my wife to wake me up. Only with great exertion can I finally break over the raised motor threshold and make a sound, and, even when this happens, the sound rarely comes out with any degree of clarity. More likely, it is a gasp or a garbled murmuring that is sufficiently disturbing to wake my wife, who in turn wakes me so that I can make midnight notes on my dream experiences. The motor involvement that commonly occurs in dreams is, according to this view, not only an outcome of the raised motor threshold, but also at least partly due to the loss of volitional control that occurs when the influence of the autocept is blocked by the raised sensory threshold.

There is a much greater similarity between night dreams and daydreams than is commonly supposed. As will be recalled from our discussion of imagining, although complete randomness and flexibility never actually take place, there is still a remarkable freedom for image juxtaposition and recombination, which is similar to that of a night dream sequence. Although novel combinations are also commonly achieved in daydreams, the surprise aspect, which is prevalent in night dreams, is missing, probably because the autocept either enters into the direction and continuity of the daydream, or serves as continuous backdrop for its progress.

Summary and Discussion

Viewing issues from a different perspective, one that may be so familiar that its importance remains unrecognized, can sometimes provide insights into areas that have long retained status as an enduring puzzle. This, I believe, is true of dreams. An examination of dreams provides insight into freedom and volition, the influence of needs and emotions upon the ongoing thought process, and even into the relative adequacy of particular theories, such as Cannon's Thalamic theory versus the James-Lange theory of emotion. These insights are provided because during sleep, for very adaptive reasons, both the sensory and motor thresholds are raised. As a result, the processes that are derived from such inputs during the waking state are affected, allowing a basis for inferring their origin and function. Thus dreams are usually volition-free, emotionally bland, haphazard juxtapositions of images following the vagaries of incidental associations. An analysis of dreams clarifies the functions of the self in ongoing behavior and thought and suggests that the experience of both freedom and volition is derived from the sensory feedback from actual and intended movement. Dreams may be influenced by profound conflicts, pathological states, or both, but only in so far as these conditions are sufficiently prepotent to produce inputs that cross the raised thresholds.

In this consideration of dreams I have placed heavy explanatory burden on the changes that take place as a result of the raised sensory and motor thresholds that happens during sleep. To determine the degree to which others had taken the same approach, a survey of recent publications in dream theory was undertaken. Most of the studies examined involved an elaboration of psychoanalytic dream theory or an examination of the contribution of other writers, including Jung, Coleridge, Shakespeare, Charcot, Janet, Craft-Ebbing, Mesmer, and Hall. There was some effort to indicate the function of dreams: As a means for organizing data pertinent to a particular problem (Fosshage, 1997), as a source for the revitalization of self (McDougal, 1993), as an attempt to transform the unconscious into three-dimensional space and Aristotelian logic (Jimenez, 1991), as efforts to purge irrelevant information and associations (Crick & Mitcheson, 1983), as a method for guarding sleep by producing mimic gratification of impulses (Stein, 1996) and, of course, the Freudian notion of wish fulfillment received wide consideration. Hobson (1988, 1990) has posited that dreams are the outcome of the brain's effort to make sense out of random excitation: from noises, pains, need to urinate, etc., and from spontaneous activity in lower centers. Smogor (June 2001, personal communication) suggested that some dreams may be initiated by the same kind of neural activity that accounts for the twitches and jerks that occasionally occur.

My own view is that dreams per se have no particular function, a curious position perhaps for an evolutionist on something as universally represented as dreams. In my view dreams consist of the apperception of encodes that sometimes occur during sleep when the constraints and directionality imposed during the waking state are removed by the raised sensory and motor thresholds.

Because of the raised thresholds, many of the sensory inputs that influence mentation during the waking state are reduced and dreams are often haphazard, disconnected, and free-wandering, in many cases following incidental, often superficial associations, which upon examination are either nonsensical or trivial in implications. However not always. Once I had a dream that was both original and creative and, many years ago when it occurred, seemed quite amusing. Just before going to sleep I had seen a political cartoon in a newspaper. The time was during the early sixties, just after Castro had nationalized American corporation properties in Cuba. The cartoon showed the American Eagle sitting on a branch of a scrubby tree. He was some miserable looking bird, forlorn, and bereft of all feathers except for one straggler sticking askew out of his tail. The caption read: "That old chicken plucker has been at it again," referring to an event that had actually occurred in a hotel in New York when Castro and his entourage, true to the austerity dictates of the revolution, prepared and cooked chicken in their hotel rooms when they presented their case at the United Nations in 1962.

I was much amused by this cartoon and more than a little envious at its cleverness. When I went to sleep that night I had the following dream, in color. A caricature of Uncle Sam, complete with striped top hat and tuxedo, was lying on an examination table, looking extremely ill and bedraggled. A caricature doctor with stethoscope in hand was standing over him, looking supportive and reassuring. The caption read, "Nothing to worry about, just a slight case of Castro-intestinal disorder."

This dream was so funny that I laughed about it when I woke, and wrote it down for future reference. Such creativity is rare in dreams, at least in my dreams. Yet there is such a multitude of loosely associated images in dreams that, on occasion, they may be productive. As one might predict, and as is often the case when we go to bed with problems, some aspects of these problems often arise in dream activity, following superficial and tangential associations perhaps, but sometimes pertinent dream outcomes may occur. On the rare occasions when this happens, they are remembered. The rest are simply forgotten.

Although the capacity for dreams did not evolve because they made an adaptive contribution they may influence future behavior: Bits of originality may surface as in my Castro dream that might be useful; back when dreams were thought to be prognostic, people sometimes made decisions based upon them; I, myself, have had sex dreams that when recalled later, aroused me considerably. Yet it is unlikely that such positive feedback shunting between dream and waking state had anything to do with the evolutionary emergence of dreaming.

However dreams do have a function—an important one that they do not accomplish in themselves, but through what they tell us about mentation during the waking state. With the constraints imposed during the waking state removed, the influence of such constraints can be determined. Thus since the raised sensory and motor thresholds attenuate the inputs from movement and affect, the issues of freedom and volition can be better understood, and the nature of the autocept can be clarified.

Reasoning

I am often struck by the manner in which words occasionally reflect the underlying process, and this is certainly true of the word reasoning, which has to do with an elucidation of the reason for things, that is, the causes that lead to certain happenings or outcomes. Thus the questions, why did you do that?, or, why did that happen?, are immediately met with a martialing of the circumstances that led up to the particular event. That is, if B happened, reasoning involves an effort to determine the A cause or causes that produced it. The reader will immediately note that this kind of mentation is a direct reflection of the if A then B contingencies that characterize the environment, which in turn suggests the selection pressures that led to the evolutionary development of this capacity. It also, with some extrapolation, allows us to infer the basis for the hypothetico-deductive method commonly stated to be fundamental to scientific discovery and theory development; that is, the process of induction is simply the determination of the particulars that lead to a given outcome, and the process of deduction involves the determination of whether a new particular is an expression of the outcome.

Meaning

The meaning of any given input (sensory or memory) is derived from the kind and quality of components that are fused into, associated, or both with it. These loadings may vary from the essentially neutral to affects of great intensity, associational webs of remarkable inclusiveness, or both, all of which may exist as a halo to a basic core, a core that may consist of an encode of an actual encounter—say a face or a representative symbol, such as a word.

Poetry, at times, gives some indication of the communicative power of such haloes, that add richness and nuance to many aspects of language expression. Consider the following line from Coleridge:

> Twas as lonely and enchanted place as ever beneath a waning moon was haunted by woman wailing for her demon lover.

Not much of literal significance here, but the lines have vivid and haunting affective overtones, which are immediately experienced by the reader. This shared affective reverberation is central to the universal appeal of poetry, and is immediately experienced as each word is encountered. There are also literal implications that emerge as apperception fluctuates over the various words of the poem. What is it with this "demon lover"?

This example suggests two characteristics of encode loading. At times the halo that carries the meaning may become separated from the core; a given input, say a face or event, may remain momentarily strange and unfamiliar until recognition and meaning rush in. Or we may apperceive the halo and then search for the core, as when we think of a person but cannot remember his or her name. Our analysis also suggests that affect may accumulate as the loadings of various symbols are sequentially encountered. The emotional reaction to Coleridge's lines previously given increases as the affects from each symbol accumulate and fuse together. Accumulation is a common characteristic of emotional responses and probably accounts for the observation that emotional feedback and escalation is a common precursor to panic. Such accumulation and escalation are to some degree dependent upon the fact that emotional responses are diffuse and enduring, since they are derived, at least in part, from visceral activity rather than from the blips of ordinary images.

PLEASURE AND MENTATION

The hedonists have long assumed that pleasure either is or should be a basic motivation for human thought and action. For some of them behavior was assumed to be rational; that is, after careful appraisal of the repercussions of many different acts, the choice made was supposed to be the one that provided the greatest net amount of pleasure. This notion of behavior as the outcome of the subtractions and additions of pains and pleasures seems reminiscent of the homeostatic negative-feedback position in which apperception is assumed to be the outcome of many different variables. Yet the similarity is only superficial. It is my view that behavior is always equilibrium-trending, but that pleasure per se never instigates activity except secondarily, as will be described later.

Is it not pleasurable to think about food when one is hungry? Indeed, whenever any particular system is out of balance, apperception tends to be centered on items and activities that have reduced or alleviated the imbalance in the past. Human beings often appear to savor certain encodes such as those representing an excellent meal, a notable sexual experience, or, perhaps, some unusual achievement. The recall of such experiences appears to add an aspect of pleasure even as the recall is taking place. Yet does pleasure per se motivate behavior?

It seems more likely to me that the loss of pleasure when an activity is terminated provides the activating input that instigates the organism to repeat it. Thus, an individual may continue thinking about food when he is hungry, not because of the pleasure that is generated by the encode, but by the loss of pleasure that occurs when the encode is no longer being focalized. This view of the manner in which pleasure may sustain activity can best be appreciated by examining situations that are presumed to provide the ultimate in pleasure input: The sexual orgasm, and the high that results from taking certain drugs. Certain people may spend a good part of their time in pursuit of a better orgasm. If this is the case, it is the memory of pleasure having been experienced rather than actual pleasure being experienced that provides the motivation for action. Pleasure as it relates to drug addiction apparently follows a similar pattern. Jules Trop, a former cocaine addict, succinctly states the essence of his motivation: "I was chasing the memory of a high."

Recently, I asked five of my male colleagues whether it was pleasurable to have an erection. After they finally took me seriously and thought about it for a while, they all agreed that it was not. Such preparatory processes are not in themselves pleasurable, but they do provide the basis for pleasure when appropriate stimulation takes place.

Much of our seemingly self-indulgent internal locomotion is probably much more complex in terms of motivational nuances than the previous discussion would suggest. There are imbalances that are simple expressions of physiological systems. There are others that relate to autocept emergence and integrity. In the latter case, familiar and comforting daydreams, or even familiar and comforting songs, may have the function of the little child's blanket when it is clutched to its bosom as it goes to sleep at night.

BUFFER REVERBERATIONS AND MENTATIONS

When a sense modality is stimulated above threshold the resulting sensation endures for a brief period, approximately one-twentieth of a second after the stimulus is removed. This interval is apparently derived from an extremely transitory perseveration of the same process that produced the sensation in the first place. These reverberations account for positive afterimages, flicker fusion, the lower threshold for pitch, and are fundamental in the perception of movement (see chaps. 4 & 5). Negative afterimages also arise from the recruitment of the receptors (particularly in vision), and these images may last for a number of minutes. Such negative afterimages appear to have no function; indeed they probably make a negative contribution, since they may interfere with ordinary experience. Apparently they are intrinsic nonfunctional artifacts of the recruitment process.

There are two other kinds of short-term residual derived from the activation of transmitting mechanisms, which are quite different from the peripheral receptor effects just mentioned. First, as has been demonstrated many times, beginning with the work of Wundt (1886) and finding its most precise and impressive measurement with the experiments of Sperling (1960), there is a brief experiential residual that lasts for approximately 750 ms each time a sensory input is imposed. As argued at length in previous sections, this input constitutes the moving belt of our reality, the specious present James talked about, and the field of awareness, as I use this term. It is preapperceptual. It is a multiparallel system wherein all sensations are simultaneously and irresistibly displayed.

The second process (which lasts approximately 15 seconds) has been given such labels as primary memory (Waugh & Norman, 1965) and short-term retention (Hebb, 1949; Hellyer, 1962). Certain theorists (Broadbent, 1957; Waugh & Norman, 1965) were convinced that the process is different from that accounting for ordinary memory, while others (Melton, 1963) argued that there is but one memory system, and that distinctions between short-term and long-term encoding mechanisms are completely arbitrary. I am in agreement with the two-mechanism theorists, but prefer the now commonly used term short-term memory to represent this more extended example of buffer delay.

It has been commonly recognized that short-term memory has a buffer function. Yet short-term memory fulfills at least three analytically distinguishable (although the operations involved overlap and interact in complex ways) information-processing functions. First, and most obvious, buffers keep inputs (sensations, memories, or both) in abeyance until the single channel of apperception can be cleared to receive them. Second, buffers allow activities, particularly sequential activities such as speech, to be monitored so that continuity can be achieved and corrections can be made. Third, overlapping this latter contribution, buffers enter into speech in a more complex manner. Before meaningful speech can occur, the pertinent encodes, affects and symbols must be scanned. Without such scanning, speech would not only be meaningless, but literally could not take place.

For example, if we were asked to talk about *Billibid*, that term would have little meaning for most people, and they could have very little to say about it. However for me the term is pregnant with meaning, since it is a prison in Manila where I spent time as a prisoner of war. For me, the term keys into an intricate web of associations with deeply embedded affective overtones. Language requires that encodes contained within this web be extricated and arranged sequentially. Here the associations representing my experiences with Billibid, and the rules of logic implicit in language, come into play. When I begin to talk about Billibid several processing stages ensue. First comes the set to respond with speech. Second, the stimulus word is apperceived.

Third, the total affective-associational blob that Billibid constitutes for me is apperceived. Fourth, apperceptual scanning of this material and the focalization of particular encodes occur, followed by the emission of sounds (as discussed at length in chap. 9). It might be concluded from this analysis that these stages are separate and discrete. This is certainly not the case. Apperception is in rapid and continuing flux over inputs and processes as they occur, and both inputs and processes typically result in affective reverberations that at any time may alter what is happening.

SUB-AWARENESS INFORMATION PROCESSING

Is information processing without awareness possible? Yes, it is. The activities under the control of the autonomic nervous system continue to occur, even when the person is asleep or unconscious, ensuring that a tolerable balance is maintained in a number of vital systems. The defensive, metabolic, and healing systems of the body proceed without the intrusion of awareness, and in certain reflexes, immediate and appropriate response to particular stimuli is made with awareness entering in, if at all, after the fact. Almost paradoxically, it would seem, the ordering and structuring of the field of awareness, which plays such a vital role in information processing, may take place without awareness of what is happening. Thus, experience, particularly in the visual modality, becomes ordered and deployed so that we perceive phenomena within the context of three-dimensional space. There is little if any awareness of the processing that results in such ordering and structuring. As Helmholtz maintained (see chap. 5), it is an unconscious inference that is automatically and irresistibly imposed.

In the sequence of events that culminates in consciousness (awareness, experience, etc.), there are typically a number of intervening factors (contributors, accentuators, etc.) that help ensure that vital information about the external environment, internal processes, or both, are registered. Thus much information processing takes place before consciousness is finally achieved. In hearing, for instance, the external ear accentuates pressure fronts, bones of the middle ear give emphasis and attenuation, the physical characteristics of pressure fronts, frequency, amplitude, and complexity, become automatically represented as pitch, loudness, and timbre; and these specific qualities (see Mullers' doctrine of specific nerve energies) are automatically cast upon the input. So it is with all information that is fed into the field of awareness, whether it represents changes in the external environment or shifts in the biological systems of the body; the categories of consciousness are determined by unconscious processes before the culminating event of consciousness occurs. Also, as Helmholtz insisted, the organizing principles of perception are automatically and unconsciously imposed.

Sub-awareness (unconscious) information processing is also implicit in the dynamic interactions that are the direct precursors of apperception. The priority resolutions that underlie the lawful course of apperceptual flux are below the level of awareness and determine that components within the field of awareness will be focalized.

Thus, in the categorization of the input from the various modalities, their perceptual structuring and their focalization, automatic, and unconscious processing is involved. The processes that are foundational to consciousness are unconscious.

This discussion leads again to the legendary question implicit in the mind-body problem: If such an extended amount of information processing can occur prior to consciousness (experience, awareness), why have consciousness at all? What is its special function? Our treatment given in numerous places in the previous pages suggests that the contribution of consciousness comes where complex, highly variable and novel resolutions are required. Where homeostacing activity is regular and repetitive, unconscious mechanisms, such as those controlled by the automatic nervous system and those present in reflexivity, suffice; but where greater variability and novelty prevails, experiential involvement becomes required. This conclusion is supported by the observation that highly practiced activities, that initially may require conscious monitoring, may become automatic as practice continues and proceed without conscious involvement.

However we may decide about this intractible issue, there is little disagreement about the participation of unconscious processing in the various sequential stages that lead up to and finally culminate in consciousness. Recently, both automaticity in behavior and the sources of our experiences of will have been analyzed in treatments that are congruent with my own (see Bargh & Chartrand, 1999; Wegner & Wheatley, 1999). Yet certain researchers posit the existence of a system that runs parallel to consciousness, while others suppose a system that works at some deeper or more subtle level. There have been many studies designed to explore these possibilities. The following is but a sample of the research and thinking in this somewhat ephemeral involvement.

Is processing of material below the limen of awareness possible? Leibnitz apparently thought so. According to Boring (1929/1950), Leibnitz held the view that "The sound of the breakers on the beach is a summation of all the falling drops of water, no one of which is conscious alone" (p. 167) and then Boring concluded "The sound of the single falling drop of water may be regarded as an unconscious perception" (p. 168). Fechner (1860) also appeared to hold the view that sensations below the limen of awareness were a possibility. Indeed, he suggested that negative sensations were a universal condition in all modalities. Upon examination, however, Fechner's evidence becomes suspect, derived as it was from his somewhat arbitrary extension of the psychophysics curve to the baseline. Both *petite perceptions* (Leibnitz' term) and Fechner's negative sensations are logically paradoxical because they imply the existence of a sensation below the limen of sensation.

More recently, evidence from three related sources—perceptual defense, subliminal perception, and the Poetzl phenomenon—have resurrected the possibility that some sub-awareness processing may occur if the conditions are appropriate.

As an example of perceptual defense, McGinnies (1949) used a tachistoscope to present two lists of words (from below the limen to above the limen), one containing emotionally laden terms such as penis and whore and the other consisting of neutral terms. McGinnies found that thresholds were higher, psychogalvanic responses greater, and number of distortions more numerous for the loaded than for the neutral words. This study has been replicated many times with positive results being reported in certain cases (Aronfreed, Messick, & Diggary, 1953; McCleary & Lazarus, 1949).

The critics have been even more numerous than the supporters. Bruner and Postman (1949) believed that unfamiliarity may have been responsible for the increase in threshold for the loaded words. Greater effort was required to recognize them, and this may have elevated the galvanic skin response. Howes and Solomon (1950) agreed, pointing out that loaded words appear less frequently, producing the indicators of perceptual defense. They also suggested that there is an obvious difference between recognition and report threshold, and that subjects would tend to delay response to the loaded words until they were sure of what they were. Casting even greater doubt on the McGinnies finding, Postman, Bronson, and Gropper (1953) replicated the study with controls for the variable of reportability. Their results were just the opposite of those obtained by McGinnies.

Other critics have focused on problems of methodology and analysis. Murdock (1954) emphasized the difficulty of finding the threshold, suggesting that direct or partial recognition is always a possibility, while Eriksen (1960) suggested that statistical artifacts may be present in all studies that have obtained positive results. Obviously this is a difficult area in which to achieve adequate control over the relevant variables.

Subliminal perception, or subception, as it is sometimes called, is a related area, haloed with similar difficulties of definition and control. Since the pioneer study by Baker (1937) many articles reported positive results. Dixon (1981) reviewed these experiments and concluded that the evidence, although conflicting, favors the conclusion that stimuli presented below the limen of awareness can influence behavior. Once again, as in the perceptual defense studies previously mentioned, critics have questioned whether there has been adequate control of the relevant variables. In a trenchant analysis Goldiamond, (1958) questioned the logic and design of all sub-awareness processing studies. He suggested that the results, whether pro or con, were contaminated by the type of response indicator used by the researcher. Some experimenters had used accuracy indicators, where the subject was given a score for correctly identifying a stimulus, while others had used semantic indicators, which required the subject to respond with a subjective judgment of how strong or how weak the stimulus is.

Interest in sub-awareness studies was highest in the 1940's and 1950's. It began to wane, probably influenced by Goldiamond's analysis, during the 1960's, and virtually disappeared during the 1970's. Dixon's (1981) book revived a fleeting involvement with the topic and brought attention to a finding called the Poetzl effect, named after the person who first reported it, that the content of dreams may be influenced by stimuli that are presented below the sensory limen. Poetzl (1917/1960) subliminally presented pictures of such items as an asparagus fern protruding from a bunch of roses, a color slide of the temple ruins at Thebes, etc., and then found evidence that representations of such pictures appeared in the dreams of his subjects. Shevrin and Luborsky (1958) replicated the Poetzl procedure and reported positive results. They subliminally presented the same picture to all subjects and then found that parts of the picture, which were not reported during a recall period as being seen, later appeared in dreams.

An impressive number of experimenters have been critical of the results supporting the Poetzl effect. Johnson and Eriksen (1961) suggested that all such experiments are methodologically weak because of the subjective nature of comparing elements in dreams with elements in pictures. Any pointed object in a dream, for instance, might be identified with practically any nonreported pointed object in the picture. Another problem emphasized by Johnson and Eriksen (1961) has to do with the fact that the stimulus pictures often depict commonly experienced scenes that, because of past learning, might be expected to be represented in dreams. They designed an experiment similar to the Shevin and Luborsky (1958) study, cited previously, but took care to control for two possible sources of error. First, a control group that was not exposed to the stimulus picture was used to establish a base rate of reporting certain items in dreams. Second, a method for assuring the independence of judges' ratings was applied. Under these conditions, no evidence for the Poetzl effect was obtained.

There has also been an effort to relate sub-awareness stimulation to personality variables and therapeutic strategies. Silverman has made the most persistent contribution to this area. Beginning in the 1960's and continuing into the 1980's, Silverman and his colleagues published many studies dealing with the influence of subliminal stimulation on such diverse topics and disorders as psycho-dynamic activation, pathological thinking in schizophrenic patients, relationships between manifest psychopathology and unconscious conflict, the speech of stutterers, the treatment of obesity, Oedipal stimuli on dart-throwing ability, cognitive functioning of schizophrenics, and ego functioning of schizophrenics. Many of these studies were given detailed and sympathetic treatment by Dixon (1981) in his book on preconscious processing. Dixon argued that subliminal applications could be used both to reduce anxiety and to provide desensitization, since such stimuli would bypass logical rejection and consciously activate anxiety.

Several efforts to replicate certain of Silverman's experiments have produced conflicting results. Hospel and Hams (1982) tried to replicate Silverman's (1978) experiment on the effect of tachistoscopic stimulation of sub-conscious Oedipal wishes on dart-throwing scores. They found no evidence to support Silverman's conclusion that the subliminal presentation of the stimulus "beating daddy is ok" produced significant improvement in dart-throwing ability.

However results supporting Silverman's hypotheses and methods have also been reported. Ariam and Siller (1982) presented 10th-grade Israeli students with four subliminal Hebrew translations of verbal stimuli: "mommy and I are one" (two versions), "my teacher and I are one" (one version), and a neutral stimulus, "people are walking in the street." Each subject was exposed to the subliminally presented stimuli four times a week for six weeks. Achievement tests showed that groups exposed to the oneness stimuli made significantly higher scores than the neutral groups.

Also following the Silverman approach, Oliver and Burkham (1982) exposed hospitalized female patients suffering from chronic depression to three different subliminal stimuli. One, called the symbiotic stimulus, consisted of "mommy and I are one," the second, called the rapprochment stimulus, consisted of "mommy loves me as I am," while the control stimulus consisted of people talking. Dependent variables consisted of scores on a number of psychological tests and scales. Oliver and Burkham found no significant effects, either taking the group as a whole or dividing the group into different levels of depression.

However Jackson (1983) obtained positive results. Extending the therapeutic implications of subliminal stimulation, he exposed schizophrenic men and women to subliminal messages relating to their relationship with their parents. He found that "mommy and I are one" stimuli reduced pathological behavior in males but not in females, while "daddy and I are one" stimuli reduced pathological behavior in females but not in males.

Several functions have been assigned to sub-awareness processing by various thinkers. Dixon (1981), for instance, insisted that sub-awareness processing (or preconscious processing as he terms it) would make an important adaptive contribution. He maintains that both conscious and unconscious processing systems evolved to serve different biologically useful functions. According to him, the unconscious system has a very large capacity that can receive sensory inflow continuously from all modalities. It is particularly responsive to weak stimuli and is resistant to inhibitory mechanisms that might distort information at the conscious level. This ensures that any information that might be useful for survival will be registered. He then stated that "the prime function of preconscious (unconscious) processes is the ... analysis, appraisal, and selection of data to be brought to conscious awareness or to be stored in memory" (Dixon, 1981, p. 31). In a later section, he postulated that the adaptive priority of the unconscious system arises from the fact that there is a "very limited channel capacity for conscious apprehension of visual inputs and a much larger channel for the assimilation and storage of information which is denied representation in consciousness" (Dixon, 1981, pp. 92–93).

It is, of course, possible that sub-awareness material may be processed, but to plead a larger channel for such processing and an adaptive (biological) function for such processing seems fanciful in the extreme. In my view, consciousness (i.e., the field of awareness) constitutes the major display for information processing, and apperception within this display simplifies, emphasizes, and singularizes pertinent components so that adaptive behavior can take place. The representation and encoding of the total display, or of material not even registered in this display, would not make an adaptive contribution. A situation of information and encoding overload would exist with disastrous behavioral repercussions. There may be some leak of information above and beyond the regular channels of processing, but to give it adaptive priority is unjustified.

I am reminded of a statement made by Carol C. Pratt to the effect that "the more unusual the claim, the more demanding we should be in terms of logical and empirical supports." It is true that on occasion harsh criticism and personal invective may stifle the innovator, as happened with Mesmer. It is even possible that the "necessary conservatism of science," as Boring called it, may result in the loss of an important discovery, but a redundancy principle typically prevails so that someone else will resurrect, perhaps at a more auspicious moment, the same finding. There may be some loss of innovation, but the elimination of hundreds of fruitless notions that would otherwise clutter, if not inundate, legitimate scientific inquiry, more than makes up for it.

CHAPTER SUMMARY

Apperception as a process is another manifestation of the evolutionary imperative that increasing complexity must always be compensated for by greater adaptability. Complexity may bring with it increased survival potential, but only when its detriments are compensated for by counter mechanisms. Apperception provides simplification in organisms burdened by increasing information overload, and was made necessary by the requirement that behavior must be singular in its direction if it is to be adaptive. The sheer fact that organisms are structured as units makes it impossible to respond simultaneously to multiple inputs.

Apperception is the most critical operation in the information-processing capacities of complex organisms. Inputs from sensation, memory, or both become momentarily focalized so that (as we see in the following chapter) adaptive behavior can take place. It is completely lawful in its activity, a focalizer rather than a selector, and works according to homeostatic negative-feedback principles, as do all of the balance-maintaining systems of the body.

Apperception takes place within the field of awareness, a multiparallel system wherein all of the various sensory inputs (as modified and structured by the perceptual organizing principles) are irresistibly deployed and displayed. The myriad encodes representing the accumulated experiences of the individual's life are also available for focalization. Such memories are not simultaneously represented in the field of awareness, but exist as preawareness potentials that become automatically available when appropriate associative conditions are present. The course of apperception is influenced by those variables that, in terms of the history of the species and the individual, have the greatest significance for adaptive behavior at each moment.

Even as the movement of a microorganism is in that direction that best resolves the tropistic influence of the chemicals being imposed upon it, the fluctuation of apperception is always the most equilibrating resolution of those forces momentarily present. Thus the mental activity of the individual, whether thinking or sensing, is completely lawful. There is no homunculus selecting or directing the course of apperceptual flux.

The tendency for dynamically interacting systems to regain and maintain equilibrium was stated by Newton (1647) in physics, by Le Chatelier (1884) in chemistry and by Helmholtz (1924) in his refinement of the law of the conservation of energy. It later was expressed as a characteristic of living systems, first by Bernard (1865) and then by Cannon (1932) in his concept of homeostasis. This concept, which assumes such central importance in my own treatment, thus has a long history as a restatement of the physical laws that determine that compensatory balances will result from the resolution of interacting fields of force. Although a number of transformations (mechanical, chemical, electrical, etc.) may be involved, such shifting balances are in perpetual process in organisms whether in the single cell or in all organ systems working in dynamic synchrony. Thus, laws that prevail in inanimate nature (as previously indicted in chap. 2 with the inverse square law) are represented in all aspects of mentation; most dramatically in that critically important-information processing operation, which we call apperception.

Apperception is the ultimate intermediating process. Without such focalization nothing (except for perceptual structuring) can be learned, and, unless a component that has been encoded reenters apperception, it can neither become a part of the thought process nor influence overt behavior. Between the input and the output lies the apperception. It is the highest refinement and the most functional outcome of the selection pressures of evolution, the attribute that underlies the operation of man's most subtle reasoning and the development of his rudest skill. In the following chapter, we see how apperception is the direct antecedent of most behavior, and is the process that makes such behavior both efficient and adaptive.

8 Reaction

The developments in the previous chapters have pointed to and culminated in apperception as the most important contributor to the information-processing capacities of complex organisms, such as ourselves; it is involved in the focalization of critical inputs, the encoding of such inputs, and the retrieval of this information.

If we were to stop here, we would have a fairly sophisticated perspective on the manner in which the selection pressures of evolution have gradually achieved living systems wherein many interdependent operations are monitored and regulated so that the balances necessary for life can be maintained. But to stop at this juncture would leave the picture far from completed. Behavior must now be integrated into our development. Behavior, whether a reaction of one part of the body or an act of the total organism, whether effective or ineffective, has been the operative mechanism in balance maintenance, the relative efficiency of which has provided the selecting in and out that marks the evolutionary history of every species. Behavior cannot just be tacked on at the end of my development but must be integrated with the information-processing capacities examined in the previous chapters so that it becomes a logical culmination of the total system.

CONTEMPORANEOUS CAUSATION AND PAST LEARNING

An evaluation of the labors of a number of theorists in experimental psychology might lead one to conclude that learning is the process that accounts for most behavior. This is not the case. Certainly much of the behavior that is emitted by most mammals has been learned, but, once such learning has occurred, the process itself ceases to play an important role in action. During any given day of a human being's life, the behavior that occurs often involves the use of learning that has already taken place, much of it during infancy and childhood. Thus, the problem of understanding behavior does not center on the learning process per se, but on the variables that determine which of the behaviors already in the individual's repertory will be used.

This leads to the affirmation of a principle that has been implicit throughout the preceding chapters: the view that action, whatever its particular nature, is always an outcome of the variables that are interacting at each given moment. This point, it is true, seems obvious, but it has been much discussed and more misunderstood. Contemporaneous causation does not negate the contribution of historical variables. A historical approach—at least at the present time—is one of the most feasible methods for determining which variables are influencing activity at any given moment; the effects of past learning are contained in the present, and as such may influence present activity. Thus, when I insist that much behavior has little to do with learning I do not mean to imply that learning makes no contribution to action, but that it is the result of the process and not the process itself that is important.

EXPERIENCE AND ACTION

It seems paradoxical that the aspects of the organism that seem most readily available for assessment are also those that have most tenaciously defied integration into a cognitive perspective. Overt behavior, although traditionally accepted as the source of objective data in psychology, is viewed as the outcome of the cognitive processes that were discussed in the previous chapters. It is here, of course, that the mind-body problem discussed at length in chapter 4 again comes into focus.

Is such cognitive activity based in complex neural networks and their interactions? This is the working assumption of most cognitive psychologists. If it takes place in some transcendental, metaphysical limbo we have all been wasting our time. Yet except for the very simple neural circuits described by Frankel (1927) and von Holst (1973) as described by Galistel (1980; to be reviewed later), we know almost nothing about neural circuits or how they activate and sustain behavior. Data on the complex neural circuits that are corollary with cognitive processing are not yet available. However data from research on cognition, as outlined in the previous chapters, is available. We can infer a great deal about the manner in which information processing takes place in complex organisms if we examine the cognitive side of the mind-brain coin. The cognitive side is an open window on the total physio-experiential process. When we combine the research and theory from cognitive psychology with the inferences derived from an evolutionary perspective, the manner in which the total physio-experiential machinery participates in the information processing and action necessary for regaining and maintaining the balances essential for life becomes understandable.

ACTION SYSTEMS

Hardwired Neural Circuits

A principle that is central to the development in the previous chapters should be emphasized again. Where conditions, whether within the environment or within the body, are relatively similar and recurrent, automatic mechanisms, free from, corollary with, or involving cognitive processing have developed. The sense organs of all of us are automatically tuned to the same kinds of energy; the rules of perceptual ordering and structuring determine that we will all be reacting within similar three-dimensional worlds; and the autonomic nervous system ensures that recurring cycles in our physiology will be monitored and effectively regulated.

Reaction shares in such automatically demonstrated representations, often without cognitive involvement. When similar problems of adaptation are repeatedly encountered during the evolution of a species, innate linkages may emerge which ensure that effective behavior will occur automatically. Reflexes, as discussed in chapter 3, provide immediate and effective reaction when certain stimulus constancies are imposed; such behaviors as sucking, withdrawal, or coughing occur immediately and automatically without cluttering the cognitive system. There are a number of rhythmic behaviors, such as the beating heart and the fin movements in fish, representing crucial adaptive conditions, that proceed automatically without control from higher levels in the brain. von Holst (1973) found that the rhythmic waving of fins continues in fish after the spinal cord has been severed just above the medulla. The eye of a marine slug (Aplysia) continues to respond, as does the heart of a turtle, for days after being removed from the respective animal. "These neural metronomes may put out their neural ticks and tocks in the absence of any sensory input" (Gallistel, 1981, p. 611).

Relatively simple neural linkages are involved in even more complex behaviors—behaviors in which cognition plays no direct role. Pearson (1976) used a coupled-oscillator model to explain cockroach locomotion. Each leg has its own oscillator that controls stepping: Feedback compensators alter the strength of muscle reactions of each leg to ensure smooth coordination in spite of terrain and load shifts

(Gallistel 1980). It should be noted that movement over varying terrain, while compensating for the ever-shifting tug of gravity, is a recurrent invariant that most animals face.

In the summary of his account of coupled oscillators in locomotion, Wilson emphasized how such low level circuitry can govern the execution of complex behavior:

> The rich variety of gaits generated by this central patterning circuitry is further enriched by the load-compensation circuitry and the trigger-delaying circuitry. The result is the bewildering variety of finely adapted stepping movements that we actually observe in the roach as it walks at varying speeds through fluctuating circumstances. (Wilson, 1980, p. 142)

Gallistel (1981) reviewed one of the most remarkable and certainly the earliest experiment demonstrating, not only the additive effects of two corollary servomechanisms (negative-feedback compensators), but also the manner in which a servomechanism can reverse its operation as a function of shifting conditions, thus demonstrating that even in a seemingly hardwired process a degree of plasticity may be allowed. This experiment by Frankel (1927), as translated by Gallistel (1980), showed that the coastal snail is both negatively phototaxic and geotaxic (i.e., that it moves away from both light and the pull of gravity). When it is placed in a vertical situation with light coming in from one side, the snail moves in an oblique direction, which effectively summates the action of the two energies being imposed. When the snail is upside down and underwater, the phototaxis becomes reversed, but not the geotaxis; that is, the snail now moves toward light.

Frankel argued convincingly that the manner in which the taxes interact, and the reversal of the phototaxis, both reflect the snail's adaptation to its particular environmental niche. The snail thus can move from the sea floor to dark crevices a few meters above high tide, where it lives most of its life cycle. I quote Frankel here:

> Assume that a Littorian neritodes finds itself on the bottom in shallow water near the shore. Littorians crawl toward dark objects. The rocks on the shore must appear dark relative to open water. Therefore the snail turns toward the shore (negative phototaxis). Having reached the rocks, it crawls upward (negative geotaxis). On the way up, it encounters a crack that cuts horizontally across the rock wall. The snail crawls without hesitation into the depths of the crevice (negative phototaxis). Arriving at the depth of the crevice, it crawls back upside down along the roof of the crevice out into the open (*positive phototaxis*), italics mine. The obstacle is overcome. Were the phototaxis incapable of changing its sign, the snail would remain hanging at the innermost part of the crack and never reach the surface of the water. The change in phototaxis has, in other words, an important biological significance. (Frankel, 1927, in Gallistel, 1980, p. 160)

The repolarization of a single neuron is the simplest and best known example of the manner in which a negative-feedback compensator (or servomechanism) works, and the irritability curve provides the clearest and most invariable record of a negative feedback system in action. In the interacting taxes with reversibility, documented by Frankel, we observe how two such systems work in conjunction to help achieve and maintain the equilibrium necessary for the life of the snail. These, along with the contribution of oscillators, previously discussed, provide a rudimentary explanation of behavior in relatively simple organisms. However as organisms became divided into organ systems with varied functions, more refined and greater numbers of homeostatic negative-feedback compensators (servomechanisms) became necessary for the monitoring and maintenance of optimal balance in these systems and in the organism as a whole. Thus, we see the contribution of the autonomic nervous system in mammals, which serves to both maintain or regain an essential balance in a number of organs and processes, and to keep them working in efficient dynamic synchrony.

Complex Behavior: Hardwired, Cognitive, or Both. Consider the salmon. It proceeds upstream, accurately selecting among the multitude of branches encountered, until it finally arrives at the particular place where it was spawned and developed as a fingerling. How does it find its way? Apparently the mineral content of the critical branch becomes encoded to provide an effective program for reaction. In some remarkable experiments, (Hasler, Scholz, & Horall, 1978), salmon fingerlings during the smolting state (a period of hormonal changes) were exposed to minute amounts of certain chemicals—one group to morpholine, another to phenethyl alcohol, and a control group to untreated water. The fish from all three groups were fin tagged and released midway between two tributary systems. At spawning time the two chemicals were released into particular streams. True to the experimenter's predictions, the chemically treated fish predominantly entered the tributaries baited with their respective chemicals, while the control fish randomly distributed among the various streams. Hasler and his colleagues not only clearly established that salmon are guided by chemical sensitization, but also found that only four brief exposures during the critical smolting period were sufficient to establish the chemical representation, and that this representation may persist for at least three years to provide the basis for the homing behavior.

The rapid development of internal representations of critical environment circumstances is best known and probably most clearly demonstrated in the imprinting studies of Lorenz (1981). He and his colleagues showed that a brief visual exposure during a critical maturational window was sufficient to create an internal representation of a particular stimulus complex, often consisting of Lorenz himself. After the imprinting had taken place, wherever Lorenz went his ducks or geese (his most common subjects) were sure to follow.

The development of visual representations is not restricted to birds. Holldobler (1980) demonstrated that a species of ant, P. Tarsatus, is capable of developing a representation of the forest canopy. This representation provides a basis for orientation in the home range, and regardless of where the ant is located within this range, it can orient itself appropriately as long as the canopy is visible.

It should be emphasized that all such examples of imprinting allow adaptive behavior to circumstances that are highly variable. There are many different stimulus configurations representing different canopies, tributaries, people, or animals. Imprinting involves a very rapid and stable kind of learning, a kind of learning that depends on the unique adaptation problems of each species, and that in all cases can occur during a very specific and adaptively appropriate set of physiological conditions representing a particular developmental window.

Do the action systems that develop in conjunction with imprinting suggest the existence of cognitive maps? Probably not in the salmon. The brief exposure during the smolting stage may simply sensitize the salmon to orient tropistically when the critical mineral (chemical) solution strikes its body. In Lorenz's geese and ducks and in canopy orienting by ants the situation is less clear, possibly because the stimulus configurations are more variable and complex. Yet the detour behavior that is mentioned by Tolman (1948) as a critical indicator of a cognitive representation is present in both of these latter examples.

There are other types of action systems that guide adaptive behavior, where the role that learning plays, if indeed it plays any role, is minimal. They are widely represented in species, and may involve a restricted set of stimuli, or in certain cases may include a number of overlapping cues from the same or different sense modalities. Thus organisms may orient themselves (i.e., behave adaptively) in terms of information derived from the position of the sun, lines of magnetic force from the earth's magnetic field, polarized light patterns, or star clusters, as well as many others.

The adaptive contributions of such action systems are almost as diverse as the survival problems of the species that utilize them, but all have certain features in common. In all cases, critical inputs representing internal or external changes becomes tied to behaviors that have had survival consequences for the species in question for an extended period during its biological and behavioral evolution. In simple reflexes, oscillators, and servomechanisms, the problem of reaction does not exist; it is implicit in the action system. However where cognition enters the picture, reaction becomes a critical issue in need of explanation. There is little doubt that cognitive processes, discussed at length in previous chapters and in this one, play a role in reaction. But how? Up until this point we have been (this is similar to Guthrie's complaint about Tolman's rat) encapsulated in the mental, if not lost in thought. We must come to grips with events and processes that will allow behavior to be integrated into our theory; the problem is to develop a cognitive explanation of which particular behavior will occur and when it will occur.

Cognition and Behavior

In chapter 7 the view was defended that the thought processes of the human being are as lawful (although the outcome of many subtle variables and their interactions) as the tropistic behavior of the moth, the vector summation reactions of an amoeba, or, as in our most recent example, the vector summation behavior of the snail. In this section, we explore the manner in which the cognitive processes (subsumed by the term cognition), discussed in the previous chapters, culminate in behavior that serves to regain and maintain the balances vital for life, and how cognition regulates such behavior in spite of the variable conditions that are continually being encountered, both on the outside and on the inside of the organism.

As we proceed we must keep in mind that it will not be acceptable to play the naming game—to give names to hypothetical processes and then use such names in the service of explanation. Nor will it be acceptable to invent convenient demons, as has often happened in modern cognitive psychology (see chap. 1)—some all purpose homunculus that pulls the strings involved in action. Nor will it be legitimate to use small explanatory elves that are themselves left unexplained, as Galistel (1980) did with the concepts of positive and negative potentiation in his efforts to show how motivation energizes action. The issue is not whether shifts in motivation may energize different and appropriate behaviors. This is certainly so. The issue is how it happens.

The research and theory on reflexes, oscillators, and servomechanisms demonstrate that the foundations of behavior, at least where simple organisms and similar recurrent adaptive circumstances are involved, may be inherited and represented in hardwired neural circuits. Research and theory at this level also have metaphorical implications for a cognitive-processing model; certainly the commitment to a nonteleological, completely lawful, interpretation is implied. The complex interactions and resolutions present in cognitive processing are just as lawful as the behavior of Frankel's snail or Wilson's cockroach. However between the two areas there is a quantum leap in complexity.

In cognitive processing and the behavior that is an outcome of such processing, we must examine a system that is immediately responsive to hundreds (perhaps thousands) of information components and their interactions—a system that moment by moment instigates behavior that is appropriate to each adaptation circumstance.

The postulate of process states that all behavior, whether overt or covert, is immediately equilibrium trending. Thus, overt behavior is viewed as homeostatic, even as thought was so considered in the last chapter. Overt behavior, except that controlled by hardwired neural circuits (as discussed above) and highly practiced skills (to be discussed), is a direct outcome of apperception. Yet at what point does overt action take place? It is one thing to surmise that apperception is the functional antecedent to most overt behavior, and quite another to give a plausible explanation of which apperceived component will occasion a particular overt action and when this action will occur. For example, a human being may indulge in a considerable amount of thought relative to a given problem before any discernible behavior takes place. At what point in this process of internal locomotion will overt behavior be initiated? Further, an individual may sequentially apperceive a large number of sensory components without overt response. To which one will response be made, and at what point will the reaction occur? Before these questions are considered, a number of factors upon which overt behavior depends must be examined.

Fusion and the Emergence of Composite Action Systems (CASs)

In the newborn infant, behavior is diffuse and undifferentiated. As time passes, maturation and learning coalesce to allow the development of coordination and the refinement of specific motor skills. The integration of body movements that is necessary for the performance of such skills is dependent upon the fusion and constancy mechanisms previously described (see chap. 5). Each repeated activity produces discrete sensory inputs that are similar and recurrent, conditions postulated as necessary and sufficient for the occurrence of fusion. As a function of such fusion, a unified skill experience emerges in much the same fashion as a phenomenal self or a phenomenal tree comes into being, and this unity, which I call the motocept, underlies all integrated and coordinated movement. Thus, the sensory feedback components peculiar to any skill, typing for instance, are subsumed in the totality and are not apperceived as long as there is no interference with coordinated action. Only at the outset are other auxiliary inputs (as from vision or touch) required for effective performance. As the motocept emerges, the skill becomes automatic; the crutch cues from other modalities are no longer necessary.

A majority of the motor skills that will be used during a lifetime have already developed by the time the child is three or four years old. Sensory inputs derived from activities such as eating, walking, grasping, and talking soon become fused into motocepts comprised of the feedback components unique to each particular set of movements.

Even as various motor skills such as walking, typing, and driving a car are becoming fused into motocepts, cognitive maps representing the environmental context wherein such motocepts are demonstrated also emerge to become integrated into a CAS. After a particular CAS has been perfected through repeated usage, the individual may move through a very complex series of activities with only occasional apperceptual monitoring. However sometimes the execution is not perfect. Take my coffee-making CAS. This morning I placed the coffee filter in the appropriate receptacle of my coffee maker, poured in the water, and turned on the switch. I returned to find, much to my astonishment and chagrin, that I had produced hot water, having failed to put coffee in the filter. Yet prior to this occasion I had automatically produced excellent coffee many different times under the auspices of an operative CAS with only occasional apperceptual monitoring, indeed often while still half asleep.

Recently I went into my kitchen to get some food for my cat Brucie, who was occasionally meowing and scratching at the door. The cat food was located on a shelf in one of the cabinets. I went into the kitchen, opened the refrigerator door, and stood there, wondering why I was there. I had forgotten all about Brucie and remained there at the refrigerator staring inside puzzling about it all until I was reoriented by some more scratching and meowing from Brucie. This sort of addled behavior is not unusual in humans. When we perform many tasks in the same general location, a remarkable amount of overlap occurs in the CASs that develop; essential aspects of the same CAS are operative for different goals (incentives, intentions, needs, etc.). If, in the process of executing one of these overlapping behavior sequences, we forget the goal, the most commonly used aspect of the inclusive CAS will tend to occur, suggesting that I often go to the refrigerator in the kitchen.

The fusion of motocepts and cognitive maps into CASs would seem to provide an adequate basis for effective reaction. Yet, without exception, other previously discussed phenomena feed into these representations and become fused with them. Cognitive maps have typically referred to internal representations of environmental features, and this usage is certainly acceptable. However something more than replicas arising from ordinary learning is involved. Perceptual learning, as discussed in chapter 5, invariably makes a contribution. Indeed, our perceptual world is the most literal and stable cognitive representation (cognitive map) of the environment that occurs in human beings. It provides the stage within which the flux of our more variable experiences is displayed and within which the cognitive maps that develop as a function of ordinary learning are always imbedded. Thus, interacting with the environment commonly involves a remarkable composite, one in which perceptual organization, motocepts, and traditional cognitive maps have become incorporated into a unifying totality, wherein the pertinent features of both behavior and environment are fused into a CAS that can operate efficiently with minimal apperceptual involvement.

This discussion emphasizes a relationship between organism and environment that is characteristic of all species. Every individual confronts essentially the same environment on many different occasions. Human beings move again and again through the same rooms of their homes, the same streets on their way to work, the same circumstances of their work place, etc. Animals repeatedly confront the same features of their territory, the same watering holes, the same foraging areas, and the same resting places. Thus much behavior becomes a functional reiteration of previous behaviors. As the same "if A then B" concurrences and the same environmental circumstances are repeatedly encountered, a CAS emerges that is peculiar to each individual's daily round of activities. The CAS is remarkably functional. Efficient and adaptive behavior can proceed with only occasional apperceptual monitoring, freeing the apperceptual mechanism for other traffic. It also serves as a relatively constant reference backdrop. If some novel happening, say the appearance of predator or prey, suddenly intrudes into the field of awareness, the CAS allows such intrusions to be more readily discriminated so that apperceptual involvement can trigger effective behavior.

The utilization of an explanatory homunculus is not necessary or even implied in the use of CASs as explanatory constructs. Maier's (1929) early experiments on cognitive maps in rats and those by Thorpe (1950), Tinbergen and Kruyt (1938), and van Iersel and van dem Aassen (1965) on the digger wasp, make an environmental representation (cognitive map) highly probable in these animals. It seems unnecessary to assume a homunculus pressing the buttons in order to reach the prescribed goals. The recent invention and demonstration of terrain-recognizing and error-compensating cruise missiles clearly indicates that environment-duplicating maps and corrective feedback based on such maps can exist without assuming an homunculus or some other metaphysical intrusion.

It is both interesting and ironic that an early version of how a homunculus might work to bring about compensatory changes in behavior was provided by Skinner (1960) during the early months of

World War II. At that time missiles were available, but there were no guidance systems to control them. In the "pigeon in a pelican" (pelican was the name of the missile), pigeons were trained to peck at the image of a ship on a television screen (the television camera was located in the nose of the pelican). The pigeon's pecks on the ship's image brought about compensatory changes in the guidance systems of the missile so that it would home in on the actual ship being pictured. We no longer need an avianunculus to bring about compensatory changes in missiles, nor do we need an homunculus for similar purposes in the human being.

Overt Behavior and the Motor Threshold

The notion of a motor threshold certainly did not originate with me. Hull (1943) developed the concept of reaction threshold and made it a central component of his behavior theory. It was an important construct in psychology during the 1950s and 1960s, largely because of Hull's influence, and it still emphasizes the fact that a lot can be happening in an organism before overt behavior is triggered.

In the human being, and probably in most mammals, three thresholds are implicit: (a) the sensory threshold, which is crossed when a neural process within a receptor results in the introduction of sensory input into the field of awareness, has been universally recognized since Fechner (1860); (b) the apperceptual threshold, which is crossed when one of the many components within the field of awareness or memory becomes focalized, has been accepted since Herbart (1882); and (c) the motor threshold, which is crossed when an apperceived input becomes the occasion for overt behavior, was clearly defined by Hull (1943) and is of central importance as an explanatory construct.

A consideration of what happens when an organism goes to sleep provides insight into the nature and functions of the thresholds under discussion. In sleep there is a raising of both the sensory and motor thresholds. The input from the various receptor systems of the body is essentially cut out so that apperception relative to such material (unless it is extremely intense) simply does not occur. Likewise, the apperception of encodes during sleep (as in dreaming) ordinarily does not result in behavior because of the elevated motor threshold (see the section on dreams in chap. 7).

Learning and the Lowering of the Motor Threshold. In the human infant, except for reflexive behavior, the motor threshold appears to be at approximately the same level regardless of the source of the apperceived input. Intensity alone seems to be the major determinant of action, and, once the motor threshold has been crossed, behavior, though undifferentiated, invariably results. Yet as the infant matures, learning plays an increasingly important role in action; motocepts emerge and become associated with apperceived encodes and sensory components. The development of such associations involves the lowering of the motor threshold between input and output, as described in the following statement:

> The lowering of the motor threshold is a direct function of the number of times a given apperceived input has been followed by a particular overt response.

Such input may be sensory, (i.e., from one of the sense modalities, such as hearing or vision); it may represent emotional states (such as fear or anger); it may represent biological imbalances (such as hunger or thirst); or it may consist of encodes or encode constellations. In all cases, when the motor threshold between input and action has been sufficiently lowered the apperception of that input will automatically trigger the behavior in question.

Most organisms are born with a relatively low motor threshold between certain inputs and particular motor responses. The various reflexes, animal communication signals, and instincts fall into this category (also see chap. 3). Such inherited stimulus-response sequences represent the ultimate in automaticity, but little more so than many learned sequences that are present in highly practiced skills. It is in terms of this automaticity that the similarity between learned and innately contributed stimulus-response linkages becomes most evident, the essential difference being that in the one case the lowering of the motor threshold occurs as a function of practice, while in the other the threshold is low to begin with.

When learned skills become highly refined, apperception as a precursor to action becomes less important. The motor threshold becomes lowered to such a degree that a complex series of reactions may unfold with only occasional apperceptual monitoring. The extent to which the motor threshold has been lowered, and the degree of automaticity that the behavior has come to assume, can be determined experimentally.

> The more a given motor skill has been practiced, the greater will be a subject's ability to indulge in another activity while simultaneously performing that motor skill.

Accordingly, a skilled typist would find it easier to learn a list of nonsense syllables while typing than one less practiced in the art, and individuals forced to sort cards while learning nonsense syllables would find the latter task less difficult as they become more skilled with the first.

Such developing automaticity has been examined in depth by Logan (1991), who demonstrated that as a task is practiced it becomes more and more automatic, until very little apperceptual monitoring is required. This process of automatizing has been studied using a divided attention task. Subjects are asked to perform two tasks simultaneously, such as playing Mozart on a piano while trying to remember an unrelated list of words. Automaticity is demonstrated when the automatic task no longer interferes with the other task (Hasher & Zacks, 1979; Shiffrin & Schneider, 1977).

Input Intensity and Response Magnitude. Is intensity of input a variable that directly influences the latency, magnitude of response, or both? The many experiments on the issue, particularly those cited by Hull (1951) in support of his concept, *stimulus intensity dynamism*, seem to provide clear evidence that this is the case. This would also seem a reasonable inference in very simple animals and even in more complex ones, such as mammals, where homeostatic regulation is achieved without conscious involvement. Yet, in one of the most elementary of all reactions, the depolarization of the neuron, an all-or-none principle prevails; that is, if the firing threshold of the neuron is reached it fires at its maximum, and an increase in the intensity of stimulations beyond this point has no influence on response magnitude.

When cognition plays an active role in reaction, the issue becomes more complicated. Certainly intensity is an important variable (along with a number of others) influencing apperception, and apperception is the critical antecedent to most behavior (see chap. 7). But, if apperception were held constant, would intensity directly influence behavior? From an adaptive standpoint it would seem so. We would predict (from the fundamental corollary, chap. 2) that as activation input (thirst, hunger, fear, etc.) becomes more intense, the magnitude of response should increase. Many experiments (reviewed by Hull, 1951, Deese and Hulse, 1967, and Hall, 1976) support this prediction. But does the more intense activation input influence behavior directly or does it just ensure that such input will be apperceived more readily, thus triggering greater and more sustained effort?

The influence of intensity on response magnitude, while holding apperception constant, is open to experimental investigation with humans. Subjects could be placed before a response panel containing three different colored buttons along with a stimulus panel containing three different colored lights corresponding to the colored buttons. A measuring device would be used to record the response latency and pressure applied to each button. The subjects would be told to respond to each separately presented colored light by pressing the button of the same color. The lights would be systematically varied in intensity for different groups and/or on different blocks of trials.

The Effect of Conflict on the Motor Threshold. The discussion thus far has suggested that overt behavior will occur automatically once the motor threshold has been crossed, and that intensity may influence the latency and magnitude of response. However, reaction is a function not only of the number of times a given input has been associated with a particular response, but also of the presence of competing inputs that have been associated with other behaviors. During practically all waking moments many different inputs are being sequentially apperceived, and certain of these have been previously associated with responses that are incompatible. For instance, an individual might sequentially apperceive two different inputs, one of which had previously been associated with toward behavior and the other which has been associated with away from responses. When apperception is fluctuating between inputs that have been associated with incompatible responses, conflict results and behavioral oscillation may occur.

One of the simplest examples of conflict can be demonstrated in a newborn baby. If the baby is hungry, the stimulation of one cheek will occasion head movement toward the side that is being stimulated. Yet if both cheeks are alternately stimulated, the child may exhibit oscillatory behavior. The oscillation, of course, is dependent upon both the timing and the intensity of the stimulations; that is, as the child moves left, the right cheek must be stimulated with greater intensity to initiate apperception and the opposite movement.

Oscillation is also dependent upon the relative amount of sensory adaption that has taken place. Nafe and Wagoner (1941) not only demonstrated the rate of adaptation, but also provided insight into the mechanism that is responsible for the introduction of sensory input. They found that a weight placed upon the skin's surface does not sink immediately; it sinks gradually, moving more slowly as it descends, until a critical rate of sinking is reached. As long as movement is above the critical rate, the subject feels pressure, but when the rate of sinking is less than the critical rate, the subject feels no pressure. Adaptation has taken place. In like fashion, many different kinds of persistent stimuli may cease introducing sensory components into the field of awareness, thus freeing the apperceptual process from the coercion of inputs that no longer have adaptive significance.

Nafe and Wagoner's (1941) classic study emphasized the stimulus side of neuronal (receptor) adaptation. More recently Kandel and Schwartz (1982) looked at the response side of the adaptation-habituation issue. They not only isolated the specific neurons activated in both habituation and sensitization, but also have determined that a specific transmitter substance, *serotonin*, secreted by the neurons themselves, is involved. As a function of repeated stimulation the amount of serotonin secreted is reduced, resulting in response dimenution (habituation), while the presentation of a noxious stimulus to the tail of the organism results in an increase in the amount of serotonin secreted and sensitization, that is, an increase in the amplitude of responses when other parts of the body are stimulated. It is true that these findings have been derived from a very simple organism, the shell-less marine mollusk *Aplysi californica*, but they mark a breakthrough into the manner in which chemicals produced by neurons may reciprocally influence neurons so that the magnitude of reaction may be either reduced or increased.

All approach-approach conflict situations follow the general pattern outlined in the previous cheek-stimulation example, but learned rather than innate responses are typically involved. Such approach-approach conflict is quite unstable. Since apperception of input varies as a function of intensity, and since there is a relationship between distance and intensity, the relative stasis demonstrable in an approach-approach conflict situation is easily resolved.

What happens when behavior is triggered by two different inputs arising from the same circumstance, one of which occasions away from and the other toward behavior? Under these conditions of approach-avoidance conflict, as apperception fluctuates, the organism is first impelled toward, and then away from, the situation. An animal eating food that burns its mouth would be an elementary example of such conflict. A thirsty individual who is finally driven to drink salt water would be a more complex demonstration of the same phenomenon. In both cases, inputs that occasion two responses that are incompatible are derived from the same circumstance.

The approach-avoidance conflict explicit in the previous examples is typical of most overt behavior. As apperception fluctuates, many of the focalized components will have been associated with responses that are above threshold but incompatible. Thus during any particular period a fair number of approach-avoidance conflicts should arise from the ongoing process of apperception itself. Since this is the case, to which of the apperceived components will response be made?

> When motor thresholds are equal, behavior will be occasioned by the most intense of various apperceived inputs.

In the classic approach-avoidance conflict studies by Brown (1948), oscillatory behavior was noted where the approach and avoidance gradients crossed. This took place when, as a function of moving further from the critical situation, the fear input became momentarily less intense than the hunger input. As was experimentally demonstrated, if either input is increased, the point at which the oscillation takes place will vary correspondingly. If both inputs are increased proportionately, the oscillation that occurs will be more rapid, indicating that the amount of dynamic tension within the system has also been increased. It should be noted that the oscillation that can be demonstrated in the approach-avoidance situation is another example of the fulcrum phenomenon (see chap. 3), which is characteristic of behavior once a dynamic equilibrium relative to antithetical inputs has been achieved.

In this discussion I have been assuming that the motor thresholds relative to the precipitating inputs have been equal. Obviously this rarely occurs in actual behavior, for the thresholds between inputs and behaviors vary as a function of the number of contiguous pairings that have taken place. If we could hold the intensity of two inputs constant, the behavior with the lower threshold would always occur, as summarized in the following hypothesis:

> When the inputs associated with two incompatible responses are equal in intensity, the input-behavior sequence with the greater number of past contiguous pairings (or lower motor threshold) will predominate.

If rats were given 50 trials to the left arm of a T-maze while under water deprivation, and one 100 trials to the right while under food deprivation, they would be predicted to go right when the two inputs (thirst and hunger) are equal in intensity. This hypothesis, of course, would be difficult to test, because very little research has been done on the equation of activation input intensity, and even less on input interaction effects. Yet, the general implications of the previous hypothesis could be determined without too much difficulty; all that would be necessary would be to demonstrate that either (a) the level of deprivation or (b) the number of training trials could differentially affect behavior, while holding the other constant.

The previous experiment involves two different primary activation inputs, thirst and hunger, and as such emphasizes a complication that has often been implicit in the preceding discussion. The variable of input pervasiveness, as well as that of input intensity, is present and complicates all such designs. It will require much ingenious research before we can say, even with minimal certainty, how much hunger is equal to how much pain, etc.

The Autocept and Overt Behavior. In previous chapters, the autocept emerged as a central contributor to behavior. As will be recalled (see in particular chaps. 4 & 6), the autocept is the outcome of the fusion of the most consistent and recurrent experiences encountered during an individual's life, particularly during his or her early childhood. It has its nucleus in the visceral rhythms of the mother, expands to include the recurrent experiences of infancy and childhood, and ultimately includes the principles and precepts of one's socialization. In essence, we become the most consistent and recurrent experiences of our life, as these are fused into phenomenological unity.

It was surmised that the autocept has made and continues to make a vital contribution to survival, and that this contribution, though perhaps rudimentary in the beginning, was the critical factor that occasioned its emergence and refinement in evolution. The survival contribution provided by the autocept is in part derived from its implications for rapid and appropriate response; it allows an organism to respond as a totality when different parts of the body are threatened. Thus, when a particular hazard is encountered, whether a general physiological imbalance or a specific trauma, the threat is to the total animal. Since the most universal of all adaptive behaviors is the movement of the entire body either toward or away from a critical circumstance, the autocept makes a considerable adaptive contribution.

As human beings evolved, the tendency for autocept emergence expanded to include the rules and regulations imperative for group (and individual) survival. The individual survived to the extent that he or she followed the dictates of the group; aberrant individuals, those who were too individualistic, selfish, or rebellious (and here we see the selection pressures involved in the refinement of the predisposition for autocept inclusion) were ejected from the group, and they took their unruly genes with them into the oblivion of isolation.

When the predisposition for autocept emergence expanded to include the principles and precepts of society, the individual, as he or she developed, in essence became his or her society; religion, family, club, and nation. As such, both the manner in which the autocept abets group cohesion (at least it did at an earlier time in our history) and enters into individual motivation becomes clear. If I am my society and I break its rules—rules vital for individual as well as group survival—a powerful activation input, guilt, coerces me from the inside to reaffirm the rules of the group and to make amends for my folly. When the integrity that the autocept has come to assume is threatened, activation input in terms of anxiety or guilt (or both) intrudes into the field of awareness as a pervasive background input that instigates and sustains behavior, even as intense hunger or thirst provides the sustaining occasion for appropriate response.

Some of the most profound conflicts in the human being arise from threats to the organization of the autocept. If an input precipitates covert behavior (thought) that threatens the integrity of the autocept, anxiety and guilt will result. If such precipitating input is so intense that overt behavior actually occurs, the anxiety or guilt may produce disruptive or even bizarre repercussions. Apperception may be flooded, jamming out all other inputs, so that behavior becomes discordant and inappropriate, as may be the case with neurotics or individuals who are being tortured.

Intensity and prepotence of input are major variables that determine whether behavior (either covert or overt) that runs counter to the organization of a particular autocept will occur. Yet autocept disruption may also be affected both by the kinds of material that have been fused into the autocept in the past and by the degree of integration that has taken place. Thus, behavior that might be extremely disruptive to one autocept could be incidental or even corollary to another.

We may surmise the conditions (selection pressures) that gradually resulted in the capacity for guilt. The survival of our early ancestors was intimately tied to the principles of cooperation and concerted activities, both of which required that each individual follow the rules and regulations of the group. External coercion (overt punishment) is a way to achieve conformity, but it brings with it the probability of violence with all its lethal implications. Guilt drives from the inside to provide individual conformity, making external and hazardous coercion unnecessary. With guilt the individual conforms to rules and regulations, not because of external and hazardous coercion, but because as a function of autocept inclusion the rules and regulations of the group have become essentially what he or she is, and to break them is to violate an essential aspect of what he or she has become.

Since humans often have the experience of freedom when we choose between alternatives, or when we persist in some patently odious activity, does some agency of freedom determine action or nonaction in such cases? The following analysis may be helpful in clarifying the freedom issue. Let us begin by asking the opposite question: Under what circumstances do we have the feeling that our behavior is caused or non-free? Certainly this is the case when our behavior is affected by external coercion, or, for that matter, when it is dominated by an obvious need system. In these examples we would answer the question, "Why did you do that?" with such answers as "because I was forced to" or "because I was hungry." In such cases the causal relationship between circumstance and behavior is apparent to the individual in question.

However when behavior is motivated by the autocept the situation is quite different. Even as the percept tree subsumes sensory particulars that have become lost in the global phenomenon, so in a much more complex fashion have the many recurrent experiences that constitute the autocept become lost in the larger representation.

Thus, the extent to which the autocept enters directly into behavior is difficult to determine in any particular case. Self-awareness is a continuous process; the encode constellations, visceral input, and other sensory components that have become fused into the autocept are perpetually being monitored as apperception fluctuates. However many subtle motivations coalesce to activate and sustain behavior. Because the source of such motivation is often obscure and because the inputs that represent the autocept are perpetually present, the individual may incorrectly assume that such behavior is self-directed when it is actually the outcome of different and more amorphous inputs.

The situation is further complicated by the fact that many activating inputs, some representing obvious imbalances, such as thirst and hunger, and others representing more subtle homeostatic systems, have become integrated into the autocept. When this happens and the individual is motivated by such inputs, the resulting behavior may be judged as arising from the autocept, which under these circumstances is indeed the case. In like fashion, the inputs from many activities, particularly those that occur repeatedly, such as brushing teeth, eating, making coffee, driving to work, etc., may become a part of the autocept. The inclusion of such CASs may explain why much behavior that is judged by the individual as arising from his or her own will can proceed so smoothly and efficiently, with little if any perceptible motivation. It may also explain why people often experience that they are determining and directing the course of action in such cases. To the degree that the autocept and the CAS have become fused into unity, they are.

Is there something about the behavior attributed to the autocept that is different from behavior initiated and sustained by other activation systems? When "I" decide to pick up a pencil and then do so, is there something that sets this behavior apart from the typical activities that occur during an individual's day? One factor is the continuing presence of an intention that can be monitored as a behavior unfolds: that is, when we intend to do something we are repeatedly apperceiving that we have this intention. Since the inputs representing intentions have recurred many times, they are an integral part of the autocept. Thus the apperception of an intention becomes a basis for the individual's judgment that an act is instigated and sustained by the autocept.

There is evidence from dreams (see chap. 7) and from studies on paralysis (see chap. 5) that the experience of freedom, which is a corollary of much behavior, may be derived from the inputs arising from anticipatory and actual movement. During sleep the sensory and motor threshold are raised, so that the inputs normally monitored to provide our experience of freedom and volition are not available. The nonvolitional character of dreams is a universal outcome. The dreamer becomes a passive bystander, somehow within the dream but unable to determine what happens. When the inputs from anticipatory and actual movements are cut out by paralysis, the feeling of volitional control over the affected area ceases.

This analysis gives some hint of the difficulties that have made the issue of volitional behavior so intractable. Everyone has the experience that certain choices arise from his or her own volition. Yet, to conclude that there is some kind of willing agency that enters into such choices relegates a critical aspect of human motivation to the infinite regress of homunculus ex machina.

Inclusiveness of Composite Action Systems (CASs). The most global homeostatic response any organism can make involves the movement of the entire body. Organisms move as a totality away from situations that account for such inputs as pain, excessive heat, and excessive cold, and toward the cues that signify food, warmth, and water. The input from such circumstances, regardless of its apparent diversity, has become tied to the same general response. On the other hand, there are many discrete motor reactions that will occur only when quite specific input is focalized. A hungry organism will tend to exhibit eating responses only when food is in its mouth, and drinking responses only when both thirst and water are present. Thus, the motor threshold involved in gross away from and toward movements may be lowered relative to a number of different inputs, while the threshold pertinent to such discrete responses as eating and drinking is crossed only when certain specific and appropriate inputs are focalized. This makes sense from an evolutionary standpoint. A multitude of different environmental situations (those comprising hazard or food sources) can be reacted to adequately only by the movement of the entire organism, while specific responses to discrete stimuli allow precision of adjustment to particular, but often no less vital, adaptation circumstances.

Many animals live in particular home ranges, where critical features of that range soon become subsumed into an inclusive, all-purpose CAS, which can be utilized in a wide variety of adaption circumstances. Thus, animals can behave appropriately to different activation inputs; they can go to one place to get water and to another to get food, depending on which activating input is being introduced.

The autocept is the most inclusive and all encompassing CAS. It emerges as the inputs derived from often encountered situations and activities become fused into a phenomenological unity—a unity that constitutes the identity of each person and occasions behavior appropriate to the welfare of each person and his or her group when the integrity of that unity is threatened.

Summary and Discussion

Innate reflexes, at least in one respect, are prototypes of all highly practiced learned behaviors, from the simplest discrete reaction to the most complex series of responses; a high degree of automaticity prevails. In such reflexes as the eye-blink and knee-jerk the behavior occurs when the stimulus is presented without awareness or apperceptual involvement, except perhaps during and after the response has occurred. The entire information processing and reaction sequence is below the level of consciousness.

When behaviors are learned the components are initially within the field of awareness and require apperception for their establishment, but as practice continues conscious awareness becomes correspondingly less involved. With practice, CASs emerge and the information-reaction process that takes place occurs automatically except for occasional apperceptual monitoring. Thus, when driving a car we not only automatically stop at red lights, but also make all the appropriate reactions to the many variables encountered without any particular awareness that we are doing so. In like fashion such complex behaviors as typing, piano playing, walking, speaking, and any other highly practiced behavior sequences can proceed efficiently.

The adaptive contribution of such automaticity is considerable; the single channel of apperception is cleared for other business; hunters can be alert to signs of game as they walk along, typists can scan other office demands while typing, and professors driving to their morning classes, can plan the coming lecture.

Briefly stated, to the extent that a vital system deviates from optimal balance, activation input derived from this system, such as thirst (or anxiety and guilt, as in the case of the autocept), will intrude into the field of awareness and dominate in apperception in direct relation to its intensity, duration and prepotence. To the extent that information inputs (sensations and memories) have been associated with such activation input they will tend to be focalized as apperception fluctuates. To the degree that the motor threshold has been lowered, either through innate mechanisms or through practice, the apperception of such information input will trigger appropriate behaviors. When such behavior occurs (whether moving toward or away from, or ingesting, or pleading for forgiveness) the imbalance in question will be reduced and the apperceptual mechanism will be freed for other service. When we tie this analysis to the observation (see chap. 4) that the various activation inputs directly represent the status (i.e., the relative balance) of many of the critical systems of the body, the developments in the previous chapters and in this one coalesce to provide a unified theory of the organism, not just as an information processor, but as a dynamic system encompassing both information processing and reaction so that the balances necessary for living can be maintained within tolerable limits.

RESPONSE CESSATION

There is general agreement that two factors are involved in every circumstance of response cessation: reduction in performance and loss of learning (see Mackintosh, 1974 for a review of pertinent research and theory). A simple example should clarify the problem, and also allow me to present an evolutionary interpretation. If a rat is trained in a Skinner box, the motor threshold between critical sensory input (the activating input of hunger and the sensory discriminatives from the box) and the bar press will become lowered. During extinction the rat will continue to respond for a relatively prolonged period after the magazine has ceased to deliver a pellet of food.

Why does it cease responding? If the activating (hunger) input is still present and is being apperceived, response should continue. Indeed this would happen if other components did not intrude into apperception and occasion incompatible behaviors. As the rat continues to press the bar, fatigue input will increase and become focalized in direct proportion to its intensity. When it is apperceived, the motocepts that have been associated with such input in the past will be triggered. These resting responses are incompatible with the bar press response, and, thus, behavior at the bar will be reduced in frequency. Yet fatigue is not the only interfering input to intrude during an extinction circumstance. If a barrier interferes with equilibrating activity (in this case, eating), secondary activation input in terms of anger will be introduced. When this input is apperceived, new behaviors in terms of attack responses will compete with smooth performance at the bar. Since attack typically involves moving toward behavior, the anger input may for a time actually bring about an increase in activity at the bar, but as it becomes more intense, specific aggression behaviors such as biting and scratching will tend to predominate, and these responses are almost directly incompatible with the bar press.

This may explain response cessation, but how about unlearning? In my view, whenever responses other than the bar press occur, these responses rather than the bar press become associated with the situation. Since to be apperceived is to be encoded (see chap. 6), the visual inputs from the experimental apparatus become associated with behaviors that interfere with the bar press; the sight of the bar no long occasions the bar press, but, rather, occasions the previously mentioned incompatible responses. Thus a loss in performance is caused by the apperception of inputs that instigates competing behaviors, while a loss of learning is brought about by the development of interfering associations.

However why does the animal finally leave the bar situation and move to some other area? Because, as the prior associations are weakened through interference, the anger that produced some of the abortive and interfering behavior will be reduced. Further, after sufficient rest, the fatigue will also be reduced until it likewise ceases to influence behavior. Since these extraneous inputs are no longer in competition, the hunger once again will predominate, and the rat will respond to such input with the next most appropriate behavior in its repertoire. We may thus hypothesize that:

> The rate of extinction will vary inversely with the number of pertinent alternative responses that are available.

Therefore, it is predicted that a rat will extinguish faster to a given goal box if another one in which it has been given a certain amount of training is accessible.

A similar explanation seems reasonable for extinction in the classical conditioning situation. It will be recalled that repeated presentation of the metronome without the meat powder brought about a cessation of salivation on the part of Pavlov's dog. In my view, the response itself is not extinguished; it is simply no longer made to the metronome. Once again interference is assumed to be the crucial factor. Classical conditioning (see chap. 6) takes place when the sound of the metronome occasions the apperception of the food encode, and as long as the sound and this encode remain associated the conditioning will remain intact. Yet when the meat powder is no longer presented with the metronome, other inputs that interfere with the food encode become associated with the sound. When the metronome is sounded, these inputs, rather than the food encode, become apperceived, causing extinction to take place.

Does the account of response cessation outlined in the last few pages add much that is different and original? Not a great deal. Miller and Dollard (1941) and Mowrer and Jones (1943) gave similar interpretations—interpretations that Hull (1951) accepted as a critical part of his behavior theory. Hull's development, in turn, was expanded by Amsel (1958), as well as by others, to include the interfering effects of emotional responses arising from the frustrating effects of reward withdrawal. My own view is similar, except that, rather than using such concepts as reactive inhibition (Hull's term) and frustration (Amsel's term), I emphasize a cognitive viewpoint that included experienced fatigue and anger as the proximal determiners of response change.

It is likely that my interpretation is much too limited and indeed may be pertinent only for subjects that are placed in response-restricting experimental procedures and apparatus. For animals in an open habitat, other factors, as well as the previously mentioned ones, probably contribute to changes in behavior. These principles, a number of which are congruent with an evolutionary interpretation, are supported by experimental results, both from psychology and from ethology; they relate to conditions that are universally imposed on animals as they forage for incentives.

For example, as many field studies have confirmed, if an animal is able to observe that its particular food, whether game or grass, is not present in a given place, it will not go there. Predators, whether raptors or large cats, scan their hunting range, and if nothing is observed they move on.

Laboratory studies have also confirmed that animals will cease going to a place where they have observed that no food is present. Seward and Levy (1949) demonstrated that the behavior of rats trained to run down a straight alley for food in a goal box could be extinguished by placing them in the goal box without any food in it. These results have been replicated many times, perhaps most recently by Premack (1983). In experiments designed to demonstrate reasoning, but which also give clear evidence for observation learning and strategies in foraging, Premack let chimps observe him put two different types of food in two widely separated containers. Out of sight of the chimps, he removed one of the foods and then let the chimps see him eating it. Afterwards the chimps went predominantly to the presumably intact container, with an accuracy comparable to 4- and 5-year-old children. In another study Premack let chimps observe him put one piece of food into a container in a field. Later, after seeing Premack remove the food, the chimp made no movement toward the container. However if the chimp saw Premack put in two pieces of food and remove only one, it went immediately to the container.

When we consider that the items necessary for maintaining balance in the various physiological systems of the body are typically distributed in the environment, such search or shift or stay strategies are probably deeply imbedded in many species. If we try to understand response cessation (or continuation) without taking such strategies into account, our insights will be restricted to artificial experimental situations where subjects behave within the straightjackets imposed by the apparatus, typically mazes or Skinner boxes. Interestingly, several experiments have been performed in which the design and apparatus have approximated a field situation. Hermann, Bahr, Bremmer, and Ellen (1982), Maier (1932), and Stahl and Ellen (1974) have all reported that rats can learn shift or stay strategies if they are trained in open situations that allow them to observe where food is or is not present.

Animals, whether hunters or grazers, follow foraging strategies. Some strategies are learned, but others are reflections of the evolutionary history of the species. For instance, animals rarely exhaust a particular resource before moving on to other areas. This tendency, whether inherited or learned, depends upon the fact that overgrazing or overhunting will decimate the resource to the point where it will be unable to renew itself, with the consequent demise of the foragers. Thus selection pressures have ensured that a kind of mutuality will develop between species and its food source, for the same reason that parasites seldom kill off their hosts; and the survival imperative before the resource is exhausted move on is imbedded in the response repertoire of many animals and manifests itself in the alternation tendencies observed in both laboratory experiments and field studies.

Dember and Fowler (1958) found a marked tendency in rats to alternate (and alternation always involves the cessation of one response as another is initiated) in a T-maze when equal rewards were presented in each goal box. In another study McSweeney (1975) reinforced pigeons at two different treadles located at opposite ends of a rectangular chamber. She found that alternation was implicit; the pigeons allocated their time at the two treadles in proportion to the amount (number, quantity, and quality) of the reinforcements available at the respective treadles. They did not fixate exclusively on the treadle that provided the greatest amount of food. This study then suggests that, although alternation is a basic tendency in pigeons, the patterns of such alternation may be learned, and accurately reflect the relative amount of food at different locations.

In field studies paralleling the experimental situations just discussed, Kamil (1978) observed that a nectar-feeding bird, the Amakihi of Hawaii, rarely visits a flower after taking the nectar from it, and does not return to it until sufficient time has passed for the nectar supply to be replenished. In an effort to emulate foraging in an experimental situation, Olton, Walker, Gage, and Johnson (1977) allowed rats to first explore an area that included three towers, with a number of food pellets on the top of each tower. The experimenters found that rats distributed their behavior among the three towers. If they obtained food at one tower they would then tend to visit another tower, rather than go back to the one where they had just received food. In a related study Haig, Rawlins, Olton, Mead, and Taylor (1983) demonstrated that rats can more easily learn a shift than a stay strategy. One group of rats was rewarded on the second trial only if it shifted to the other arm of a T-maze, while another group was rewarded on the second trial only if it went to the same arm of the maze. The shift rats learned significantly faster than the stay rats. Apparently the tendency to alternate overrode immediate reinforcement effects.

Thus, results from both field observations and laboratory experiments suggest that the patterns of an animal's behavior are influenced by distal directives from its evolutionary history and proximal directives from individual learning. Both directives enter into each reinforcement situation and are apparent in the food-finding strategies of each species. Alternation tendencies are simply one of these strategies. It should appear in every circumstance where the consumption of food at a particular place reduces the probability that food will be located at that place until the resource in question has had an opportunity to be replenished.

This all relates to what Mayr (1982) has termed a cost-benefit principle. This notion suggests that foraging strategies reflect both the evolutionary history of the species and the learning of each individual and will tend to achieve an optimal balance between the energy gained by a particular behavior and the energy expenditure required to perform that behavior. A number of laboratory experiments (Collier, 1983; Collier & Rovee-Collier, 1981; Kaufman, 1979; Krebs, 1978) have substantiated the cost-benefit principle. In one study Mellgren, Misasi, and Brown (1984) placed different numbers of pellets on six towers that varied in height from one to seven feet. Rats were allowed continued access to the towers until their behavior stabilized. In general it was found that, as effort increased, as determined by tower height, the rats consumed more food at that tower before alternating to other towers. In line with a foraging hypothesis, rats continued to alternate between towers until the number of pellets left unconsumed were approximately equal, supporting Charnov's (1976) notion that animals will forage at a particular place until the food available at that spot is equal to the amount available on average in the environment.

Thus, research data and theory from both ethology and experimental psychology coalesce to suggest that cessation (or initiation) of response at a given place is far more complex than the simple analysis given at the beginning of this section would indicate. Animals appear to bring into each foraging situation a number of response cessation sets that may reflect inherited, learned, or both patterns of alternation. When an animal leaves one place to go to another it is apparently influenced by a number of coalescing variables, some of which are just now being isolated and examined.

In the beginning of this chapter, it was stated that past, rather than present, learning is involved in most behavior. This statement can be expanded. Present behavior is influenced not only by past learning, but also by inherited intrusions, as represented in both morphology and the neural underpinnings of action, which set limits and constraints (as discussed in chap. 6) upon the learning that takes place.

Response cessation in humans probably differs only quantitatively from the same process in other animals, but it is much more complex. In human beings, observation learning becomes much more predominant. Most human learning (see chap. 6) is a direct and necessary outcome of the process of apperception, resulting in the establishment of the myriad encodes that serve as a functional internalized accumulation of the individual's experiences.

It will be recalled that in human beings, and probably in many other animals as well, at least four steps are involved in the encoding process. First, there is the development of percepts arising from the influence of fusion and constancy mechanisms. Next, there is the emergence of even more inclusive encodes, that is, concepts, which represent a summarizing memory residual derived from recurrent exposure to similar percepts; then, as a function of the sequential and repeated presentation of many percepts, encode associations, and encode constellations (cognitive maps) are established.

Finally, to the extent that sequential behaviors are repeated (practiced), CASs emerge, allowing such behavior, whether speaking, typing, or walking, to proceed easily and efficiently with only occasional apperceptual monitoring. The information processing-reaction sequence occurs automatically and unconsciously, freeing the limited capacity of the single-channeled apperceptual system for other inputs.

This automaticity highlights a central aspect of response cessation, which has been both obscured and neglected because of the fixation on traditional extinction studies in animals, where both the activation for action and the nature of the response are restricted by the demands of experimental design. Most response cessation, such as in speech, occurs when the behaviors involved shift back and forth between alternatives in reaction to the myriad circumstances encountered by the individual. In classical extinction studies the instigating circumstances, hunger, thirst, pain, etc., have been imposed by the experimenter. In the flux of daily human encounters, such basic demands are minimal. In speech, which involves a predominant amount of behavior onset and offset, the basic motivation (proximal cause of activity) is to respond, or not respond, appropriately to each particular situation, social or individual. In speech, response alternation achieves its most facile and rapid demonstration, and suggests again that most response cessation is related to a shift in performance, not to a loss of learning.

As discussed in an earlier section of this chapter, relatively few behaviors must suffice for literally thousands of different potential inputs, both sensory and memory. When response cessation takes place, a shift in performance, rather than a loss of learning, is usually involved; if a particular input-response sequence does not result in the reduction of an imbalance (i.e., activation input), the next behavior in what may be a very complex behavioral hierarchy will take place. Thus, if a background input such as hunger instigates the apperception of associated encodes, which in turn triggers behavior, if that behavior does not result in finding food, the next response in the hierarchy will appear. This hierarchical model works at all levels and is particularly appropriate when discussing the effective use of language. If after a few moments of apperceptual scanning a certain word is not focalized, the next word in the hierarchy may be focalized, triggering an appropriate verbal motocept.

Thus, response cessation rarely involves a loss of learning, but much more often simply demonstrates a shift in apperception that triggers different behaviors. Yet loss of learning does occur; it happens when prior fusions, associations, or both become disrupted by the interference of new learning. If a subject practices mirror drawing for only two hours (see chap. 5), normal perceptual organization may become so altered that he or she automatically writes the mirror image of his or her signature. New learning may also alter the subsumptive character of concepts, either to include or exclude material. Encode associations may be disrupted by new contiguous pairings that represent environmental concurrences that are different from those previously encountered. Lowered thresholds between apperceived inputs (whether encodes or sensory components) and motocepts are also capable of being altered as different input-response sequences, which interfere with old associations, occur.

Interference or Decay?

What is the basic principle of extinction or forgetting? This question has generated a great deal of research and much theoretical speculation (see Hall, 1976 for an excellent review of this material). In general, two different processes have been held accountable: decay and interference. From an evolutionary frame of reference, decay (except for buffers, see chap. 6) would have little survival value. Ideally, any association that has been functional in the past history of the organism should remain intact as long as there is a possibility that the situation will be encountered again. Thus, to the extent that decay might take place, the processes accounting for it would be detrimental to survival and would probably reflect limitations of the encoding system, or decrements from aging, disease, variations during cell replacement, or perhaps to changes in neurotransmitters.

On the other hand, interference would appear to be a functional type of unlearning mechanism. It will be recalled (see chaps. 3 & 6) that learning provided a major increase in survival potential in that it allowed adaptive responses to short-term concurrences in each organism's particular environment. As long as such concurrences were present, the associations that represented them would have great survival significance. However the same conditions that account for the associations in the first place should also account for their disruption when new concurrences are encountered; to the extent that such new concurrences involve some aspect of the old one, interference or facilitation (depending on the nature of the similarity) should occur.

In short, as far as extinction and forgetting are concerned, interference would have profound adaptive repercussions. It would provide the basis for the development of new associations reflecting changes in the short-term concurrences in the individual's own environment, and the cancellation of associations reflecting the concurrences that were present prior to these changes. It was the survival value contributed by effective response to short-term environmental concurrences that provided the basis for the emergence of learning in the first place, and which accounts for its present functionality. Yet, it would appear not only reasonable, but also parsimonious, that new learning, canceling out the old, would take place when environmental circumstances are altered. In a sense there is no such thing as extinction or forgetting, there is simply the development of new associations that effectively interfere with those that have become detrimental, since they would lead to inappropriate, perhaps lethal, behavior.

In this treatment of forgetting through interference I am dealing with long-term memory. Short-term buffers are a different story. In short-term memory (lasting approximately 15 seconds) forgetting is primarily due to decay (see chap. 6). This makes sense from an adaptive point of view. Short-term memory provides a holding display (a buffer) that allows time for the limited-capacity apperceptual mechanism to be cleared so that the focalization of relevant material within the display can take place. When apperception occurs, the focalized component becomes transferred to long-term memory, while nonfocalized components become rapidly lost, effectively clearing the buffer for new material. The adaptive contribution of such rapid decay is derived primarily from a reduction in complication, that is, the rapid loss of inconsequential clutter that otherwise might inundate the information-processing system.

The principle of reaction summarizes the basic notions in this chapter, and is, with only slight modifications, a restatement of the postulate of process that was given in chapter 2. **The postulate of process** will be recalled: *All activity, whether overt or covert, serves the immediate function of moving the organism toward equilibrium.* The similarity between this postulate and the statement that follows is evident.

Principle of reaction *The overt act that takes place at any given moment will always be the one that, in terms of the organism's structure (both as derived from heredity and altered through learning) provides the greatest probability of regaining equilibrium.*

This statement is the **postulate of process** restricted to overt behavior. Yet it is a reflection of the **postulate of inference** as well. This postulate, it will be recalled, states that *every attribute that has been within a species for an enduring period abets the survival potential of the organism*. Thus the principle of reaction demonstrates what the reader may have noticed all along, that the two fundamental postulates are interlocked; the process or attribute that abets survival must make its contribution by adding to the facility with which an organism is able to maintain or regain the equilibrium necessary for life.

CHAPTER SUMMARY

All organisms (this is particularly true of human beings) bring into each situation a complex array of inherited and learned modes for sensing, integrating, retrieving, and behaving. These modes serve as the context within which all specifics of learning and reaction must reside, and they inevitably influence the specifics that take place. Thus even in the most controlled experiment there are no naive subjects; each subject brings with it an abundant heritage from its past, both as individual and as species, that colors everything it does and everything it learns.

Much of the heritage from the past comes in terms of CASs that are syntheses of physiological and behavioral operations (sometimes including and sometimes excluding cognitive contributions) that provide the basis for efficient reaction to commonly encountered situations. Such CASs may depend upon inherited hardwired neural circuits, may involve a combination of inherited circuitry and learned modifications, or may be almost entirely derived from learning. They may vary from specific reactions to environmental particulars, to total reactions to general environmental circumstances. Such reactions may occur with rigid and inflexible automaticity or, in their more complex expressions, they may be intimately tied to plastic cognitive processes.

Overt reaction depends upon two processes: the development of motocepts and the lowering of the motor threshold between such motocepts and some apperceived input. In inherited associations the motor threshold between specific inputs and particular behaviors is low to begin with, but in learned associations, a series of contiguous occurrences between apperceived input and response is required. The lower the threshold between input and response, the greater the automaticity of the behavior. This is most apparent in highly practiced habits and in actions attributed to the autocept.

Conflict occurs when two or more inputs that have been associated with incompatible responses are sequentially apperceived. Although apperception is single-channeled and many input components are always in competition for focalization, extreme conflict is relatively uncommon, since—and this is fortunate for adaptive behavior—the same circumstance rarely introduces two different inputs that are associated with incompatible responses.

Response cessation occurs when inputs that have been associated with alternate responses predominate in apperception, and loss from long-term memory (whether associations between encodes or the alteration of percepts or concepts) takes place through interference. The loss of material from short-term memory is primarily caused by decay, although the introduction of extraneous inputs may interfere with recall.

9 Language

The logic of our development is now complete. A development that suggests how organisms have evolved to be appropriately reactive to internal and external changes, so that the balances necessary for vulnerable life can be maintained. Reaction, whether the rudimentary tropisms of amoebas or the refined adjustments of human drivers in heavy traffic, has been integrated. Now we come to human language. It is appropriate that its treatment comes toward the end of the book. It is the final step (see chap. 3) in the emergence of the adaptation attributes facilitating differential response, the most facile method for implementing individual and group interactions that evolution has produced, and the most refined demonstration of reaction that has appeared. As such, it will be examined here as an exemplar of the developments in the previous chapters, and as a demonstration of the appropriateness and subsumptive power of the theory. This chapter also includes an update on some of the history and research in the area, and some of the controversy.

THE NATURE-NURTURE CONTROVERSY

As with many of the issues having to do with human development, the nature-nurture controversy intrudes to add contention and inspiration in our efforts to understand language. Such behaviorists as Watson (1913) and Skinner (1957) agreed that the child is born with an innate capacity for language learning but insisted, in the tradition of John Locke, that this capacity is a clean slate (tabula rasa, etc.) upon which the experiences of the child are gradually represented. Skinner's position is the clearest example of this approach. The child, during the babbling stage, emits a vast number of behaviors with their corresponding sounds. Certain of these sound-behavior emissions are differentially reinforced by the reactions of others, primarily the mother, until words and then sentences emerge in the child's repertoire. Language development is shaped by the sculpting action of reinforcement in a manner reminiscent of the shaping action of the selection pressures that account for the emergence and function of all biological attributes.

The use of the conditioning paradigm (both classical and instrumental) marked the culmination of pure empiricism, (i.e., learning, nurture, etc.) in the explanation of language.

It seems curious that observation learning never became a serious theoretical contender in explaining language development. The child, although probably not a good listener, is continually bombarded with language cajolings from solicitous and curious adults. Certainly it is capable of apperceiving and, according to our treatment of observation learning (chap. 6), the apperception of a given input is sufficient for the establishment of an encode. The observation learning paradigm was probably too late in appearance for it to have much impact; the dramatic introduction of nativistic interpretations, as represented by Chomsky and certain present-day evolutionists, captured contemporary interest.

In my view there is little doubt that observation learning accounts for much of the origin and development of human language. Once, in discussing this issue with Melvin H. Marx, I asked my two-year-old son to say a word *karophatalapinx* that he had never heard before. He immediately said the word clearly and with all the proper spacings and enunciations; then he looked around beaming with satisfaction as he observed Marx's bemused thoughtfulness and my gratification. Yes, sounds can be observed, as can odors and tastes, etc., and they can be encoded as a direct function of apperception. "Monkey see, monkey do, child hear, child speak."

When Chomsky critically evaluated Skinner's 1957 book on verbal behavior, a powerful trend toward nativism in language development was initiated. There are several nativistic arguments. The babbling stage that takes place from about the fifth to the 15th month, with minor exceptions, appears in all children everywhere at approximately the same time, even in deaf children. Language usage begins universally at around the 18th month and proceeds rapidly, suggesting the presence of a developmental window. Adults, of course, can learn new languages, but it takes them much longer, and there is typically an interfering accent (Chomsky, 1959; Lenneberg, 1967).

Chomsky (1965) believes that children have an innate language acquisition device (LAD) that not only accounts for this rapid learning but also imposes an intrinsic grammar on the child's utterances. Such LADs seem remarkably similar to the contemporary evolutionists' notion of modules that are assumed to reflect particular selection pressures.

All children, regardless of culture, appear to go through the same stages of language development, in the same sequence (Pinker, 1994; Slobin, 1970). There is a flexibility and creativity in language usage that the nativists believe obviates the stereotyped rigidity they perceive in the Skinnerian explanation, as when my son said karophatalapinx. They also posit the existence of universals that coerce grammar into particular patterns. Certain of the evolutionists apparently agree:

> All languages are complex computational systems employing the same basic kinds of rules and representations with no notable correlation with technological progress; the grammars of industrial societies are no more complex than the grammars of hunter-gatherers; Modern English is not an advance over Old English. Within societies, individual humans are proficient language users regardless of intelligence, social status, or level of education. Children are fluent speakers of complex grammatical sentences by the age of three, without benefit of formal education. (Pinker & Bloom, 1992, p. 451)

Such statements would seem to indicate that there is a rapprochement between Chomsky and the evolutionists, but this is only superficially the case. Chomsky, and indeed the evolutionist Gould, insisted that human language cannot be explained as the outcome of natural selection, but must be an emergent exaptation (Gould) or be the result of unknown, innate, rational capacities (Chomsky, who follows Descartes). In this regard Chomsky made the curious statement: "It is perfectly safe to attribute this development (of innate mental structure) to 'natural selection,' so long as we realize that there is no substance to this assertion and that it amounts to nothing more than a belief that there is some naturalistic explanation for these phenomena" (1972, p. 97).

There seems to be no basis for substantive conflict between the evolutionists and Chomsky; both adhere to a nativistic interpretation. The modules of contemporary evolutionists and the LADs of those who follow Chomsky seem remarkably similar in their segmentation of a general domain interpretation. The conflict is at a more basic level. Most Darwinians view language as a natural expression of evolutionary variables, while Chomsky (and Gould) see language as a transcendent emergent. This fundamental difference remains. Yet perhaps I am being picky here. Calvin and Bickerton (1998) defend the view that Chomsky and Darwin are fully reconcilable, and, in terms of the practical outcomes, perhaps they are. It is possible for completely divergent approaches to arrive at essentially similar conclusions, but sooner or later the empirical data should help us determine the relative legitimacy of the positions. Whether a given attribute is an outcome of variables that fit into a framework that can be empirically evaluated or is the result of forces that, by their very nature, defy such evaluation seems a fundamental distinction that the optimistic efforts to conciliate by Calvin and Bickerton cannot hide.

To sum up, the nativists whether evolutionists (Darwin, Gould) or linguists (Pinker, Chomsky and his followers) not only hold that the general capacity for language is inherited, but that there are innate particulars (modules, LADs) that influence the pacing and pattern that language will take both in its acquisition and in its use. The disagreement among the nativists has to do with the origin of these capacities. The evolutionists (Gould excepting) hold that the capacities in question, both general and specific, are the outcomes of natural selection, while Chomsky and his adherents hold that they are the result of innate ways of knowing of transcendent origin.

The empiricists, of course, have counter arguments. They agree with the view that all humans have an innate capacity for language development, but suggest that innate particulars are not necessarily involved. All children do babble, but they also make all kinds of movements with their hands and feet. The occurrence of babbling drift (Weir, 1966) indicates that such babbling rapidly becomes altered to fit the language that the child hears. For instance, Chinese children begin to express tonality in their babbling at a very early age (eight or 10 months). Mothers can identify which language group a child is being raised in by listening to their babbles. Both babbling drift and mother identification suggest that learning makes rapid inroads into whatever innate capacities there are.

The developmental window argument is countered by the observation that the child learns language on a clean slate, while adult language learning invariably involves interference from the language already in use, which could easily account for the difference in acquisition. The universals that the nativists posit may occur, but they are the outcome of common circumstances, both in the environment and in the socialization of all people everywhere, rather than being imbedded in innate predispositions. The empiricists also point out that assumptions of nativism invariably move the issue into the inaccessible, a safe position since it is invulnerable to objective exploration. They also suggest that nativistic criticisms are typically made against experimental findings and theoretical statements, as in Skinner's *Ver-*

bal Behavior, which are given with sufficient clarity to allow nativistic assumptions to achieve dubious credibility through invidious (as in Chomsky's criticism of Skinner) comparison.

The nativists, of course, suggest that simplicity and clarity bring little guarantee of truth, that important problems are often complicated and that science progresses only when superficial notions are replaced with ones with greater integrating power and deeper meaning.

The nativists suggest that language involves novel and creative combinations and relationships that are far removed from the stereotyped sequencing that is characteristic of learned behaviors, as expressed in Skinner's position. The empiricists counter that, although objective sequencing may be required in any well-controlled experiment, this does not rule out flexibility in the learning that has taken place, that seemingly stereotyped behavior is always variable, as when a bird approaches its nest from many directions, or a spider spins its web. Originality and creativity in language are likely a reflection of the flexibility implicit in all thinking and thought-driven activity, (as suggested in the sections on daydreaming and dreaming), which can be checked by anyone who relaxes in his or her easy chair and lets their thoughts roam around. Thoughts, although following intricate lines of association, can be remarkably flexible and creative, as can language when it is triggered by such thoughts.

In passing it may be noted that the empiricists have a cogent, although peripheral, counter against the stereotyped and rigid argument. When we examine nativistic behavior sequences, whether reflexes, instincts, migration, or web spinning, flexibility may be present, but it is invariably reduced to a minimum. Learning typically provides the potential for flexibility and creativity, while nativistic propensities typically impose encapsulation and rigidity.

Of course, in terms of general acceptance, the complexity of presentation may have a lot to do with the palatability of an argument. I have always been impressed with the simplicity and clarity of Skinner's treatment of language in *Verbal Behavior*, and have often been baffled as I try to explore the ramifications of Chomsky's position. One appreciates again the cycles of human effort relative to complex issues. Occasionally someone comes along with a clear, if somewhat simplistic, position. The very clarity of the statement is an open invitation, as it should be in science, for attack and criticism. In Skinner's treatment of language the difficulties are exposed, and the failings become apparent. The warts show. In Chomsky's position the attack may be, as I think it is, totally appropriate and justified, but it is couched at an abstruse level, with such esoteric terms and arcane analytical methods, that it achieves a degree of invulnerability simply because it lies beyond the competence of many investigators.

The give and take in any nature-nurture controversy typically moves from an objective and testable statement by the empiricists to corrective and at times embarrassingly obvious criticisms by the nativists. This was the case with Kant's analysis of Hume's position, which was so clear and simple that the flaws were sufficient to wake Kant from his dogmatic slumber, making him renounce pure empiricism in favor of nativistic categories. It was true of Bridgman's operationism, stated with such simplicity and clarity that its practical limitations were exposed, making it apparent that much of scientific fact or truth is inferential. It was true of the sensationalism of Mach whose view that the stick in the water is bent, and if there be any illusion it is that it is still straight exposed the hazards of becoming encapsulated in the immediate and made an acceptance of inference in the search for fact an obvious requirement.

Those who present a clear, even though simplistic, view provide an essential service as a focal point for argument, as a tangible marker against which problems can be exposed and clarified. This is the way that science works. Yet there are particular dangers where language is concerned. As with all withdrawals into nativism, the retreat is often into obscurity, where speculations can run rampant, where hairsplitting tangents can proceed apace, where explanatory efforts can become so convoluted that they become invulnerable. Nativists often discern the problems exposed by clear, objective, though simplistic, explanations, but they seldom solve them.

In my own treatment I will adhere to the commitment that nativistic principles will not be assumed until empirical positions have been examined and found wanting. There is little benefit in retreating into the obscure and often intractable until the failure of empiricistic (learning, nurture, etc.) arguments make it necessary.

LANGUAGE DEVELOPMENT

Language involves one of the clearest examples of the emergence of motocepts and the lowering of the motor threshold. As has often been observed (Locke, 1983), a child's first words are made up from sounds that have been highly practiced during the prelingual babbling phase. As a function of babbling, the inputs derived from tongue, lip, and mouth movements become fused into motocepts long before the child has begun to use language in a meaningful fashion. Even as such motocepts are being established, the process of motor threshold lowering is taking place. After much practice the motor threshold between a sensory input and the response that produces a sound becomes lowered to such a degree that the child may automatically make the sound when the input is presented.

Each time the child's visual image dog is followed by the word dog, not only does the motor threshold between the image and the word become lowered, but also the encodes representing both become more firmly established. Once established, the apperception of either encode (or the fused encode) will occasion the motor responses essential to the utterance of the word. Thus the child may say the word dog even when neither the visual image nor the sound is presented. The apperception of the encode alone becomes sufficient for the utterance of the word.

Speech then involves the emission of the sounds produced by the motocepts that have previously become associated with particular encodes. In effective speech the encode must always precede the motocept that occasions the vocal sound. Thought and speech are remarkably similar; thought involves the sequential apperception of encodes, while speech includes the occurrence of the motocepts that have been previously associated with such encodes. When speech has been highly practiced it may proceed rapidly and efficiently, each utterance being momentarily preceded by the apperception of the appropriate encode.

Speech Initiation

Once speech is initiated it typically proceeds smoothly with minimal hesitation. As the language composite action system (LCAS) is demonstrated, the speech sequence unravels with only occasional apperceptual monitoring. Yet what initiates the sequence? A long series of thoughts may precede speech involvement. It's as if there is some kind of speech switch that is suddenly turned on to begin and sustain the speech process—some event in the nervous system that no doubt will someday be isolated and measured. Are there any hints about where we may begin our search? A few perhaps. One of the fondest memories I have of my father was of him sitting in his easy chair lost in deep thought with his lips moving! Apparently his thoughts and speech had become so intimately fused that the occurrence of one was almost sufficient for the occurrence of the other. So much so that on occasion he would utter garbled sounds that, as I view it now, were probably indications of a spill over, a leakage through the speech switch.

There are other indications of the intimate relationship between thought and speech. Watson (1913, 1998), for theoretical reasons, believed that thought was implicit speech, and suggested that thoughts were correlated with minute but measurable movements of the vocal cords. It turned out that he was correct. A similar relationship between thought and behavior is present in deaf people, where thinking has been found to be related to implicit movements in the hands.

As a function of practice, the off-on switch between thought and speech becomes automatic and effortless. Consider the nursery rhyme "Mary Had a Little Lamb." A person can think it through without speaking and then speak it. Or he or she can think it and speak only parts of it; the switching back and forth is achieved automatically and with minimal effort. The encode-sound segments may have become fused, but they can be separated. We can think without speaking, but we typically do not speak without thinking, though on occasion everyone blurts out things they did not intend.

It should be emphasized again that a thought-speech sequence, regardless of length or complexity, is a lawful outcome of the dynamic interaction of many variables that coalesce to momentarily determine the processes that are taking place. There is no homunculus turning the off-on switch.

The "Mary Had a Little Lamb" example is typical of all language sequences, though the degree of automaticity is usually not as great as that found in highly practiced nursery rhymes or in often-used words, sentences, or both. As previously noted, LCASs may proceed automatically, depending upon the extent of fusion (practice) unless some input, whether incongruous thought or sensation, disrupts the process.

Speech is always undergirded by many layers of automatically imposed influences, both innate and learned: physiological structures (mouth, tongue, vocal cords, and correlated neurological machinery, etc.); the particular language with its complex of accents, intonations, rhythms, etc.; and the operative words and sentences (slogans, cliches, aphorisms, etc.) that achieve their automaticity through practice. Speech, whenever and wherever it occurs, is couched within these layered influences, all of which may be viewed as indications of the unconscious information processing that serves as the precursor of speech.

The immediate precursor of speech, of course, is consciousness itself, that is, the dynamic matrix that includes the field of awareness and the apperception of components within this field. However, as just discussed, with practice and resulting fusion the sequential behavior implicit in speech can occur automatically with minimal apperceptual monitoring.

Here again we encounter the adaptive function of fusion, simplification, and the relegation of practiced material to unconscious processing. Such subordination helps free the single-channeled apperceptual system from clutter, determines the priority of components vying for focalization, and makes the monitoring of other inputs, some of which may be important for adaptive behavior, more easily accomplished.

To sum up, with practice more and more of the information processing that takes place is relegated to the unconscious level; the dynamic interactions that determine which components within the field of awareness, memory, or both will be apperceived are below the level of consciousness. I do not mean that it occurs in some kind of unconscious mind, but simply that the dynamic interactions that are precursors to apperception are below the level of awareness. Is this quibbling? Perhaps, but consider this. The term unconscious mind has picked up a misleading burden; it has been invested with all kinds of attributes, powers, and functions, as a repository for unusual experiences and motives, and as the source for bizarre behavior.

In retrospect, the off-on switch metaphor, except in situations where people are shifting back and forth between languages, is misleading. What happens in typical mentation is more akin to multiple competition for a restricted area. Apperception is a single-channeled system, with myriads of components dynamically interacting to determine the focalization that occurs at each moment. Shifting from one language to another is not essentially different from shifting from one CAS to another, or from simply thinking about different topics.

LANGUAGE AS SYMBOLIC REPLICA

All living things, from the simplest to the most complex, are made of vulnerable material residing in a context of energy changes, some degrees of which can harm or destroy them. Selection pressures in this ever-shifting display have ensured that organisms are reflections of those aspects of the environment that were significant in maintaining the balances necessary for life to continue until replication could take place. This has ensured that the structures and processes of organisms represent, in truncated form, certain critical aspects of the physical world. A further isomorphic transduction occurred when the operative aspects of the physical world were internalized to constitute the surrogate world of memory. When language emerged, a further truncation and simplification occurred, comprising a functional symbolic summary of the events and relationships implicit in this surrogate world. Thus, each human being, in his or her structure and processes, is a reflection of the circumstances in which he or she evolved: his or her internalized world of memory is an internalized representation of the actual world, and language in turn, is a symbolic duplicate of this representation.

In all humans certain universals in the environment, in our intricate social interactions and our personal internal processes, are represented, regardless of particular modes of expression in the languages of all of us. Language is a truncated symbolic reflection of the physical world, the conditions present in each person, and the myriad interactions between members of the group. Indeed it was from the communication and reaction requirements of group cooperation and concerted action that the refinements of human language became honed to their remarkable degree of subtlety and flexibility.

If memory is the physical world internalized, language is that internalized world symbolized. In both cases of transduction there is remarkable simplification and unification. The items, events, and relationships of our experience become funneled into fused encodes or concepts; the "if A then B," concurrences become represented by associative webs and the three-dimensional characters of the physical environment become transformed into the perceptual context that provides the stage for the deployment and display of such experiences. All of this flux and variability becomes represented by the symbols and syntax of language. Thus, words and their relationships, though two steps removed, are functional reflections of the physical world, the processes of each person, and the interrelationships between things in the world and the individuals who populate it. Accurate communication is provided to the extent that this isomorphic system is representational. Effective communication is ensured to the extent that people have common internalized worlds and to the degree that language reflects the crucial features of these worlds.

This is the case with most symbolic systems; to be functional they must be representational. Mathematicians may postulate purely symbolic systems, where the relationships are restricted to the symbols alone without reference to anything outside the system, but the outcome of exploring the ramifications of such a system are inevitably tautological. All practical mathematics are function because the symbols and their relationships represent things and happenings in the actual world. When such is the case, complex questions about the workings of the world can be resolved by exploring the ramifications of the symbolic system. Calculus is a language that has practical applications to the degree that its symbols and their relationships truly represent the way things are in the physical world. The same is true of scientific theories. However abstract, they achieve their functionality to the extent that they are valid representations of some aspect of nature. Euclidian geometry is a good example of a symbolic system that fulfills the representational requirements of language, mathematics and theory.

Thus, even as thought is an expression of things and relationships in the physical world internalized, language is an externalized reflection of thought. Language represents, as does thought, central features of man's experience with the physical world, his own processes, and other human beings. It takes the multiplicity of these happenings and reduces them into simplicity so that the critical adaptive interactions with others in the group can take place. Thought is the world transformed for planning and foresight; language is thought transformed for communication and concerted action.

WORDS AS MIRRORS OF PROCESS

Words are not only the fundamental units of communication, but also windows into the nature and function of the processes they represent. Words, whether guilt or jealousy, stand for something, and the relative durability and utility of such words are indications both of the existence of such a state and of the contribution to adaptive behavior that it makes. The words rock and water in English, and comparable terms in all languages, have had a long history of usage because they stand for something common in human experience. The same is true for such abstruse words as awe, jealousy, remorse and the many other terms that represent subtle emotions. All languages have such terms because they represent states, seemingly internal and private, that are common to the experience of all of us. When you say you feel guilty, anxious, or remorseful, etc., I can understand you because, as a member of the same species subjected to similar circumstances, I have had such experiences. Considered from this frame of reference, words are literal instruments exposing the nature and function of internal states.

BODY LANGUAGE AND EMPATHY

Words are the primary communicators of such internal states, but there is an auxiliary language that adds precision and sometimes intensity to our understanding of them: facial expressions, body postures, laughter and tears, conjoin to provide this auxiliary language. When we say that we are sad, others can understand us, but if we are weeping the communication is much more vivid and more deeply understood.

All of such communication is abetted by a remarkable capacity that is highly developed, if not unique, in humans—the capacity for empathy; that is, the knowledge that others are experiencing the same thing we are. When we look at something and others are looking in the same direction, we automatically assume that they are experiencing the same thing, and the same is true for sounds, temperatures, and smells. Such an assumption of common experience has long been and yet remains vital to concerted action toward external things, such as hunting or preparing shelter. Yet it is even more important in helping us to comprehend the internal states of others, which are central to group cohesion and motivation. When we are exposed to the indicators of such states as sadness or joy we not only have words to designate such states, but also literally experience such states in ourselves. To see tears or hear weeping in others brings sadness, to see or hear laughter brings joy. Panic is contagious, and so is the blind rage that sometimes results in the cruelties of mob action. However the indicators of rage or fear in others may instigate and sustain behavior essential for the survival of the group, as may happen in war or massive threat. Indicators of such internal states trigger empathetic responses, a kind of social communication that implements group cohesion and group action.

Great actors are deceivers, able, at a moment's notice, to express the entire panoply of signals that trigger emotional states, from anguish to awe. They are experts in empathy. They are able to project with their posture, voices, and facial expressions all the nuances of such subtle internal states, and they can depend on the shared capacity for empathy in their audiences for a predictable response. Yet in the continuing social interplay that has been central to our evolution and still remains fundamental to our nature, we are all deceivers when the demands of a situation require that we be so. Rare is the person who has not shed crocodile tears.

How universal are these hidden, covert experiences? To the extent that the words designating such processes and experiences have endured through time and across cultures they are common in the experience of all of us, and, as such, they represent fundamental, innate, evolutionarily derived characteristics of our species. Thus, research should examine the longevity of such terms in each language and the universal representation of such terms in all languages. It can be inferred that jealousy is a fundamental attribute of human beings because it has been represented by a designating term for hundreds of generations in the English language and because there is a corresponding term in all languages. Words, when combined with the auxiliary indicators previously mentioned and the capacity for empathy that human beings possess in marked degree, become avenues into the nature and function of processes that are seemingly private and covert.

Language, as often noted, is not unique to humans, the Rubicon that separates us from other animals. All animals have forms of communication, sexual signals, signals of territoriality, infant-mother interactions, cooperation signals in the hunt, the warning cries of the vervet monkey, and the waggle dance of the honey bee. It is not my task here to consider the vast research that has taken place on animal communication, certainly a worthy involvement both for its own sake and for any light it may shed on human language origins and functions. Rather, I look at the possibility of universals, constants within the welter of particular sounds, intonations, and combinations that distinguish all languages and persist in spite of the word drift that takes place through time. I am not at this point concerned with whether such universals, if they exist, represent circumstances that are common to all cultures or innate modules or algorithms that coerce particulars into common molds. I am primarily interested in the manner in which the common features of human experiences, as myriad as they are, become simplified and ordered so that efficient communication can take place. In this effort the nature of universals may be clarified.

Against the backdrop of this discussion we take a cursory look at the English language, not only to shed light on the possibility of universals, but also to examine the manner in which the essential features of mankind's physical and social environments and enduring aspects of his own nature become reflected in both thought and language. The process termed transduction—the functional reduction of complexity into simplicity, the funneling of myriads of gradations into functional and tractable singularity—should be kept in mind as we proceed. The cardinal rule in this reduction of the many into the few, as we shall see, is that where things, processes, or relationships are similar and recurrent, they become fused into unity so that intractable multiplicity is reduced to functional simplicity. The development and function of language follow the evolutionary imperative of getting the most functional use from the least process complication.

LANGUAGE TRANSDUCTION

Every language represents only a small portion of the inputs that an individual may experience. Any word, regardless of its specificity, is always a summarization, and regardless of how many qualifying terms we may apply, distortion is an inevitable outcome of language usage. People vary on a continuum of pulchritude, but we have no continuum of terms to represent such variation. There are ugly people and beautiful people and very ugly and very beautiful people. We can, of course, make finer language differentiations, but not a great many. So it is with all of our experiences; an almost infinite number of inputs must be represented by a limited number of sounds. Any human being is capable of billions of discriminations, as indicated by apperceptual shift, but even the most learned person uses something less than 100,000 words.

Thus, in the development and use of a language there is a tendency to represent the greatest possible amount of experience with the smallest possible number of sounds. Consider the suffix "ing" in the

English language and marvel at what a broad category of experience is subsumed by this single sound. Hook it to the end of certain other sounds and it stands for process taking place whether we are speaking of talking, walking, killing, or loving.

"Ing" indeed has a mighty subsumptive and generalizing power. However no more so than many other sounds that condense human experience into the categories that make communication possible. The familiar eight parts of speech are remarkable examples of the manner in which complexity is reduced to order and simplicity. In each case the principle of this transduction is to represent a wide band of experience (which, regardless of its variety, has something in common) by a single sound. Consider nouns. A noun, says the grammar book, is the name of a person, place or thing. There are common nouns and proper nouns, singular nouns and plural nouns, and most of the latter, in English, are created by adding an "s." So with a muted hissing sound we subsume the vast category of the more than one. Pronouns, too, participate in this functional transduction. Take "it" for instance, and consider again the remarkable generalizing power implicit in a sound. "It" can stand for any idea or place or thing.

Since much of our experience is undergoing change, every language needs sounds to represent such flux. Verbs not only fulfill this function, but also tell us when this process happened. Again we see the inclusive power of a sound. Place "will" before certain other words and we subsume the future, add "ed" and we represent the past. Nouns and pronouns stand for static inputs, but verbs place them in process, and the particular kind of verb tells us whether this process was or is or will be.

Certain sounds even have the function of partially alleviating the bluntness and imprecision inherent in language usage. Both adverbs and adjectives represent, even if only to a slight degree, the infinite shadings and gradations of our experience. Adjectives break things down into finer categories. They tell us what kind, which one, or how many. They place nouns and pronouns in perspective; they add concreteness and specificity. They tell us that there are several reasons why the blue eyes of that woman are beautiful. Adverbs do for verbs what adjectives do for nouns and pronouns. They represent the gradations that are taking place in process; the how, when, where, and extent of the action of the verb.

Components of our experience never occur in isolation. They are always related to other items in time and space. Though there exists a multiplicity of ways in which things can be related, a few terms, the prepositions, represent them all. Think of above or in or upon and see again how a few words represent vast segments of our experience. I leave conjunctions to the grammarians, but cannot terminate this brief survey without a nod to interjections; in these we find the human effort to represent different emotions with certain sounds.

LANGUAGE UNIVERSALS

This brief examination suggests that all human languages, in spite of the many different dialects, have certain features in common, and this possibility has inspired much constructive research and speculation during the past few years. Such universals may exist, but when we search for them it should be kept in mind that speech is a derivative of thought; that if there are specific innate representations in language such modules (whatever we call them) may be reflections of innate predispositions in thought processes, as these have arisen from long-term selection pressures. The comprehensive review by Cosmides and Tooby (1994) suggested that humans appear to have a number of innately represented thought modules, such as cheater detection mechanisms, that facilitate the capacity to solve problems implicit in social interchange. Where such influences are found, they may be reflected in speech, and when they are found in all languages they may be indicators of innate language universals.

Does the existence of universals in language necessarily imply specific innate propensities? Perhaps. Certain circumstances have been of such enduring importance during the evolution of human mental capacities that there may be reflections of these conditions in thought and speech. For instance, humans may have an innate predilection for learning and expressing the "if A then B" concurrences that have been critical in their evolution, ensuring the development of certain kinds of associations and the means of their expression, that is, we may tend to learn and think and speak in causal categories. Or there may be maturational windows of hypersensitivity, which may account for lingual precocity in children.

Certainly the general capacity for language emergence and expression is innate, the common heritage of humans everywhere. Whether innate particulars within this domain-general capacity exist remains an open question. Innate particulars derived from differences in the speech apparatus are present in men, women, and children, but overlap is invariably present, which should make us sensitive to the probability that particulars in other areas of expression, if found, will also demonstrate overlap, adding to the difficulty of extracting such specifics from domain-general capacities.

What then, if any, are the universals in language? They include the morphological and neural structure underlying speech. They may be present in propensities for particular modes of expression, as in Chomsky's LADs; they may reside in certain sounds or sound combinations that symbolize enduring aspects of the environment, relationships between people, or fundamentals of each person. There may also be sounds or sound combinations that are particular to each language that have a universal meaning. Such universality of meaning may also hold for expressions of emotion and the capacity for empathy. Do smiles say the same thing in all languages? Darwin thought so.

If there are universals they will reflect experiences common to all people—certain critical and fundamental recurrences during the evolution of language such as location, distance, concurrence, number, night, day, past, present, and future. They will manifest themselves in sounds or phrases that, although different for different languages, have the same meaning in all languages.

A friend, Louis Smogor, can speak five languages fluently, switching back and forth between them with ease. The languages, according to him, seem to be contained in separate but easily accessible areas, each with its own encode matrix. My grandson, Andrew, was from the earliest months of his life exposed to two languages, French and English, so that by age three he could converse fluently in both languages. On a number of occasions when he was absorbed in playing, I asked him questions in both languages. Without appearing to think about it, he would respond in the corresponding language, seemingly unaware that he was switching back and forth between them. The set to respond was apparently sufficient to occasion the unconscious processing that allowed the activation of the appropriate LCAS.

An examination of what happens when multilingual individuals shift from one language to another may help expose the commonalities that exist between them. That is, an examination of the processes that allow interpreters to extract the meaning of a statement in one language and then search and find the same meaning in another language may expose the universals commonly assumed by contemporary linguistic and evolutionary theorists. Interpreters must delve deeper than ideosyncratic particulars to the shared features of language and, in so doing, may be exposing the fundamentals common to all languages.

Although universals may eventually be found, such exposure will not necessarily support the existence of nativistic propensities or modules. All humans everywhere share common biological processes and social conditions. They also share environments that may vary in certain particulars, but all environments have fundamentals in common. The universals, if found, may reflect the learning resulting from encountering such common conditions, rather than demonstrating innate capacities. Much more research, under conditions of precise control, must be pursued before we can settle the nature-nurture issue in language development.

The window for language development that is commonly assumed by nativists does not yet have adequate empirical support. Certainly, young children have a remarkable capacity for language, but the evidence used to support an innate evolutionarily derived propensity, although I believe that such may exist, is suspect because the learning situations for children and adults are not comparable. That is, children learn with a clean slate, while adults invariably learn a new language in the context of another language, so that the differences found may be due to interference in the adult, rather than to precocity in the child.

In passing it should be noted that LCASs are not exceptional. They are simply more dramatic examples of the behavior sequences (CASs) that characterize all activity, in animals as well as humans. All behavior involves series of acts, most of which have become fused into functional unity as a function of many repetitions. Such action systems may vary from discrete particulars to multiple inclusions, where as a function of continuing practice they become relegated to unconscious processing, and where, as such, they require occasional apperceptual overscan to insure efficient operation. LCASs are thus not appreciably different from other highly practiced behavior sequences, whether driving to the grocery store, or preparing coffee. The particular acts, encodes, and performance feedback become fused into a functional unity that is relegated to unconscious processing, where it only requires apperceptual monitoring for efficient operation.

Such monitoring or overscan is probably not essentially different from the vigilance activity that takes place in most organisms. I placed grains of rice on one of the wooden railings of my deck and observed birds eating. Without exception, pecking would continue for a brief period, for a few seconds, then the bird's head would cock in all directions, apparently scanning the environment for hazard, reminiscent of the manner in which humans automatically scan for hazards as they drive a car down a busy street. The scanning occurs automatically, and the behavior sequence proceeds automatically, to be interrupted only if some anomaly intrudes, allowing corrective action to take place. Incidentally, perhaps because they are more vulnerable, smaller birds demonstrate more rapid oscillations between peck and scan than larger birds.

The relegation of highly practiced behavior sequences to unconscious processing requiring only occasional monitoring abets rapid and efficient operation, but, more importantly, it reduces the clutter that might otherwise inundate the field of awareness and overload the limited capacity of the apperceptual system. This dual contribution to adaptation facility was responsible for the evolutionary emergence of the capacity for CAS development and suggests their present function.

The critical reader may have noticed that the previous treatment neglects at least one fundamental attribute of speech—its volitional character: The fact that the speaker is almost always aware that he or she is both emitting the sounds and giving direction to their progress. This view, I believe, is essentially correct. The LCASs that are responsible for speech have been associated with the autocept on so many different occasions (beginning with the babbling phase of language development) that the inputs involved become fused, the threshold between autocept and motocept becomes lowered to a marked degree. These LCASs ensure that most, if not all, language emissions, even though they may proceed quite rapidly and with seeming automaticity, still appear to the speaker as volitional. If one talks in one's sleep or talks aloud to one's self when immersed in some problem or is driven into verbalization by extreme anger, pain, or grief, the volitional overlay is missing, and when the person remembers that this has happened, it may seem strange indeed.

The responses that occasion language sounds may seem exceptional. Yet this is hardly the case. The movements (of the vocal cords and mouth) that produce such sounds are different from other responses in only three respects. Such movements may be varied rapidly, sound is typically emitted, and the instrument (the vocal cords) that produces the sound is always handy. However even as the motor threshold between particular inputs (sensations or encodes) and certain vocal cord movements can be lowered, so can the threshold between such components and any other response. Indeed, if a sufficient variety of signals is emitted, any behavior can take over the function of the vocal cords. Thus human beings can learn to communicate with drums, telegraph keys, or musical instruments, or they can dispense with sound entirely and converse fluently with sign language.

BUFFER REVERBERATIONS AND SPEECH

Buffer reverberations play a critical role in speech as well as in the development and demonstration of most motor skills. The ongoing progress of speaking produces reverberations that are continually and automatically monitored by apperceptual overscan. When most speech emissions occur the words involved have typically been so often practiced that only occasional monitoring takes place; the more often practiced, as with any motor activity, the less overscan is required. As long as the speech sequence proceeds normally the monitoring is minimal but if a word or phrase is uttered that is inconsistent with the LCAS in operation, that incongruity assumes apperceptual dominance triggering corrective measures.

This is the case with all highly practiced motor activities. Vigilance can be viewed as a set for the detection of incongruity. Driving a car through a complicated route to a particular destination requires extensive overscan (vigilance) in the beginning, but as many trips over the same route are made the total complex of cues become melded into an inclusive CAS so that less and less overscan is required and the trip can proceed automatically with only occasional monitoring. If some incongruity, say the sound of a siren, a flashing light, an obstruction in steering or missing a turn occurs, such an incongruity immediately triggers apperceptual dominance and corrective measures. These examples suggest that the incongruity can intrude from any sense modality demonstrating the remarkable inclusiveness of the CAS, and also that the CAS may serve as a composite relative to which incongruities stand out for easier discrimination and reaction. The unity implicit in the CASs semi-automatic operation not only frees apperception for other mental operations but by contrast provides the basis for immediate focalization upon incongruity. Incongruities may vary from incidental intrusions to these of great disruptive import. A piece of paper fluttering from a car would occasion momentary focalization while a child falling from a car would be totally disruptive.

It sometimes happens that buffer reverberations are jammed out by some distraction so that the activity in progress may continue for a while without the overscan. When this happens in speech a rapid rescan is required to insure that what was said was what the person intended to say. Usually such rescan confirms that the utterance was congruent but it sometimes reveals that it was so at variance that corrections must be made.

Such lapses in overscan and the emission of inappropriate responses are potential in all sequential activities, and occasionally the incongruity does not register for a while. In typing, for instance, the movements required occur smoothly and automatically with only occasional acceptual monitoring. But in my case (I am a very bad typist) I sometimes get my fingers on the wrong keys and produce a number of lines of nonsense before the incongruity seeps in.

One of the most common examples of incongruity in speech occurs when a single word appears out of place or out of context, as occurs when one misuses the name of a family member. Last night while talking to my son Brad on the phone I called him Boyd, the name of one of my other sons. I immediately knew that I did it but not before he huffily reminded me that he was Brad. Confusing names of family members would be predicted. All family members, particularly one's children, share a vast web of overlapping memories. So in talking about one or to one of them the utterance of another's name would be highly potential.

The functions and nature of the apperceptual overscan of buffer reverberations can be examined experimentally. One approach could involve the introduction of sounds through earphones that overlap sounds being uttered. Such earphone introduced sounds (typically words) could be varied in a number of ways. As example, subjects might be asked to say the words "red", "green" or "blue" aloud while these same words were being presented through earphones. The earphone words could be varied for congruence, time delay, intensity, etc. and the behavior of the subject, reaction times, emotional response and verbal reports recorded.

Subjects might be asked to tell a story, say to describe what they did the previous day, while simultaneously performing some other task or being subjected to distracting circumstances, such as learning a visually presented list of nonsense syllables, being subjected to shock, being presented with emotionally arousing situations, etc., and determining the extent that the continuity of the subject's story was disrupted.

Experiments could also be designed to study the influence of variables upon the development of automaticity and upon the disruption of automaticity once it had been established. The research of Logan (1991) and his colleagues point the way. While training toward automaticity was progressing various kinds of distractors such as the learning of other material, the solving of puzzles or emotion producing stimuli imposed could be introduced and their influence determined. Once automaticity has been achieved such distractors and the performance decrement occasioned by their introduction measured.

Such research would not only be important for determining the contribution that the monitoring of buffer reverberations makes to speech (and other sequential behaviors) but it might have implications for understanding a number of inappropriate behaviors. Although dealing with breakdown or malfunction (the subject matter of abnormal psychology) is beyond the scope of this book. Brad Goodson (personal communication) suggests such research might have implications for the understanding and treatment of such anomalies as stuttering and stammering, tics and tourette syndrome. This discussion may also have implications for an understanding of Freudian slips. When an encode constellation (CAS) relating to some need or problem (whatever the issue) is present ion the background while one is talking about something else, a term related to that CAS may intrude into the progress of what is being said. If one is hungry or angry or has some ongoing problem related words may be uttered before overscan reveals the incongruity and corrections can be made. There is, according to my view, no unconscious mind but there are vast numbers of overlapping CASs with shifting potential for demonstration.

The lessening requirement of apperceptual overscan in the learning and performance of sequential activities suggests the function that consciousness may play in the total adaptive economy of the organism. This has been touched upon a number of times in the previous pages: Where conditions are simple and recurrent unconscious processing such as that found in reflexes, metabolic activity and the automatic nervous system suffice: When activities which initially required constant conscious involvement are highly practiced the participation of consciousness in that activity becomes diminished. These two mutually reinforcing observations suggest that consciousness evolved and reached its refined level of adaptive utility because of the selection pressures imposed from encountering novel and complex situations.

CHAPTER SUMMARY

Language involves the utilization of LCASs in which the motocepts have become fused into functional unity, where the motor threshold has been lowered through repetition to such a degree that a given sequence can proceed with only occasional conscious involvement, that is, apperceptual overscan. It demonstrates a remarkable transduction wherein myriads of experiences are funneled into summarizing encodes and their relationships, and where these are reduced into sounds and their relationships. As such, language is an isomorphic reflection of one's environment, social interactions, and inner and outer processes. To the extent that many of these circumstances are shared by all humans, regardless of particular culture, common modes of expression or universals may be implicit in the meanings which accrue to particular sounds. Such universals may be determined by studying the history of various dialects, modes of expression common to all cultures, and the efforts of translators as they extract meaning from one language and translate it into another. Where found, universals may indicate the existence of long-term circumstances present during the evolution of language as these are demonstrated by inherited modules, or they may simply be evidence for common learning circumstances shared by all people everywhere. Much research will be required before it can be determined whether the general domain capacity for language is segmented into innate particulars. For instance, one of the most commonly assumed innate particulars, the language facility window in children, may be due to inappropriate comparison conditions: Adults always learn a new language in an interference situation while children learn with a clean slate.

The language precosity of the little child that has been of such interest may be only one aspect of a more general facility for rapid learning that is present at that early age. Little children are precocious in learning language but they learn many other activities at a rapid pace as well. The language decrement found in feral children and those raised in isolation (Curtiss, 1977) may not be caused by missing out during a critical developmental language window but because language was simply one aspect of a more inclusive incapacity induced by lack of appropriate stimulation or the development of interfering habits.

Before we accept the existence of evolutionarily derived modules or predispositions let us always be sure that we have ruled out the influence of learning (cultural) overlay. It is fittingly ironic, perhaps, that one of our greatest evolutionary achievements—our facile capacity for learning—should obscure and complicate our search for the evolutionary roots of other attributes.

It was suggested that buffer reverberations play a critical role in the learning and continuity of serial activities. They provide the basis for monitoring such activities and for correction when incongruities are encountered. It was surmised that consciousness gradually evolved and became a central aspect of information processing because of the selection pressures imposed from encountering novel and complex situations.

Epilogue

Does the elevation of the principle of association to such an all-inclusive status add anything new to its long history as an explanatory construct? Yes, it does. In the early history of the principle, from Locke to Hartley, association was restricted to sensations and memory images, with the cementing (connecting) said to be attributable to such variables as contiguity, similarity, frequency, and contrast. With the advent of psychology, very little changed. Wundt accepted J. S. Mill's doctrine of mental chemistry without alteration but continued to limit the application of the principle of association to the same mental categories isolated by the English empiricists. With the advent of behaviorism, association was still widely used, except instead of mental components, the elements being cemented consisted of stimulus-response conjunctions (Watson, Guthrie, Hull, and Thorndike) or stimulus-stimulus concurrences (Tolman), with the cementing principle variously consisting of contiguity, need reduction, or both.

Although the use of association as an explanatory principle is as old as Aristotle, its use has never, to my knowledge, emphasized those inputs, such as pain, hunger, fear, and anger, that activate the organism. To include activation input within the associational matrix, let us view the organism as a unified system; one where information, activation and reaction are continuing reciprocal processes that moment-by-moment insure that the balances necessary for life can be maintained.

Associations, it should be remembered, halo our efforts in a broader and more inclusive sense. They are represented in the cycles and the "if A then B" concurrences of the environment in which we evolved, are reflected in the dynamic interdependencies that obtain both within and between all organisms, and are required by our efforts to explain and communicate (see intrinsic dilemma, chap. 2). We reside within an associational web that occasioned our existence, constitutes a central aspect of our nature, and dictates the methods of explanation and communication that are available.

Is the theory that this book comprises an accurate statement about the manner in which information processing and reaction actually occur in complex organisms? I believe that it is. The research and theory from experimental psychology, cognitive psychology, and ethology coalesce to provide a unified rationale in support of this claim. When combined with the research and theoretical underpinnings provided by an evolutionary perspective, the theory achieves considerable subsumptive and explanatory power. The legitimacy of theory, of course, must depend on how well it fulfills the requirements of a theory, as these requirements are interpreted by workers in the field (see Goodson & Morgan, 1976b, for an analysis of the criteria used in the evaluation of theory).

Does the theory provide a complete integration? No, it does not. In particular, much of the neural machinery implicit in cognitive processing remains unknown. We know very little about the events in the nervous system that account for the transformation of excitation into sensation, or thought into action. We know even less about the neurological processes involved in perceptual organization, apperception, and encoding. In 1976, Goodson and Morgan made the observation that psychologists have been concerned with three levels of data—the experiential, behavioral, and physiological—and then they went on to say, "None of these types of data are the exclusively appropriate one for psychology. Data of all three levels and particularly that from across levels, should be emphasized" (Goodson & Morgan, 1976a, p. 406).

I still agree with this statement. The theory outlined in this book provides direction in our search for the neurophysiological particulars and the behavioral indicators that are corollary with the cognitive processes that have been discussed. It also serves as a conceptual structure into which both physiological and behavioral data, once they are isolated, can be fitted, and in relation to which their function can be understood. As the data from the three levels, and that derived from research across levels, comes in, the theory will provide a simplifying inclusive statement of how information processing and reaction function interdependently and dynamically to ensure the maintenance of essential balance in vulnerable organisms in a context of perpetual change.

As a final note, let me emphasize that psychologists enjoy a remarkable vantage point, a dual perspective that other scientists are not privy to. Although our subject matter may be the most complex of all scientific pursuits, this dual perspective gives us a unique basis for unraveling the intricate workings of human beings. We cannot only observe others from the outside as they indulge in their myriad pursuits, but also, as fellow products of evolution, can observe the processes that are fundamental to such pursuits from the inside. It is as if, while puzzling about the workings of a machine, say a computer, by analyzing what it can do or what it cannot do, one were transported to the inside of the machine and became suddenly and intimately involved in the processes that made such behavior possible. We have this dual perspective as we try to unravel the working of the human being by both observing and introspecting. When we introspect we are not only examining the intricate workings of ourselves, we, as fellow products of evolution and members of the same species, can generalize our findings to others. Yes, introspection as a legitimate tool for human study is back. It was the central approach of our discipline at its inception with the involvements of Wundt and Bretano, and the contents and acts discovered and clarified by them and their disciples now fit neatly into an evolutionary frame of reference.

When psychologists defined their discipline as the psychology of the other one, as the behaviorists did during much of the last century, one side of the dual perspective was rejected and often vilified. The behaviorists insisted that their subject matter, their psychology, should be based upon experience but then paradoxically denied that experience itself was a topic worthy of consideration. With the reemergence of cognitive psychology and its grounding in evolutionary theory, this paradox is finally resolved. We can look within; we can introspect on the nature of our inner workings, on the experiences that are the defining essence of human nature, the wellsprings of our behavior, and the subject matter of our discipline.

With a cognitive psychology based on evolutionary theory, there is both a return to experience as it was treated in the early days of our discipline and a new emphasis on functionalism, not only on how each aspect of the mental machinery contributes to the immediate operations of the organism but also, feeding into and abetting this concern, on how the adaptive contribution of each such aspect insured its evolutionary emergence and is still manifested in our present nature.

According to some people, efforts, such as mine, to include experience as a lawful aspect of information processing negates what they find important in the human being. A colleague in theology inveighing against what he called efforts to subsume the transcendent with naturalistic explanations maintains that such efforts will come to nowhere. That science should know its limitations. This is not a new complaint. Some felt that Aristarchus and Copernicus diminished mankind's importance with the heliocentric theory, that Darwin demeaned humans by viewing them as participants in a web of kinship encompassing all living things, and that Newton debased the rainbow by determining that it was the outcome of a myriad of tiny prisms refracting light. But Newton did not ruin the rainbow. The rainbow is still beautiful. And when the precise role that experience played in evolution and still plays in each person's nature and conduct is finally achieved, experience will still retain its richness and variety, will still define the essence of what we are, will still be wonderful.

Glossary

Note: This glossary is limited to terms that have been given special emphasis or used in special ways and those that have been coined for particular purposes.

Anthropomorphism: The explanation of an attribute by reference to some human characteristic. Ex: Genes are selfish.

Apperception: The process whereby inputs from sensation, memory, or both become automatically and sequentially focalized. Ex: If you are thinking, looking, or listening, the ongoing process of apperception is taking place and as each item (whether memory or sensation) moves into focus, it is being apperceived.

Apperceptual flooding and Apperceptual jamming: Occurs when an input is so prepotent that it excludes other components that might ordinarily be focalized. Ex: If a hot iron were fastened to the skin, the pain would flood the apperceptual mechanism so that nothing else could be focalized.

Apperceptual threshold: That point at which one of the many components available (whether sensory or memory) moves into focalization. Ex: Who discovered America? Columbus suddenly moved from being out of awareness to being focalized and the apperceptual threshold was crossed. A faucet is dripping and you suddenly notice it; at that moment the apperceptual threshold was crossed.

Attention: See Apperception.

Autocept: A phenomenological unity arising from the fusion of basic inputs (visceral, dual stimulation, mother's face, etc.) that serves as the basis for an individual's identity and much of his motivation. Ex: When a person indulges in volitional behavior or experiences his own identity and continuity, the autocept is involved.

Behavioral fulcrum: Demonstrated when an organism moves in one direction to a given level of energy but in the opposite direction to a different level. Ex: Euglena will move towards a dim light but away from more intense levels.

Blip: The briefest period of apperceptual focalization. Approximately one-twentieth of a second.

Blob: An inclusive component resulting when particulars are fused into a unity including meanings, emotional import, or both.

Brightness constancy: Regardless of the amount of light being reflected from a particular item it is perceived as having the same brightness. Ex: A piece of coal will appear to be black regardless of the amount of reflected light.

Cheater detection mechanism: A module derived from the selective pressures involved in the refinement of reciprocal altruism. See module.

Chronocept: A fusion-constancy phenomenon derived from a synthesis of recurrent cycles and tempos. The basis for the experience of time.

Cognitive map: An encoded replica of activities, situations, or both. Ex: A representation that allows us to find our way about when we are in a familiar place.

Concept: An inclusive memory encode that subsumes many similar percepts. Ex: Although we perceive many types of houses, all of this information is encoded under a summary memory residual that consists of a generalized visual image, sound image, word image, or blob, or all three combined.

Conceptual generalization: Occurs when an intermediating encode (whether visual image, auditory image, or word image) allows a common response to two or more different inputs. Ex: A dog can be taught to salivate both to a bell and a buzzer, but it is the food encode that intervenes, and actually initiates the salivation.

Contemporaneous causation: The view that behavior is the outcome of the variables that are interacting at each moment.

Cue utilization: Is demonstrated when an organism responds differentially to an energy dimension that in itself constitutes neither hazard nor sustenance (may be either innate or learned), and by so doing avoids or approaches hazard or sustenance. Ex: A spider responds to vibrations in the web and obtains food. A child responds to the sight of a hot stove and avoids being burned.

Cultural overlay: Occurs when a biological (evolutionarily derived) predisposition is hidden or altered by cultural learning. Ex: Laughter is biologically based but its expression varies depending on social situations.

Disequilibrium: Any condition of imbalance sufficient to occasion the introduction of activation input into the field of awareness or to occasion compensatory action. Ex: That level of fluid deficit that is sufficient to introduce the experience of thirst or to initiate drinking.

Disequilibrium range: The continuum of imbalance that exists between initial reaction (minimum disequilibrium) and damage to the system (maximum disequilibrium). Ex: The state that exists in a blood cell placed in distilled water between the initial compensatory action of the membrane and the bursting of the cell.

Doctrine of specific input qualities: Each type of input (cue function, need, pleasure, emotion) must be qualitatively distinct from all the others or the organism could not respond appropriately. Ex: Hunger must be qualitatively different from thirst or the organism could not respond differentially to food and water sources.

Domain general: Refers to evolved characteristics derived from selection pressures that have been widely operative for many different species. Ex: The adaptive contribution of responding to certain wavelengths of the electromagnetic continuum has resulted in the appearance of eyes in vast numbers of different organisms.

Domain specific: The view that domain general attributes are actually segmented into numerous separate abilities, sometimes called modules, that reflect each species' unique evolutionary history or special problems encountered during certain maturational stages. Ex: Rats rapidly learn to avoid poisons. Humans have a remarkable facility for learning language during early childhood.

Edgeness: A sensory characteristic that is foundational to the perception of phenomena derived from the differential response of receptors to varying intensities of imposed energy. Ex: If you look at a chair you will observe that it is separated from other aspects of the visual field by a continuous line which marks its boundary. This line is the outcome of a differential activation of the receptors in the retina.

Equilibrium: The condition of balance that exists in a dynamic and reactive system during which no compensatory action is taking place. That condition that exists when a dynamic and reactive system is retaining its optimal steady state. Ex: A body cell that is in such optimal balance with its environment that no traffic is occurring through the membrane.

Evolutionary interdependence: Species that have interacted for extended periods during evolutionary development will demonstrate attributes that reflect this interdependency. Ex: Mice have acute hearing while certain owls have dampeners on their wings that greatly reduce the noise of flight. See Process ecology.

Exaptation: An attribute that has evolved to perform one function may at times fulfill a different function. Ex: Feathers may have evolved as protection against the cold and then become a factor involved in flight (Gould). Also see preadaptation.

Flare input: An almost instantaneous experiential repercussion arising from sudden and unexpected situations. It typically lasts only a few seconds and may be correlated with the psychogalvanic reflex and the startle reflex. Ex: The input that would be experienced by a person who suddenly sees a truck on his side of the road at the top of a hill.

Focalization: See Apperception.

Founder effect: When a number of individuals migrate they automatically have gene frequencies that are different from the population since they cannot carry the entire genetic diversity of the larger group.

Functional causality: Refers to the manner in which reaction to selection pressures has resulted in the evolutionary emergence of an adaptive attribute. Ex: Teeth evolved because of the increased facility they provided in the utilization of food.

Functionalism: A school in psychology, and the emphasis in this book, that determining or infering the adaptive contribution of an attribute can abet our understanding of its evolutionary origin and its present function.

Fusion: One of the basic preceptual processes that reduces the multiplicity of sensory input into functional unity. A simplification mechanism.

Group selection: The view that certain outcomes of group living (culture) can become of sufficient importance to displace genetic variation as the major arbiter in evolution. An issue being debated at the present time.

Homeostasis: The tendency for body systems to maintain or regain a particular steady state. Ex: Heart rate tends to remain around 75 beats per minute.

Immediate awareness: The experienced "now" lasting approximately 750 mms. See Field of Awareness, Wundt's blickfeld.

Information Processor: A system such as a living organism or computer that reacts appropriately to variable inputs. May vary in complexity from a furnace thermostat to a human being.

Input: Any sensory or memory component that becomes represented in immediate awareness.

Internal locomotion: The fluctuation of apperception relative to memory encodes; has the same meaning as the conventional term thinking. Ex: If you think about the various steps involved in the grilling of a steak you are indulging in internal locomotion.

Intrinsic attribute: A characteristic that is implicit in the material from which an organism is constructed. Ex: Whiteness of teeth.

Intrinsic dilemma: Pertains to the fact that we must use sequential categories (words, sentences, paragraphs, etc.) to explain organisms that work on contemporaneous principles; that we must deal with organs and processes separately when all aspects of the organism are actually dynamically interdependent. Ex: Learning and motivation are interdependent processes yet we must discuss them in separate sections, with words that are sequential.

Junk DNA: The proportion of DNA present in most if not all species that does not code for anything.

Language: The symbolic representation of aspects of experience with sounds, words, gestures, postures, facial expressions, etc., for the purpose of communication.

Locus specificity: Refers to the fact that each receptor (and this is particularly true of those on the skin) has a unique quality that is characteristic of a particular location on the body. Ex: If an insect is crawling on one's skin, he or she knows with some precision where it is.

Long-term memory: A cumulative record of experiences that were of sufficient pertinence to be fitted into enduring associational structures, the internalized world, or both.

Metaphor error: Occurs when a seems like or works like concept takes on a life of its own and then confuses rather than clarifies.

Module: In computer terms, a self-contained assembly of electronic components and circuitry, designed to perform a particular task. In biological systems, a highly specific functional attribute residing within a more inclusive domain general capacity. Ex: Rats are remarkably facile in learning to avoid poisons.

Motocept: A phenomenological unity arising from the recurrent presentation of inputs arising from particular movements of the body. Ex: After much practice, the various inputs derived from walking have become fused so that the process can occur almost automatically.

Motor threshold: That point at which a given apperceived input (whether sensory component or encode) occasions a motor response. Ex: A person may apperceive an input for some time before a response occurs. The point at which a response is initiated, is the motor threshold.

Negative Feedback: The condition that exists when greater deviations from balance occasions more extreme compensatory reactions. Ex: The longer a person goes without food the more rapidly he will eat.

Observation learning: The shear apperception of an input, whether sensory or memory, will establish or strengthen an encode. Ex: If one glances out a window he or she will tend to remember what they saw (Bandura).

Outlocking: If an individual cannot participate efficiently in the preliminaries (i.e., appropriate song, mating dance, etc.) of sexual activity his or her evolutionary contribution ceases.

Percept: A phenomenological unity arising from the simultaneous and contiguous presentation of sensory components that are separated from others by an edge or contour. Ex: Although a tree is a composite of many sensory components, it is still perceived as unity.

Preadaptation: See exaptation.

Prepotence: Certain activating inputs are more immediately intolerable than others. Ex: Pain is more intolerable than hunger.

Primary activation input: The input arising from critical physiological imbalances within the organism or from traumatic energy applications. Ex: Hunger activates the organism when a particular level of depletion is reached, and pain activates the organism when trauma reaches a certain degree of intensity.

Process ecology: The complimentarity of functional attributes that may develop between two species that have been involved in interdependent evolution. Ex: Bats have sonar and agility for catching moths while moths have receptors sensitive to the bats' emissions and aerobatics to avoid being caught. See Evolutionary interdependence.

Progressive step: An attribute that constitutes a major leap in adaptation facility as organisms evolved from simple to complex systems. Ex: The capacity to learn was a major advance over reflexive and instinctual behaviors.

Punctuated equilibrium: The view that evolution is marked by long periods of stasis occasionally interrupted by intervals of rapid change (Gould).

Rider: A characteristic that is genetically tied to some other more fundamental trait, such as nipples in males; though not functional in males, they are tied to the more fundamental species requirement of nursing the young.

Secondary activation input: The input, commonly called emotion, that instigates the reaction to previously neutral cues. Ex: The child is activated by fear on seeing the stove after it has been burned.

Selection pressures: Circumstances that influence the survivability of the genetic material. There are five major ones: (a) competition for resources, (b) predation, (c) the environment, (d) disease, (e) competition for mates.

Sensing: The ongoing process of apperception relative to inputs arising from the various transmitting mechanisms of the body. Ex: The process involved in listening, feeling, seeing, etc.

Sensory input: Any experiential derivative arising from the various receptors of the body (whether specific modality, specialized nerve ending or fluid). It is both the manner whereby the environment is represented and the spur that makes reaction within the environment possible. Ex: Pain, anxiety, sound, cold, grief, are all classes of sensory input.

Sensory threshold: That point at which a given energy change activates a transmitting mechanism sufficiently to introduce an experiential component into the field of awareness. Ex: If a feather lightly touches the skin, one may not feel it; but if the pressure is increased, a point will be reached at which the pressure will be felt.

Shape constancy: Regardless of the different shapes that an object projects upon the eye as a function of different perspectives it is seen as having the same shape. Ex: A window is seen as a rectangle although the shape projected on the eye is more often trapezoidal.

Short-term memory: A post apperceptual encode that lasts approximately 15 seconds. Ex: Look up a telephone number and unless the encode is strengthened with new apperceptions, it is soon gone.

Simplification mechanism: A perceptual transduction helping ensure that information overload does not inundate processing and reaction. See size, shape, and brightness constancy; fusion, and apperception.

Size constancy: Regardless of the shifting size of the retinal image as distance from an object varies it is perceived as having the same size. Ex: People do not become midgets as they move to greater distances from us.

Speciation: Occurs when as a function of many variables, most notably competition between like groups, sexual incompatibility between groups takes place.

Teleology: The view that an attribute evolved to fulfill some goal. Ex: Ears evolved so we can hear, rather than the functionalist view that better hearing provided the basis for the evolution of ears.

Thinking: See Internal locomotion.

Transduction: A coined term that represents the process taking place when components at one level are funneled (reduced, translated, represented, etc.) into a summary representation at another level. Ex: The multiple sensory components that are present in an object become fused into a unitary phenomenon called a percept.

Unidimensional differential behavior: The behavior that would be demonstrated by an organism that could respond in gradient fashion to only one energy dimension.

Unnatural selection: Occurs when certain members of a species are selected by humans in order to promote particular attributes through controlled breeding. Ex: Horses that can run fast have been selected and specially bred to produce the thoroughbred.

Volitional behavior: (a) Activity that is initiated and sustained by input arising from threat to the organizational integrity of the autocept; or (b) activity that is motivated by some other system, but which is mistakenly attributed to the autocept; or (c) activity that is motivated by an input (such as sex) that has become basic to, and indeed may comprise, an essential aspect of an autocept.

World internalization: The kind of encoding involved in observation learning whereby a functional duplicate of the external environment is represented. Ex: When we look out a window a replica of what we saw is encoded. See Cognitive map.

World externalization: The process whereby experience, which is internal, is organized so that we perceive items in environment at varying distances from us. Ex: Although the experience of what you are now reading is within the brain the material is seen as a number of inches away from you.

References

Ackerman, B. P., Abe, J. A., & Izard, C. E. (1998). Differential emotions theory and emotional development: Mindful of modularity. In M. F. Mascolo & S. Griffin (Eds.), *What develops in emotional development?* (pp. 85–106). New York: Plenum.

Adams, J. (1962). Test of hypothesis of the psychological refactory period. *Journal of Experimental Psychology, 64,* 280–287.

Alcock, J. (1984). *Animal behavior: An evolutionary approach* (3rd ed.). Sunderland, MA: Sinauer.

Alexander, R. D. (1987). *The biology of moral systems.* Hawthorne, NY: Aldine de Gruyter Press.

Allen, R. D. (1962). Amoeboid movement. *Scientific American, 206,* 112–120.

Allport, D. A., Antonis, B., & Reynolds, P. (1972). On the division of attention: A disproof of the single channel hypothesis. *The Quarterly Journal of Experimental Psychology, 24,* 225–235.

Amsel, A. (1958). The role of frustrative non–reward in noncontinuous reward situations. *Psychological Bulletin, 55,* 102–119.

Ankel-Simons, F. (1983). *A survey of living primates and their anatomy.* New York: Macmillan.

Aoki, C., & Siekevitz, P. (1988). Plasticity and brain development. *Scientific American, 259,* 56–64.

Ariam, S., & Siller, J. (1982). Effects of subliminal oneness stimuli in Hebrew on academic performance of Israeli high school students: Further evidence on the adaptation enhancing effects of symbiotic fantasies in another culture using another language. *Journal of Abnormal Psychology, 91*(5), 343–349.

Aronfreed, J. M., Messick, S. A., & Diggary, J. C. (1953). Reexamining emotionality and perceptual defense. *Journal of Personality, 21,* 517–528.

Axelrod, R. (1984). *The evolution of cooperation.* New York: Basic Books.

Axelrod, R., & Hamilton, W. D. (1981). The evolution of cooperation. *Science, 211,* 1390–1396.

Ayala, E. J. (1983). Genetic variation and evolution. *Carolina Biology Readers, 126.* Burlington, NC: Carolina Biological Supply Co.

Baker, L. E. (1937). The influence of subliminal stimuli upon verbal behavior. *Journal of Experimental Psychology, 20,* 84–100.

Bandura, A., Ross, D., & Ross, S. (1961). Transmission of aggression though imitation of aggressive models. *Journal of Abnormal and Social Psychology, 63,* 572–582.

Bandura, A., Ross, D., & Ross, S. (1963). Imitation of film–edited aggressive models. *Journal of Abnormal and Social Psychology, 66,* 3–11.

Bandura, A., & Walter, R. H. (1963). *The social learning of deviant behavior: A behavioristic approach.* New York: Holt, Rinehart & Winston.

Banks, M. S., & Shannon, E. (1993). Spatial and chromatic visual efficiency in human neonates. In C. E. Granrud (Ed.), *Carnegie-Mellon symposium on cognitive psychology* (pp. 1–46). Hillsdale, NJ: Lawrence Erlbaum Associates.

Bargh, J. A., & Chartrand, T. L. (1999). The unbearable automaticity of being. *American Psychologist, 54,* 462–479.

Barkow, J. H. (1992). Beneath new culture is old psychology: Gossip and social stratification. In J. H. Barkow, L. Cosmides, & J. Tooby (Eds.), (pp. 627–636). *The adapted mind.* New York: Oxford University Press.

Barkow, J., Cosmides, L., & Tooby, J. (Eds.) (1992). *The adapted mind: Evolutionary psychology and the generation of culture.* New York: Oxford University Press.

Bartlett, F. C. (1932). *Remembering.* Cambridge, England: Cambridge University Press.

Bartley, S. H. (1958). *Principles of perception.* New York: Harper & Row.

Becklen, R., & Cervone, D. (1983). Selective looking and the noticing of unexpected events. *Memory and Cognition, 11,* 601–608.

Bekhterev, V. M. (1932). *General principles of human reflexology.* New York: International Publishers. (Original work published 1907)

Benkman, C. W., & Lindholm, A. K. (1991). The advantages and evolution of morphological novelty. *Nature, 349,* 519–520.

Berkeley, G. (1709). *An essay toward a new theory of vision.* London: J. M. Dent & Sons.

Bernard, C. (1865). *Introduction a la medecine experimentale.* Paris: Sarbonne. (This was his masterpiece.)

Bessel, F. W. (1823). *Astronomische Beobachtungen.* Konigsberg, Germany: Academia Albertina.

Blaney, P. H. (1986). Affect and memory: A review. *Psychological Bulletin, 99*(2), 229–246.

Bliss, J., Crane, H., Mansfield, K., & Townsend, J. (1966). Information available in brief tactile presentations. *Perception & Psychophysics, I,* 273–283.

Blumenthal, A. L. (1977). *The process of cognition.* Englewood Cliffs, NJ: Prentice-Hall.

Bolles, R. C. (1975). *Learning theory.* New York: Holt, Rinehart & Winston.

Boring, E. G. (1929, 1950). *A history of experimental psychology.* New York: Appleton-Century-Crofts.

Bower, G. H. (1981). Mood and memory. *American Psychologist, 36*(2), 129–148.

Bower, G. H. (1987). Commentary on mood and memory. *Behavior: Research and Theory, 25*(6), 443–455.

Bower, G. H., Gilligan, S. G., & Monteiro, K. P. (1981). Selectivity of learning caused by affective states. *Journal of Experimental Psychology: General, 110*(4), 451–473.

Boyd, R. (1988). Is the repeated prisoner's dilemma a good model of reciprocal altruism? *Ethology and Sociobiology, 9,* 211–222.

Bransford, J. D. (1979). *Human cognition.* Belmont, CA: Wadsworth.

Breland, K., & Breland, M. (1951). A field of applied animal psychology. *American Psychologist, 6,* 202–204.

Breland, K., & Breland, M. (1961). The misbehavior of organisms. *American Psychologist, 16,* 681–684.

Brentano, F. (1924–1925). *Psychologie von empirischen standpunkt.* O. Kraus (Ed., Vol. 2). Leipzig, Germany: Mainer. (Original work published 1874)

Brett, G. S. (1921) *A history of psychology.* New York: Macmillan.

Broadbent, D. E. (1954). The role of auditory localization and attention in memory span. *Journal of Experimental Psychology, 37,* 191–196.

Broadbent, D. E. (1957). A mechanical model for human attention and immediate memory. *Psychological Review, 64,* 205–215.

Brooks-Gunn, J., & Lewis, M. (1984). The development of early visual self recognition. *Developmental Review, 4,* 215–239.

Brown, F. A., Jr. (1958). Living clocks. *Science, 130,* 1535–1544.

Brown, J. S. (1948). Gradients of approach and avoidance responses and their relation to level of motivation. *Journal of Comparative and Physiological Psychology, 41,* 450–465.

Bruell, J. H. (1962). Dominance and segregation in the inheritance of quantitative behavior in mice. In E. L. Bliss (Ed.), *Roots of Behavior.* New York: Harper & Row.

Bruner, J. S., & Goodman, C. D. (1947). Value and need as organizing factors in perception. *Journal of Abnormal and Social Psychology, 42,* 33–34.

Bruner, J. S., & Postman, L. (1949). Perception, cognition, and behavior. *Journal of Personality, 18,* 14–31.

Brussell, E. M. (1968). The speed of attentional shifting under conditions of a visual-auditory complication. Unpublished master's thesis, DePauw University, Green Castle, IN.

Bub, D. N. (2000). Methodological issues confronting PET and fMRI studies of cognitive functions. *Cognitive Neuropsychology, 17*(5), 467–485.

Bucherof, J. R. (1998). From abled to disabled: A life transition. *Topics in Stroke Rehabilitation, 5*(2), 19–29.

Bundy, A. (1975). *Analyzing mathematical proofs.* Edinburg, Scotland: University of Edinburg, Department of Artificial Intelligence, Research Report No. 2.

Buss, D. M. (1987). Sex differences in human male selection criteria: An evolutionary perspective. In C. Crawford, M. Smith, & D. Krebs (Eds.), *Sociology and psychology: Ideas, issues, and applications* (pp. 335–351). Hillsdale, NJ: Lawrence Erlbaum Associates.

Buss, D. M. (1989). Sex differences in human male preferences: Evolutionary hypothesis tested in 37 cultures. *Behavioral and Brain Sciences, 12,* 1–49.

Buss, D. M. (1996). Sex differences in jealousy: Not gone, not forgotten and not easily explained by alternative mechanisms. *Psychological Sciences, 7*(6), 273–275.

Buss, D. M., & Schmitt, D. P. (1993). Sexual strategies theory: A contextual evolutionary analysis of human mating. *Psychological Review, 100,* 204–232.

Butlin, R. K., & Tregenza, T. (1997). Is speciation no accident? *Nature, 387,* 552–553.

Buunk, B. P., Angleitner, A., Oubaid, V., & Buss, D. M. (1996). Sex differences in jealousy in evolutionary and cultural perspectives: Tests from the Netherlands, Germany and the United States. *Psychological Science, 7,* 359–363.

Calvin, W. H., & Bickerton, D. (1998). *Lingua ex machina: Reconciling Darwin and Chomsky with the human brain.* Cambridge, MA: MIT Press.

Campos, J. J., Langer, A., & Krowitz, A. (1970). Cardiac responses on the visual cliff in prelocomotor infants. *Science, 170,* 196–197.

Cannon, W. B. (1929). *Bodily changes in pain, hunger, fear, and rage.* New York: Appleton-Century.

Cannon, W. B. (1932). *The wisdom of the body.* New York: Norton.

Cantril, H. (1950). *The "why" of man's experience.* New York: Macmillan.

Carlile, M. J. (1975). *Primitive sensory and communication systems.* New York: Academic Press.

Carlson, N. R. (1981). *Physiology of behavior.* 2nd ed. Boston: Allyn & Bacon.

Carr, H. A. (1925). *Psychology: A study of mental activity.* New York: McKay.

Carroll, S. P. & Boyd, C. (1992). Host race radiation in the soapberry bug: Natural history with the history. *Evolution, 46,* 1052–1069.

Cashdan, E. (1989). Hunters and gatherers: Economic behavior in bands. In S. Plattner (Ed.), *Economic Anthropology* (pp. 21–48). Stanford, CA: Stanford University Press.

Cattell, J. M. (1885). The influence of the intensity of the stimulus on the length of the reaction time. *Brain, VIII,* 512–515.

Cattell, J. M. (1886). The time it takes to see and name objects. *Mind, XI,* 68–74.

Charnov, E. L. (1976). Optimal foraging: the marginal value theorem. *Theoretical Population Biology, 9,* 129–136.

Chase, W. G., & Ericsson, K. A. (1980). Skilled memory. In J. R. Anderson (Ed.), *Cognition skills and their acquisition.* Hillsdale, NJ: Lawrence Erlbaum Associates.

Cherry, E. C. (1953). Some experiments on the recognition of speech, with one and with two ears. *Journal of the Acoustical Society of America, 25,* 975–979.

Chomsky, N. (1959). Review of B. F. Skinner, *Verbal behavior, language, 35,* 26–58.

Chomsky, N. (1965). *Aspects of the theory of syntax.* Cambridge, MA: MIT Press.

Chomsky, N. (1972). *Language and mind.* New York: Harcourt Brace.

Chomsky, N. (1986). *Knowledge of language: Its nature, origins, and use.* New York: Praeger.

Clark, L. B., & Hess, W. N. (1943). Swarming of the Atlantic palolo worm, Leodice fucata (Ehlers). *Papers from Tortugas Laboratory, 33,* 21–70.

Collier, G. H. (1983). Life is a closed economy: The ecology of learning and motivation. In M. D. Zeiler & P. Harzem (Eds.), *Foraging behavior: Ecological, ethological, and psychological approaches* (pp. 223–274). New York: Garland STPM Press.

Collier, G. H., & Rovee-Collier, C. K. (1981). A comparative analysis of optimal foraging behavior: Laboratory simulations. In A. C. Kamil & T. D. Sargent (Eds.), *Foraging behavior: Ecological, ethological, and psychological approaches* (pp. 39–76). New York: Garland STPM Press.

Conrad, R., & Hull, A. (1966). The role of the interpolated task in short-term retention. *Quarterly Journal of Experimental Psychology, 18,* 266–269.

Cook, L. M., Mani, G. S., & Varley, M. E. (1986). Past industrial melanism in the peppered moth. *Science, 231,* 611–613.

Corballis, M. C. (1992). On the evolution of language and generativity. *Cognition, 44,* 197–225.

Cosmides, L., & Tooby, J. (1992). *Cognitive adaptations for social exchange.* In J. Barkow, L. Cosmides, & J. Tooby, *The Adapted Mind:. Evolutionary psychology and the generation of culture* (pp. 163–228). Oxford, England: Oxford University Press.

Cosmides, L. & Tooby, J. (1994). Origins of domain specificity: The evolution of functional organization. In S. Gelman & L. Hirschfeld (Eds.), *Mapping the mind: Domain specificity in cognition and culture* (pp. 85–116). New York: Cambridge University Press.

Cowan, N. (1995). *Attention and memory: An integrated framework.* New York: Oxford University Press.

Crick, F. H. C., & Mitcheson, C. (1983). The function of dream sleep. *Nature, 304,* 111–114.

Cronquist, A. (1968). *The evolution and classification of flowering plants.* Boston: Houghton-Mifflin.

Curtiss, S. (1977). *Genie: A psycholinguistic study of a modern-day "wild child."* New York: Academic Press.

Daniels, A. (1895). Memory, afterimage and attention. *American Journal of Psychology, 6,* 558–564.

Darwin, C. (1859). *On the origin of species.* London: Murray.

Darwin, C. (1871). *The descent of man.*

Darwin, C. (1897). *The variation of plants and animals under domestication.* Vols. I & II. New York: D. Appleton & Company.

Darwin, C. (1958). *The autobiography of Charles Darwin.* (N. Barlow, Ed.). London: Collins.

Darwin, C. (1987). *The expression of the emotions in man and animals.* New York: D. Appleton & Company.

Darwin, C., Turvey, M., & Crowder, R. (1972). An auditory analogue of the Sperling partial report procedure: Evidence for brief auditory storage. *Cognitive Psychology, 3,* 255–267.

Davis, R. (1957). The human operator as a single channeled system. *Quarterly Journal of Experimental Psychology,* 119–129.

Dawkins, R. (1976). *The selfish gene.* Oxford, England: Oxford University Press.

Deese, J., & Hulse, S. H. (1967). *The psychology of learning.* New York: McGraw-Hill.

Dember, W. N., & Fowler, H. (1958). Spontaneous alternation behavior. *Psychology Bulletin, 55,* 412–428.

Descartes, R. (1892). *Passions of the soul* (H. A. P. Torrey, Trans.). New York: Henry Holt. (Original work published 1650)

DeSteno, D. A., & Salovey, P. (1996). Genes, jealousy and the replication of misspecified models. *Psychological Science, 7*(6), 376–377.

Devreotes, P. N., & Zigmond, S. H. (1988). Chemotaxis in eukaryotic cells: A focus on leukocytes and dictyostelium. *Annual Review of Cell Biology, 4,* 649–686.

DeWall, F. B. M. (1989). Food sharing and reciprocal obligations among chimpanzees. *Journal of Human Evolution, 18*, 433–459.
Diamond, J. M. (1988). Founding fathers and mothers. *Natural History, 97*(6), 10–15.
Dias-de-Carvalho, S. A., Andrade, M. J., Tavares, M. A., & Sarmento-de-Freitas, J. L. (1998). Spinal cord injury and psychological response. *General Hospital Psychiatry, 20*(6), 353–359.
Dillon, L. S. (1978). *Evolution: Concepts and consequences*. St. Louis, MO: Mosby.
Dixon, N. F. (1981). *Preconscious processing*. New York: Wiley.
Dobzhansky, T. (1964). *Genetics and the origin of species* (3rd ed.). New York: Columbia University Press.
Donders, F. C. (1862). Die Schnelligkeit psychischer Processe. *Archives of Anatomy and Physiology*, 657–681.
Duzel, E., Yonelinas, A. P., Heinz, H. J., Mangum, G. R., & Tulving, E. (1997). Event-related brain potential correlates of two states of conscious awareness in memory. *Proceedings of the National Academy of Science USA, 94*, 5973–5978.
Ebbinghaus, H. (1885). *On Memory*. Leipzig, Germany: Duncker & Humboldt.
Edmunds, M. (1974). *Defense in animals*. New York: Longmans.
Egly, R., & Homa, D. (1984). Sensitization of the visual field. *Journal of Experimental Psychology: Human Perception and Performance, 10*, 778–793.
Eisenbach, M. (1985). References for motility and behavior of bacteria. In M. Eisenbach & M. Balaban (Eds.), *Sensing and response in microorganisms* (pp. 237–302). Amsterdam: Elsevier.
Ekman, P. (1992). An argument for basic emotions. *Cognition and Emotion, 6*, 169–200.
Eriksen, C. W. (1960). Discrimination and learning without awareness: A methodological survey and evaluation. *Psychological Review, 67*, 279–300.
Fantz, R. L., & Fagan, J. F. (1975). Visual attention to size and number of pattern details by term and preterm infants during the first six months. *Child development, 46*, 3–18.
Fechner, G. (1966). *Elements of psychophysics* (H. E. Adler, Trans.). New York: Holt, Rinehart & Winston. (Original work published 1860)
Fernberger, W. W. (1921). Preliminary study of the range of apprehension. *American Journal of Psychology, 32*, 121–133.
Ferrari, M., & Sternberg, R. J. (Eds.). (1998). *Self-awareness, its nature and development*. NY: Guilford Press.
Fischer, K. W., & Hogan, A. E. (1989). The big picture in infant development: Levels and variations. In J. Lockman & N. Hazan (Eds.), *Action in social context: Perspectives on early development* (pp. 175–305). NY: Plenum.
Folsome, E. F. (1979). *The origin of life: A warm little pond*. San Francisco: Freeman.
Fossage, J. L. (1997). The organizing functions of dream mentation. In *Contemporary Psychoanalysis, 33*(3), 429–458.
Frackowaik, R. S. J., Frith, C. D., Dolan, R., & Mazziotta, J. C. (Eds.). (1997). *Human brain function*. London: Academic Press.

Frankel, G. S. (1927). Beitrage zur Geotaxis und Phototaxis von Littorina. *Zeitschrift fur vergleichende Physiologie (Abt. C der Zeitschrift fur wissenschaftliche Biologie)*, *5*, 585–597.

Freud, S. (1980). *The interpretation of dreams* (J. Strachey, Ed. & Trans.). New York: Avon Books. (Original work published 1900)

Freud, S. (1935). The instincts and their vicissitudes. In *Collected papers, Vol. IV* (pp. 111–140). London: Hogarth Press.

Frith, C. D., & Friston, K. J. (1997). Studying brain function with neuroimaging. In E. S. J. Frackowaik (Eds), *Human brain function* (pp. 169–195). London: Academic Press.

Galistel, C. R. (1980). *The organization of action: A new synthesis*. Hillsdale, NJ: Lawrence Erlbaum Associates.

Galistel, C. R. (1981). Precis of Gallistel's The organization of action: A new synthesis. *The Behavioral and Brain Sciences*, *4*, 609–650.

Gallup, G. G., Jr. (1970). Chimpanzees: Self-recognition. *Science*, *167*, 86–87.

Galton, F. (1880). Statistics of mental imagery. *Mind*, *5*, 301–318.

Galton, F. (1883). *Inquiries into human faculty and its development*. London: Macmillan.

Garcia, J., & Koelling, R. A. (1966). Relation of cue to consequence in avoidance learning. *Psychonomic Science*, *4*, 123–124.

Gardner, R. W., & Long, R. (1962). Cognitive controls of attention and inhibition: A study of the individual. *British Journal of Psychology*, *62*, 381–388.

Gauthier, I., & Logothetis, N. (2000). Is face recognition not so unique after all? *Cognitive Neuropsychology*, *17*, 125–142.

Gelman, R. (1990). First principles organize attention to and learning about relevant data: Number and the animate–inanimate distinction as examples. *Brain*, *52*, 491–510.

Gibson, E. J., & Walk, R. D. (1960). The visual cliff. *Scientific American*, *202*, 64–71.

Gibson, J. J. (1966). *The senses considered as perceptual systems*. Boston: Houghton Mifflin.

Goldiamond, I. (1958). Indicators of perception: Subliminal perception, Subception, unconscious perception: An analysis in terms of psychophysical methodology. *Psychological Bulletin*, *55*(6), 373–411.

Goldstein, E. B., & Fink, S. I. (1981). Selective attention in vision: Recognition memory for superimposed line drawings. *Journal of Experimental Psychology: Human Perception and Performance*, *2*, 954–967.

Goodson, F. E. (1981). The plasticity of the perceptual process. *Bulletin of the Psychonomic Society*, *17*(1), 26.

Goodson, F. E., & Morgan, G. A. (1976). On levels of psychological data. In M. H. Marx & F. E. (Eds.), *Theories in contemporary psychology* (pp. 394–407). New York: Macmillan.

Goodson, F. E., & Morgan, G. A. (1976). Evaluation of theory. In M. H. Marx & F. E. Goodson (Eds.), *Theories in contemporary psychology* (pp. 261–286). New York: Macmillan.

Goodson, F. E., Ritter, S., & Thorpe, R. (1978). Motion parallax in depth and movement perception. *Bulletin of the Psychonomic Society*, *12*(5), 349–350.

Goodson, F. E., Snider, T. Q., & Swearingen, J. E. (1980). Motion parallax in the perception of movement by a moving subject. *Bulletin of the Psychonomic Society, 16*(2), 87–88.
Gould, J. L. (1982). *Ethology: The mechanisms and evolution of behavior.* New York: Norton.
Gould, S. J. (1994). Tempo and mode in the macroevolutionary reconstructionism of Darwinism. *Proceedings of the National Academy of Sciences, 91,* 6764–6771.
Gould, S. J., & Lewontin, R. C. (1979). The spandrels of San Marco and the Panglossian program: A critique of the adaptationist programme. *Proceedings of the Royal Society of London, 205,* 281–288.
Gould, S. J., & Vrba, E. S. (1982). Exaptation: A missing term in the sciences of form. *Paleobiology, 8,* 4–15.
Grant, P. R. (1993). Hybridization of Darwin's finches on Isla Daphne Major, Galapagos. *Philosophical Transaction of the Royal Society of London (B), 340,* 127–139.
Grant, P. R., & Grant, R. (1983). The origins of a species. *Natural History, 91*(9), 76–82.
Greeno, J. G. (1978). A study of problem solving. In R. Glaser (Ed.), *Advances in instructional psychology* (pp. 13–75). New York: Wiley.
Haig, K. A., Rawlins, J. N. P., Olton, D. S., Mead, A., & Taylor, R. (1983). Food searching strategies of rats: Variables affecting the relative strength of stay and shift strategies. *Journal of Experimental Psychology, 9,* 337–348.
Haig, R. A. (1988). *The anatomy of humor: Biopsychosocial and therapeutic perspectives.* Springfield, IL: Charles C. Thomas.
Hall, C. S. (1951). The genetics of behavior. In S. S. Stevens (Ed.), *Handbook of experimental psychology.* New York: Wiley.
Hall, J. E. (1976). *Classical conditioning and instrumental learning: A contemporary approach.* Philadelphia: Lippincott.
Hall, R. P. (1964). *Protozoa: The simplest of all animals.* New York: Holt, Rinehart & Winston.
Hamilton, T. H. (1967). *Process and pattern in evolution.* New York: Macmillan.
Hamilton, W. D. (1964). The genetic evolution of social behavior, I & II. *Journal of Theoretical Biology, 7,* 1–52.
Hardin, C. L. (1990). Why color? *Proceedings of the SPIE/SPSE symposium on electronic imaging: Science and technology,* 293–300.
Harlow, H. F. (1949). The formation of learning sets. *Physiological Review, 56,* 51–65.
Harris, C. R., & Cristenfeld, N. (1996). Jealousy and rational responses to infidelity across gender and culture. *Psychological Science, 7*(6), 378–379.
Hart, B. L. (1988). Biological basis of the behavior of sick animals. *Neuroscience and Behavioral Reviews, 12*(2), 123–137.
Harter, M. R., & Aine, C. J. (1984). Brain mechanisms of visual selective attention. In R. Parasuraman & D. R. Davies (Eds.), *Varieties of attention* (pp. 43–75). Orlando, FL: Academic Press.
Hasher, L., & Zacks, R. R. (1979). Automotive and effortful processes in memory. *Journal of Experimental Psychology: General, 108,* 356–388.
Hasler, A. D., Scholz, A. T., & Horall, R. M. (1978). Olfactory imprinting and homing in salmon. *American Scientist, 66,* 347–355.

Hazelbauer, G. L. (1988). The bacterial chemosensory system. *Canadian Journal of Microbiology, 34,* 446–473.
Hebb, D. O. (1949). *The organization of behavior.* New York: Wiley.
Hellyer, S. (1962). Supplementary report: Frequency of stimulus presentation and short term decrement in recall. *Journal of Experimental Psychology, 64,* 650.
Helson, H. (1948). Adaptation-level as a basis for a quantitative theory of frames of reference. *Psychological Review, 55,* 297–313.
Herbart, J. F. (1882). *Lehrbuch zur Psychologie.* Hamburg, Germany: G. von Hartenstein.
Hernandez-Peon, R., Scherrer, H., & Jouvet, M. (1956). Modification of electrical activity in the cochlear nucleus during "attention" in unanesthetized cats. *Science, 123,* 331–332.
Herrmann, T., Bahr, E., Bremmer, E., & Ellen, P. (1982). Problem solving in the rat: Stay versus shift solutions on the three-table task. *Analysis of learning and behavior, 10,* 39–45.
Hillix, W. A., & Marx, M. H. (1960). Response strengthening by information and effect in human learning. *Journal of Experimental Psychology, 60,* 97–102.
Hillyard, S. A., & Kutas, M. (1983). Electrophysiology of cognitive processing. *Annual Review of Psychology, 34,* 33–61.
Hobson, J. A. (1988). *The dreaming brain.* New York: Basic Books.
Hobson, J. A. (1990). Dreams and the brain. In Krippner, Stanley (Ed.), *Dreamtime and dreamwork: Decoding the language of the night.* Los Angeles: Jeremy P. Tarcher, Inc.
Hogan, J. A. (1973). How young chicks learn to recognize food. In R. A. Hinde & J. Stevenson-Hinde (Eds.), *Constraints on learning.* New York: Academic Press.
Holldobler, B. (1980). Canopy orientation: A new kind of orientation in ants. *Science, 210,* 8–88.
Horn, G. (1960). Electrical activity of the cerebral cortex of the unanesthetized cat during attention behavior. *Brain, 83,* 57–76.
Hospel, K. C., & Hams, R. S. (1982). Effect of tachistoscopic stimulation of subconscious Oedipal wishes on competitive performance: A failure to replicate. *Journal of Abnormal Psychology, 91*(6), 437–443.
Howes, D., & Solomon, R. L. (1950). A note on McGinnie's "Emotionality and perceptual defense." *Psychological Review, 57,* 235–240.
Hubel, D. H., & Wiesel, T. N. (1962). Receptive fields, binocular interaction and functional architecture in the cat's visual cortex. *Journal of Physiology, 160,* 106–154.
Hugelin, A., Dumont, S., & Paillas, N. (1960). Tympanic muscles and control of auditory input during arousal. *Science, 131,* 1371–1372.
Hull, C. L. (1943). *Principles of behavior: An introduction to behavior theory.* New York: Appleton-Century-Crofts.
Hull, C. L. (1951). *Essentials of behavior.* New Haven, CT: Yale University Press.
Hull, C. L. (1940). *Mathematico-deductive theory of rote learning.* New Haven, CT: Yale University Press.
Hunt, E., Frost, N., & Lunnebork, C. (1973). Individual differences in cognition: A new approach to intelligence. In G. Bower (Ed.), *The psychology of learning and motivation* (Vol. 7; pp. 87–122). New York: Academic Press.

Hunter, W. S., & Sigler, M. (1940). The span of visual discrimination as a function of time and intensity of stimulation. *Journal of Experimental Psychology, 26,* 160–179.
Hurford, J. R. (1991). The evolution of the critical period for language acquisition. *Cognition, 40,* 159–201.
Isen, A. M. (1984). Toward understanding the role of affect in cognition. *Social cognition.* Hillsdale, NJ: Lawrence Erlbaum Associates.
Ittelson, W. H., & Cantril, H. (1954). *Perception: A transactional approach.* Garden City, NY: Doubleday.
Jackson, J. M. (1983). Effects of subliminal stimulation of oneness fantasies on manifest pathology in male vs. female schizophrenics. *The Journal of Nervous and Mental Diseases, 171*(5), 280–288.
James, W. (1890). *The principles of psychology.* New York: Holt.
Jennings, H. S. (1906). *Behavior of the lower organisms.* New York: Columbia University Press. (Reprinted in 1962 by Indiana University Press)
Jevons, W. S. (1871). The power of numerical discrimination. *Nature, 3,* 281–282.
Johnson, C. T., & Eriksen, C. W. (1961). Preconscious perception: A reexamination of the Poetzl phenomenon. *Journal of Abnormal & Social Psychology, 62,* 497–503.
Johnston, R. F., & Selander, R. K. (1964). Correlation between severe winter temperatures and mean body weight for fourteen populations of the house sparrow. *Passer Domesticus. Science, 144,* 548–550.
Johnston, W. A., & Dark, V. J. (1986). Selective attention. *Annual Review of Psychology, 37,* 43–75.
Kamil, A. C. (1978). Systematic foraging by a nectar-feeding bird, the Amakihi *(loxop vivens). Journal of Comparative and Physiological Psychology, 92,* 388–396.
Kandel, E. R., & Schwartz, J. H. (1982). Molecular biology of learning: Modulation of transmitter releases. *Science, 218,* 433–443.
Kant, I. (1781). *Critique of pure reason.* (J. Watson, Trans.) New York: St. Martin's Press.
Kaplan, S. (1987). Aesthetics, affect, and cognition: Environmental preference from an evolutionary perspective. *Environment and Behavior, 19–1,* 3–32.
Kapur, S. (1995). Functional role of the prefrontal cortex in retrieval of memories. A PET study. *Neuroreport, 6*(14), 1880–1884.
Kaufman, E. L., Lord, M. W., Reese, T. W., & Volkman, J. (1949). The discrimination of visual number. *American Journal of Psychology, 62,* 498–525.
Kaufman, L. W. (1979). *Foraging strategies: Laboratory simulations.* Unpublished doctoral dissertation, Rutgers University, New Brunswick, NJ.
Kawai, M. (1965). Newly acquired precultural behavior of the natural troop of Japanese monkeys on Koshima Islet. *Primates, 6,* 1–30.
Keller, F. S., & Schoenfeld, W. N. (1950). *Principles of psychology.* New York: Appleton-Century-Crofts.
Kettlewell, H. B. D. (1959). Darwin's missing evidence. *Scientific American, 200,* 48–53.
Kilpatrick, F. P. (1952). *Human behavior from the transactional point of view.* Hanover, NH: Institute for Associated Research.
Kim, Jaegwon (1999). *Mind in a physical world: An essay on the mind-body problem and mental causation.* Cambridge, MA: MIT Press.

Kohler, W. (1917). *The mentality of apes.* Berlin, Germany: Royal Academy of Sciences.

Kohler, W. (1920). *Static and stationary physical Gestalts.* Braunschweig, Germany: Wieweg.

Kohler, W. (1971). Human perception. In M. Henle (Ed. & Trans.), *The selected papers of Wolfgang Kohler* (pp. 142–147). New York: Liverright. (Original work published in French, 1930)

Krebs, J. R. (1978). Optimal foraging: Decision rules for predators. In J. R. Krebs & N. B. Davies (Eds.), *Behavioral Ecology* (pp. 23–63). Sunderland, MA: Sinauer.

Krueger, W. C. F. (1929). The effect of overlearning on retention. *Journal of Experimental Psychology, 12,* 71–78.

Kudler, H. (1989). The tension between psychoanalysis and neuroscience: A perspective on *dream theory* in psychiatry. *Psychoanalysis-and-Contemporary-Thought, 12*(4), 599–617.

Kuhn, D. (1992). Cognitive development. In M. H. Bornstein & M. E. Lamb (Eds.), *Developmental psychology: An advanced textbook* (pp. 211–225). Hillsdale, NJ: Lawrence Erlbaum Associates.

Lack, D. (1944). Ecological aspects of species formation in passerine birds. *Ibis., 86,* 260–286.

Lambowitz, A. M. (1998b). Retrohoming of a bacterial group II interon: Mobility via a completely reverse splicing independent of homologous DNA recombination. *Cell, 94,* 451–462.

Larkin, J. H. (1979). *Models of strategy for solving physics problems.* Paper presented at the annual meeting of the American Educational Research Association.

Lashley, K. (1929). *Brain mechanisms and intelligence.* Chicago: University of Chicago Press.

Le Chatelier, H. L. (1884). General statement of the laws of chemical equilibrium. *Comptes Rendus, 99,* 786–789.

Lenneberg, E. H. (1967). *Biological foundations of language.* New York: Wiley.

Lettvin, J. U., Mauturans, H. C., McCulloch, W. S., & Pitts, W. H. (1959). What the frog's eye tells the frog's brain. *Proceeding from the Institute of Radio Engineers, 47,* 1940–1951.

Leuba, C., & Lucas, C. (1945). The effects of attitudes on description of pictures. *Journal of Experimental Psychology, 35,* 517–524.

Levi, W. H. (1951). *The pigeon.* Columbia, SC: Brim.

Lewis, J. L. (1970). Semantic processing of unattended messages using dichotic listening. *Journal of Experimental Psychology, 85,* 225–228.

Locke, J. L. (1983). *Phonological acquisition and change.* New York: Academic Press.

Logan, G. D. (1991). Automaticity and memory. In W. E. Hockley & S. Lewandowsky (Eds.), *Relating theory and data: Essays on human memory in honor of Bennet R. Murdock* (pp. 347–366). Hillsdale, NJ: Lawrence Erlbaum Associates.

Lorenz, K. Z. (1981). *Foundations of ethology.* New York: Springer-Verlag.

Lorenz, K. Z. (1970). *Studies on animal and human behavior.* (Vols. 1 & 2). Cambridge, MA: Harvard University Press.

Lotze, R. H. (1852). *Medicinische Psychologie.* Leipzig, Germany: Weidmann Press.
Mach, E. (1959). *The analysis of sensation.* (T. S. Szasz, Trans.) New York: Dover. (Original work published 1897)
Mackintosh, N. J. (1983). *Conditioning and association learning.* Oxford, England: Clarendon Press.
Mackintosh, N. J. (1974). *The psychology of animal learning.* New York: Academic Press.
Maier, N. R. F. (1932). The effect of cerebral destruction on reasoning and learning in rats. *Journal of Comparative Neurology, 54,* 45–75.
Malthus, T. R. (1826). *An essay on the principle of population as it affects the future improvement of society.* 6th ed. London: Murray.
Manning, A. (1961). The effects of artificial selection for mating speed in *Drosophilia Melanogaster. Animal Behavior, 9,* 82–92.
Marler, P. (1959). Developments in the study of animal communication. In P. R. Bell (Ed.), *Darwin's biological work* (pp. 150–206). New York: Cambridge.
Marler, P. (1955). Characteristics of some animal calls. *Nature, 176,* 6–8.
Mascolo, M. F., & Griffin, S. (1990). Alternative trajectories in the development of appraisals in anger. In M. F. Mascolo & S. Griffin (Eds.), *What develops in emotional development?* New York: Plenum.
Matthew, P. (1831). *On naval timber and arboriculture.* London: Longman.
Mayr, E. (1959). Where we are. *Symposia on quantitative biology* (Vol. XXXIV). New York: Biological Laboratory, Cold Springs Harbor.
Mayr, E. (1982). *The growth of biological thought.* Cambridge, MA: Harvard University Press.
McClean, J. D. (1979). Perspectives on the forest and the trees: The precedence of parts and whole in visual processing. *Dissertation Abstracts International, 39,* 6162B–6163B. (University of Michigan Microfilms No. 79-12574).
McClearn, G. E. (1959). The genetics of mouse behavior in novel situations. *Journal of Comparative and Physiological Psychology, 52,* 62–67.
McCleary, R. A., & Lazarus, R. S. (1949). Autonomic discrimination without awareness. *Journal of personality, 18,* 171–179.
McDougall, J. (1993). Of sleep and dream: A psychoanalytic essay. *International Forum of Psychoanalysis,* 2(4), 204–218.
McGinn, C. (1999). *The mysterious flame: Conscious minds in a material world.* New York: Basic Books.
McGinnies, E. (1949). Emotionality and perceptual defense. *Psychological Review, 56,* 244–251.
McGraw, K. O. (1987). *Developmental psychology.* New York: Harcourt Brace.
McNeill, D. (1979). *The conceptual basis of language.* Hillsdale, NJ: Lawrence Erlbaum Associates.
McSweeney, F. K. (1975). Matching and contrast on several concurrent treadle-press schedules. *Journal of Experimental Analysis of Behavior, 23,* 193–198.
Mellgren, R. L., Misasi, L., & Brown, S. W. (1984). Optimal foraging theory: Prey density and travel requirements in Rattus norvegicul. *Journal of Comparative Psychology, 98,* 142–153.

Melton, A. W. (1963). Implications of short-term memory for a general theory of memory. *Journal of Verbal Learning and Verbal Behavior, 2,* 1–21.

Mill, J. (1829). *Analysis of the phenomena of the human mind.* London: Baldwin & Cradock.

Mill, J. S. (1848). *System of logic, ratiocinative and inductive.* New York: Harper & Brothers.

Miller, D. (1979). Global precedence in attention and decision. *Journal of Experimental Psychology; Human Perception and Performance, 6,* 1161–1174.

Miller, G. A. (1956). The magical number seven, plus or minus two: Some limits on our capacity for processing information. *Psychological Review, 63,* 81–97.

Miller, G. A., Galenter, E., & Pribram, K. H. (1960). *Plans and the structure of behavior.* New York: Holt, Rinehart & Winston.

Miller, L. (1993). *Psychotherapy of the brain-injured patient: Reclaiming the shattered self.* New York: Norton.

Miller, N. E., & Dollard, J. (1941). *Social learning and imitation.* New Haven, CT: Yale University Press.

Minsky, M. (1975). A framework for representing knowledge. In P. H. Winston (Ed.), *The psychology of computer vision* (pp. 211–277). New York: McGraw-Hill.

Moore, B. R. (1973). The role of directed Pavlovian reactions in simple instrumental learning in the pigeon. In R. H. Hinde & J. Stevenson-Hinde (Eds.), *Constraints on learning* (pp. 159–168). New York: Academic Press.

Mowbray, G. H. (1954). The perception of short phrases presented simultaneously for visual and auditory perception. *Quarterly Journal of Experimental Psychology, 22,* 565–571.

Mowbray, G. H. (1953). Simultaneous vision and audition: The comprehension of prose passages with varying levels of difficulty. *Journal of Experimental Psychology, 46,* 365–371.

Mowrer, O. H., & Jones, H. M. (1943). Extinction and behavior variability as functions of effortfulness of task. *Journal of Experimental Psychology, 33,* 369–386.

Mueller, D. M., & Feldman, M. W. (1987). The evolution of altruism by kin selection: New Phenomena with strong selection. *Ethology and Sociobiology, 9,* 223–229.

Muller, J. (1843). *Handbuch der physiologie das menschen II* (book V). Philadelphia: Lea & Blanchard.

Munn, N. L. (1950). *Handbook of psychological research on the rat.* New York: Houghton Mifflin.

Murdock, B. B., Jr. (1954). Perceptual defense and threshold measurement. *Journal of Personality, 22,* 565–571.

Nafe, J. P., & Wagoner, K. S. (1941). The nature of pressure adaptation. *Journal of Genetic Psychology, 25,* 323–351.

Navon, D. (1977). Forest before trees: The precedence of global features in visual perception. *Cognitive Psychology, 9,* 353–383.

Navon, D. (1981). The forest revisited: More on global precedence. *Psychological Research, 43,* 1–32.

Neisser, U. (1967). *Cognitive psychology.* New York: Appleton-Century-Crofts.

Nesse, R. M. (1990). Evolutionary explanations of emotions. *Human Nature, 1,* 261–289.

Nesse, R. M. (1994). An evolutionary perspective on substance abuse. *Ethology and Sociobiology, 15,* 339–348.

Newell, A., Shaw, J. C., & Simon, H. A. (1958). Elements of a theory of general problem solving. *Psychological Review, 65,* 151–166.

Newell, A., & Simon, H. A. (1972). *Human problem-solving.* Englewood Cliffs, NJ: Prentice-Hall.

Newton, I. (1934). *Principia* (F. Cajori, Trans). Berkeley: University of California Press. (Original work published 1687)

Norman, D. (1966). Acquisition and retention in short-term memory. *Journal of Experimental Psychology, 72,* 369–381.

Norman, D. (1969). Memory while shadowing. *Quarterly Journal of Experimental Psychology, 21,* 85–94.

Nyberg, L., Cabeza, R., & Tulving, E. (1996). PET studies of encoding and retrieval: The HERA model. *Psychonomic Bulletin, 3,* 135–148.

Olds, J. (1960). Differentiation of reward systems in the brain by self-stimulation techniques. In Ramey & O'Doherty (Eds.), *Electrical studies and the unanesthetized brain.* New York: Harper & Row.

Olds, J., & Peretz, B. (1960). A motivational analysis of the reticular activating system. *Electroencephalography and Clinical Neurophysiology, 12,* 445–452.

Oliver, J. M., & Burkham, R. (1982). Subliminal psycho-dynamic activation in depression: A failure to replicate. *Journal of Abnormal Psychology, 91*(5), 337–342.

Olton, D. S., Walker, J. A., Gage, F. H., & Johnson, C. T. (1977). Choice behavior of rats searching for food. *Learning and Motivation, 8,* 315–331.

Oomura, Y. (1976). Significance of glucose, insulin, and free fatty acid on the hypothalanic feeding and satiety neurons. In D. Novin, W. Wywricka, & G. A. Bray (Eds.), *Hunger mechanisms and clinical implications* (pp. 145–157). New York: Haven.

Oparin, A. I. (1953). *The origin of life.* New York: Dover.

Oparin, A. I. (1968). *Genesis and evolutionary development of life.* New York: Academic Press.

Paivio, A. (1971). *Imagery and verbal processes.* New York: Holt, Rinehart & Winston. (Reprinted by Lawrence Erlbaum Associates, Hillsdale, NJ, 1979)

Pally, R. E. (1997). How brain development is shaped by genetic and environmental factors. *International Journal of Psycho-Analysis, 78*(3), 587–593.

Paulesu, E., Frith, C. D., & Frackowiak, R. S. J. (1993). The neural correlates of the verbal component of working memory. *Nature, 362,* 342–344.

Pavlov, I. P. (1927). *Conditioned reflexes* (G. V. Anrep, Trans.). London: Oxford University Press.

Pearson, K. G. (1976). The control of walking. *Scientific American, 235,* 72–86.

Pengelley, E. T., & Asmundson, S. J. (1974). Circannual rhythmicity in hibernating animals. In E. T. Pengelley (Ed.), *Circannual clocks.* New York: Academic Press.

Pepper, R. & Herman, L. (1970). Decay and interference effects in the short-term retention of a discrete motor act. *Journal of Experimental Psychology, Monographs, 83*(2).

Phillips, W., & Baddeley, A. (1971). Reaction time and short-term visual memory. *Psychonomic Science, 22,* 73–74.

Piaget, J. (1924). *The language and thought of the child* (M. Warden, Trans.). New York: Harcourt, Brace & World.

Piaget, J. (1929). *The child's conception of the world.* New York: Harcourt Brace.

Pinker, S. (1989). *Learnability and cognition: The acquisition of argument structure.* Cambridge, MA: MIT Press.

Pinker, S. (1994). *The language instinct.* London: Penguin.

Pinker, S., & Bloom, P. (1992). Natural language and natural selection. In J. H. Barkow, L. Cosmides, & J. Tooby (Eds.), *The Adapted Mind* (pp. 451–487). Oxford, England: Oxford University Press.

Pipp, S. K., Fischer, K. W., & Jennings, S. L. (1987). The acquisition of self and mother knowledge in infancy. *Developmental Psychology, 22,* 86–96.

Plomin, R., & DeFries, J. C. (1998). The genetics of cognitive abilities and disabilities. *Scientific American, 6,* 62–69.

Plutchik, R. (1980). *Emotion: A psychoevolutionary synthesis.* New York: Harper & Row.

Poetzl, O. (1960). The relationship between experimentally induced dream images and indirect vision. Monograph No. 7. *Psychological Issues, 2,* 41–120.

Posner, M. I., Boies, S. J., Eichelman, W. H., & Taylor, B. (1969). Retention of visual and name codes of single letters. *Journal of Experimental Psychology, 7,* 1–16.

Posner, M. I., & Cohen, Y. (1984). Components of visual orienting. In H. Bouma & D. Bowhuis (Eds.), *Attention and performance* (pp. 531–556). Hillsdale, NJ: Lawrence Erlbaum Associates.

Postman, L., Bronson, W. C., & Gropper, G. L. (1953). Is there a mechanism of perceptual defense? *Journal of Abnormal and Social Psychology, 48,* 215–224.

Postman, L., Bruner, J. S., & McGinnies, E. (1948). Personal values as selective factors in perception. *Journal of Abnormal and Social Psychology, 43,* 142–154.

Premack, D. (1983). The codes of man and beast. *The Behavioral Brain Sciences, 6,* 125–167.

Rao, K. P. (1954). Tidal rhythmicity of rate of water propulsion in mytilus and its modifiability by transplantation. *Biological Bulletin, 106,* 353–359.

Remington, R., & Pierce, L. (1984). Moving attention: Evidence for time-invariant shifts of visual selective attention. *Perceptual Psychophysics, 35,* 393–399.

Restle, F. (1975). *Learning: Animal behavior and human cognition.* New York: McGraw-Hill.

Revulsky, S. H. (1971). The role of interference over a delay. In W. K. Honig (Ed.), *Animal memory.* New York: Academic Press.

Richter, W., Georgopoulos, A. P., Ugurbil, K., & Kim, S. G. (1997). Detection of brain activity during mental rotation in a single trial by fMRI. *Neuroimage, 5,* 549.

Rock, I. (1983). *The logic of perception.* Cambridge, MA: MIT Press.

Roe, A. A. (1951). Study of imagery in research scientists. *Journal of Personality, 19,* 459–470.

Roeder, K. D. (1965). Moths and ultrasound. *Scientific American, 212,* 94–102.

Rosenzwieg, M. R., & Leiman, A. L. (1982). *Physiological psychology.* Lexington, MA: D.C. Heath & Co.

Rothbart, M. K. (1977). Psychological approaches to the study of humour. In A. J. Chapman & H. C. Foot (Eds.), *It's a funny thing humour* (pp. 87–94). New York: Pergamon.

Roy, C. S., & Sherrington, C. S. (1980). On the regulation of the blood supply of the brain. *Journal of Physiology, II,* 85–108.

Saltz, E. (1971). *The cognitive bases of human learning.* Homewood, IL: Dorsey.

Sanford, R. N. (1936). The effect of abstinence from food upon imaginal processes. *Journal of Psychology 2,* 227–241.

Schafer, R., & Murphey, G. (1943). The role of autisms in visual figureground relationship. *Journal of Experimental Psychology, 32,* 335–343.

Schopf, J. W., & Oehler, D. Z. (1976). How old are the eukaryotes? *Science, 193,* 47–49.

Scott, J. P., & Fuller, J. L. (1965). *Genetics and the social behavior of the dog.* Chicago: University of Chicago Press.

Seligman, M. E. P. (1970). On the generality of the laws of learning. *Psychological Review, 77,* 406–418.

Seward, J. P., & Levy, H. (1949). Latent extinction: Sign learning as a factor in extinction. *Journal of Experimental Psychology, 39,* 660–668.

Shepard, R. N., & Metzler, J. (1971). Mental rotation of three-dimensional objects. *Science, 171,* 701–703.

Shettleworth, S. J. (1983). Function and mechanism in learning. In M. D. Zeiler & P. Horzem (Eds.), *Advances in analysis of behavior* (pp. 1–39). New York: Wiley.

Shevrin, H., & Luborsky, L. (1958). The measurement of preconscious perception in dreams and images: An investigation of the Poetzl phenomenon. *Journal of Abnormal and Social Psychology, 56,* 28–37.

Shiffrin, R. M., & Schneider, W. (1977). Controlled and automatic human information processing II: Perceptual learning, automatic attending, and a general theory. *Psychological Review, 30,* 127–190.

Shultz, T. R. (1977). A cross-cultural study of the structure of humour. In A. J. Chapman & H. C. Foot (Eds.), *It's a funny thing, humour* (pp. 176–179). New York: Pergamon.

Siipola, E. M. (1935). A study of some effects of preparatory set. *Psychological Monographs, 46,* 28–37.

Silverman, L. H. (1975). An experimental method for the study of unconscious conflict: A progress report. *British Journal of Medical Psychology. 48*(4), 291–298.

Silverman, L. H. (1978). Simple research paradigm for demonstrating subliminal psychodynamic activation. Effects of Oedepal stimuli on dart-throwing accuracy in college males. *Journal of Abnormal Psychology. 87*(8), 341–357.

Skinner, B. F. (1938). *The behavior of organisms.* New York: Appleton-Century-Crofts.

Skinner, B. F. (1953). *Science and human behavior.* New York: Macmillan.

Skinner, B. F. (1957). *Verbal behavior.* New York: Appleton-Century-Crofts.

Skinner, B. F. (1960). Pigeons in a pelican. *American Psychologist, 15,* 28–37.

Skinner, B. F. (1981). Selection by consequences. *Science, 213,* 501–504.

Slobin, D. I. (1970). Universals of grammatical development in children. In F. D'Arcais & J. M. Levelt (Eds.), *Advances in psycholinguistics* (pp. 174–186). Amsterdam, Netherlands: North-Holland Publishing Co.

Small, W. S. (1901). Experimental study of the mental processes of the rat. II. *American Journal of Psychology, 12,* 206–239.

Solomon, R. S., & Turner, L. H. (1962). Discriminative classical conditioning in dogs paralyzed by curare can later control discriminative avoidance responses in the normal state. *Psychological Review, 69,* 202–219.

Spelke, E. S. (1991). Physical knowledge in infancy: Reflections on Piaget's theory. In S. Carey & R. Gelman (Eds.), *The epigenesis of mind: Essays on biology and cognition* (pp. 133–169). Hillsdale, NJ: Lawrence Erlbaum Associates.

Spelke, E. S., Breinlinger, K., Macomber, J., & Jacobson, K. (1992). Origins of knowledge. *Psychological Review, 99,* 605–632.

Sperling, G. (1960). The information available in brief visual presentations. *Psychological monographs, 74,* Whole No. 11.

Sperling, G., & Reeves, A. (1980). Measuring the reaction time of a shift of visual attention. In R. S. Nickerson (Ed.), *Attention and performance* (pp. 347–360). Hillsdale, NJ: Lawrence Erlbaum Associates.

Stahl, J. J., & Ellen, P. (1974). Factors in the reasoning performance of the rat. *Journal of Comparative and Physiological Psychology, 87,* 598–604.

Stebbens, G. L. (1982). *Darwin to DNA, molecules to humanity.* San Francisco: Freeman.

Stein, H. H. (1996). The dream is the guardian of sleep. *Psychoanalytic Quarterly, 64*(3), 533–550.

Steiner, J. E. (1979). Human facial expressions in response to taste and smell stimulation. In H. W. Reese & L. P. Lipsitt (Eds.), *Advances in child development and behavior* (Vol. 13). New York: Academic Press.

Stewart, W. N. (1983). *Paleobotany and the evolution of plants.* Cambridge, England: Cambridge University Press.

Stratton, G. M. (1897). Vision without inversion of the retinal image. *Psychological Review, 4,* 341–360, 463–481.

Street, P. (1971). *Animal weapons.* New York: Taplinger.

Stroop, J. R. (1935). Studies of interference in serial verbal reactions. *Journal of Experimental Psychology, 18,* 643–662.

Sullivan, M. W., & Lewis, M. (1993, March). *Determinants of anger in young infants: The effect of loss of control.* Poster presented at the 30th meeting of the Society for Research in Child Development, New Orleans, LA.

Symons, D. (1987). If we're all Darwinians, what's the fuss about? In C. B. Crawford, M. F. Smith, & D. L. Krebs (Eds.), *Sociobiology and psychology* (pp. 121–146). Hillsdale, NJ: Lawrence Erlbaum Associates.

Teale, E. W. (1949). *The insect world of J. Henri Fabre.* New York: Dodd, Mead & Co.

Thorndike, E. L. (1898). Animal intelligence. *Psychological Review Monographs, 2*(4), 304–462.

Thorpe, W. H. (1950). A note on detour behavior with *Ammophilia pubescens. Behavior, 2,* 257–264.

Thorpe, W. H., & Griffin, D. R. (1962). The lack of ultrasonic components in the flight noise of owls compared with other birds. *Ibis, 104*, 256–257.

Timberlake, W. (1983). The functional organization of appetitive behavior: Behavior systems and learning. In M. D. Zeiler & P. Harzem (Eds.), *Advances in analysis of behavior* (pp. 177–221). New York: Wiley.

Tinbergen, N. (1948). Social releasers and the experimental method. *Wilson Bulletin, 60*, 6–52.

Tinbergen, N., & Kruyt, W. (1938). Uber die Orientierung des Bienenwolfes (Philanthrus triangulum Faber). III. Die Bevarzuging bestimmeter Wegmarken. *Zeitschrift fur vergleichende Physiologie, 25*, 292–334.

Tinbergen, N., & Perdeck, A. C. (1950). On the stimulus situation releasing the begging response in the newly hatched herring gull chick (Larus argentatus Pont). *Behavior, 3*, 1–39.

Tinklepaugh, O. L. (1928). An experimental study of representative factors in monkeys. *Journal of Comparative Psychology, 8*, 197–236.

Titchener, E. B. (1909). *Lectures on the experimental psychology of the thought processes.* New York: Macmillan.

Tolman, E. C. (1948). Cognitive maps in rats and men. *Psychological Review, 55*, 189–208.

Tolman, E. C., & Honzik, C. H. (1930). Introduction and removal of reward and maze performance in rats. *University of California publications in psychology, 4*, 257–275.

Treisman, A. M. (1960). Contextual cues in selective listening. *Quarterly Journal of Experimental Psychology, 12*, 242–248.

Treisman, A. (1964). Monitoring and storing of irrelevant messages in selective attention. *Journal of Verbal Learning and Verbal Behavior, 3*, 449–459.

Tryon, R. C. (1942). Individual differences. In F. A. Moss (Ed.), *Comparative psychology.* Englewood Cliffs, NJ: Prentice-Hall.

Tsal, Y. (1983). Movements of attention across the visual field. *Journal of Experimental Psychology: Human Perception and Performance, 9*, 523–530.

Underwood, B. J. (1949). *Experimental psychology.* New York: Appleton-Century-Crofts.

von Ehrenfels, C. (1890). Ueber Gestaltqualitaten. *Psychologische Forschung, 14*, 249–292.

von Helmholtz, H. (1924). *Physiological optics* (J. P. C. Southall, Trans.). New York: Optical Society of America.

von Holst, E. (1973). Relative coordination as a phenomenon and as a method of analysis of central nervous functions. In *The behavioral physiology of animals and man. Selected papers of Eric von Holst.* Coral Gables, FL: University of Miami Press.

van Iersel, J. J. A., & van den Assen, J. J. (1965). Aspects or orientation in the digger WASP Bembix nostrada. In *Animal Behavior Supplement, 1.*

von Tchisch, W. (1885). Prior entry and attention. *Philosophical Studies, 2*, 603–634.

Wagoner, K. S., & Goodson, F. E. (1976). Does the mind matter? In M. H. Marx & F. E. Goodson (Eds.), *Theories in contemporary psychology* (pp. 201–204). New York: Macmillan.

Wagoner, K. S., Goodson, F. E., & Nunez, A. E. (1980). The shrink phenomenon. *Bulletin of the Psychonomic Society, 16*,(5), 403–404.

Wald, G. (1959). Life and light. *Scientific American, 201,* 92–100.

Wallace, A. R. (1891). *Natural selection and tropical nature.* London: Macmillan.

Wasserman, E. A. (1997). The science of animal cognition: Past, present, and future. *Journal of Experimental Psychology, 23,* 123–135.

Watson, J. B. (1913). Psychology as the behaviorist views it. *Psychological Review, 20,*158–177.

Watson, J. B. (1998). *Behaviorism.* Gregory Kimble (Ed.). New Brunswick, NJ: Transaction Publishers. (Original work published 1925)

Watson, J. S., & Ramey, C. T. (1972). Reactions to response-contingent stimulation in early infancy. *Merrill-Palmer Quarterly, 18,* 219–227.

Watson, M., & Fischer, K. W. (1977). A developmental sequence of agent use in late infancy. *Child Development, 48,* 828–835.

Watt, H. J. (1904). Experimentelle beitrage zur einer theorie des denkens. *Arch. Ges. Psycholl., 4,* 289–436.

Waugh, N. C., & Norman, D. A. (1965). Primary memory. *Psychological Review, 72,* 89–104.

Wegner, D. M., & Wheatley, T. (1999). Apparent mental causation: Sources of the experience of will. *American Psychologist, 54,* 480–492.

Weiner, J. (1995). *The beak of the finch.* New York: Vintage Books.

Weir, R. H. (1966). Some questions on the child's learning of phonology. In F. Smith & G. A. Miller (Eds.), *The genesis of language* (pp. 153–172). Cambridge, MA: MIT Press.

Weisfeld, G. E. (1993). The adaptive value of humor and laughter. *Ethology and Sociobiology, 14,* 141–169.

Werner, T. K., & Sherry, T. W. (1987). Behavioral feeding specialization in *Pinaroloxias inornata,* the 'Darwin's Finch' of Cocos Island, Costa Rica. *Proceedings of the National Academy of Sciences, 84,* 5506–5510.

Wertheimer, M. (1945). *Productive thinking.* New York: Harper & Row.

Wertheimer, M. (1950). Gestalt theory. In W. D. Ellis (Ed.), *A source book of Gestalt psychology.* London: Routledge & Kegan Paul. (Original work published in German, 1925)

Wickelgren, W., & Norman, D. (1966). Strength models and serial position in short-term recognition memory. *Journal of Mathematical Psychology, 3,* 316–347.

Wilson, D. M. (1980). The role of coupled oscillators in locomotion. In C. R. Gallistel (Ed.), *The organization of action: A new synthesis.* Hillsdale, NJ: Lawrence Erlbaum Associates.

Wilson, E. O. (1975). *Sociobiology: The new synthesis.* Cambridge, MA: Harvard University Press.

Wilson, D. S., & Sober, E. (1994). Introducing group selection to the human behavioral sciences. *Behavioral and Brain Sciences, 17,* 585–654.

Witkin, H. A., Lewis, H. B., Hertzman, M., Mackover, K., Meisner, P. B., & Wapner, S. (1954). *Personality through perception.* New York: Harper & Row.

Wood-Gush, D. G. M. (1960). A study of sex drive of two strains of cockerels through three generations. *Animal Behavior, 18,* 43–53.

Wundt, W. (1886). *Outline of psychology* (C. H. Judd, Trans.). Leipzig, Germany: Englemann.

Wundt, W. (1899). Zur kritik tachistoskopisher versuche. *Pilosophische Studien, 15,* 287–317.

Author Index

A

B

C

D

K

L

M

R

S

T

SUBJECT INDEX

A

I

J

K

L

P

Q

R

T

U

V

W

X

Y

Z